纪念华罗庚先生诞辰100周年

《华罗庚文集》编委会

王　元　万哲先　陆启铿　杨　乐

李福安　贾朝华　尚在久　周向宇

国家出版基金项目
中国科学院华罗庚数学重点实验室丛书

华罗庚文集
代数卷 II

华罗庚 著
李福安 审校

科 学 出 版 社
北 京

内 容 简 介

本书汇集了华罗庚先生 1930–1952 年关于代数和矩阵几何的代表性论文 22 篇, 以及万哲先关于华罗庚在代数和几何领域成就的一篇介绍文章. 华罗庚的论文内容深刻, 技巧性很强, 要求的预备知识并不多. 本书适合数学专业的研究生和研究人员阅读, 大学数学系的高年级学生也能读懂其中大部分内容.

图书在版编目(CIP)数据

华罗庚文集: 代数卷 II/华罗庚著, 李福安审校. —北京: 科学出版社, 2011
(中国科学院华罗庚数学重点实验室丛书)
ISBN 978-7-03-030014-0

I. ①华… II. ①华… III. ①数学–文集②代数–文集 IV. O1-53

中国版本图书馆 CIP 数据核字 (2011) 第 008066 号

责任编辑: 赵彦超 / 责任校对: 宋玲玲
责任印制: 赵　博 / 封面设计: 黄华斌

科 学 出 版 社 出版
北京东黄城根北街 16 号
邮政编码: 100717
http://www.sciencep.com

三河市春园印刷有限公司印刷
科学出版社发行　各地新华书店经销
*

2011 年 2 月第　一　版　　开本: B5(720×1000)
2025 年 8 月第四次印刷　　印张: 24 1/4
字数: 475 000

定价: 198.00 元
(如有印装质量问题, 我社负责调换)

《华罗庚文集》序言

2010年是著名数学家华罗庚先生诞辰100周年. 值此机会, 我们编辑出版《华罗庚文集》, 作为对他的美好纪念.

华罗庚先生是他那个时代的国际领袖数学家之一, 也是中国现代数学的主要奠基人和领导者. 无论是在和平建设时期, 还是在政治动荡甚至是战争年代, 他都抱定了为国家和人民服务的宗旨, 为中国数学的发展倾注了毕生精力, 受到了中国人民的广泛尊敬.

华罗庚先生最初研究数论, 后将研究兴趣拓展至代数和多复变等多个领域, 取得了一系列国际一流的成果, 引领了这些领域的学术发展, 产生了广泛持久的影响. 他从一名自学青年成长为著名数学家, 其传奇经历激励了几代中国数学家投身于数学事业.

华罗庚先生为我们留下了丰富的精神遗产, 包括大量的学术著作和研究论文. 我们认为, 认真研读这些著作和论文, 是深刻把握华罗庚学术思想精髓的最佳途径. 无论对于数学工作者还是青年学生, 其中许多内容都是很有启发和裨益的.

华罗庚先生担任中国科学院数学研究所所长30余年, 他言传身教, 培养和影响了一批国际水平的数学家, 他的学术思想和治学精神已经成为数学所文化的核心. 自2008年起以中国科学院数学研究所为基础成立的中国科学院华罗庚数学重点实验室, 旨在继承和弘扬华罗庚先生的学术思想和治学精神, 积极推动中国数学的发展. 为此, 我们选择华罗庚先生的著作和论文作为实验室的首批出版物, 今后还将陆续推出更多优秀的数学出版物.

在出版《华罗庚文集》的过程中, 我们得到了各方面的关心和支持, 包括国家出版基金的资助, 在此我们表示深深的感谢. 同时, 对于有关人员在策划、翻译和审校等方面付出的辛勤劳动, 对于科学出版社所作的大量工作, 我们表示诚挚的谢意.

中国科学院华罗庚数学重点实验室
《华罗庚文集》编委会
2010年3月

目　　录

《华罗庚文集》序言

Algebra and geometry ··· 1

苏家驹之代数的五次方程式解法不能成立之理由 ··································· 5

Geometries of matrices. I. Generalizations of von Staudt's theorem ············· 8

Geometries of matrices. I_1. Arithmetical construction ······························ 54

Orthogonal classification of Hermitian matrices ····································· 64

Geometries of matrices. II. Study of involutions in the geometry of symmetric
　　matrices ·· 82

Geometries of matrices. III. Fundamental theorems in the geometries of
　　symmetric matrices ··· 122

Some "Anzahl" theorems for groups of prime power orders ···················· 152

On the automorphisms of the symplectic group over any field ················· 166

On the existence of solutions of certatin equations in a finite field ············ 189

Characters over certain types of rings with applications to the theory of
　　equations in a finite field ··· 195

On the automorphisms of a sfield ··· 201

On the number of solutions of some trinomial equations in a finite field ········ 205

On the nature of the solutions of certain equations in a finite field ············ 210

Some properties of a sfield ·· 218

On the generators of the symplectic modular group ································ 223

Geometry of symmetric matrices over any field with characteristic other than
　　two ··· 236

On the multiplicative group of a field ··· 262

环之准同构及对射影几何的一应用 ·· 267

A theorem on matrices over a sfield and its applications ························· 273

Supplement to the paper of Dieudonné on the automorphisms of classical
　　groups ·· 320

Automorphisms of the unimodular group ·· 351

Automorphisms of the projective unimodular group ································ 370

《华罗庚文集》已出版书目 ··· 377

Algebra and geometry*

Z. X. Wan

Finite Groups. Early in 1938, while Hua taught at the Southwest Association Associated University in Kunming, he conducted a seminar on finite groups; among the topics studied were p-groups. In [46, 81], he introduced the concept of the rank of a p-group. A p-group $\mathfrak{g}$ of order p^n is said to be of rank α, if the maximum of the orders of its elements is $p^{n-\alpha}$. Using this concept he proved that a pseudo-basis exists in p-groups, i.e., if $p \geqslant 3$ and $n \geqslant 2\alpha + 1$, then every element of $\mathfrak{g}$ can be expressed uniquely as

$$G = A^\delta A_\alpha^{\delta_\alpha} A_{\alpha-1}^{\delta_{\alpha-1}} \cdots A_1^{\delta_1}, \quad 0 \leqslant \delta \leqslant p^{n-\alpha} - 1, \quad 0 \leqslant \delta_i \leqslant p - 1,$$

where A is of order $p^{n-\alpha}$ and $A_i^{p^i} = 1$. With the aid of pseudo-bases, and of a modified form of the enumeration principle of P. Hall, he proved several "Anzahl" theorems. For instance, if $\mathfrak{g}$ is a group of order p^n and rank α ($p \geqslant 3$, $n \geqslant 2\alpha+1$), then (i)$\mathfrak{g}$ contains one and only one subgroup of order p^m and rank α ($2\alpha + 1 \leqslant m \leqslant n$); (ii) $\mathfrak{g}$ contains p^α cyclic subgroups of order p^m ($\alpha < m < n - \alpha - 1$); (iii) the number of elements of order $\leqslant p^m$ ($\alpha \leqslant m \leqslant n - \alpha$) in $\mathfrak{g}$ is equal to $p^{m+\alpha}$. The second and third results improved theorems of G. A. Miller and A. A. Kulakoff respectively.

Skew Fields. Since Hamilton's first example of a non-commutative division algebra——the quaternion algebra——division algebras have received a great deal of attention. By comparison, infinite dimensional division algebras——skew fields ——were neglected; until, around 1950, with his perceptive direct algebraic method Hua proved several remarkable theorems in this area.

First, in 1949, Hua[88] proved that "every semi-automorphism of a skew field is either an automorphism or an anti-automorphism" (by a semi-automorphism of a skew field we mean a one-to-one mapping σ from the skew field into itself with the properties $(a + b)^\sigma = a^\sigma + b^\sigma$, $(aba)^\sigma = a^\sigma b^\sigma a^\sigma$ and $1^\sigma = 1$). This theorem was referred to as the beautiful theorem of Hua by E. Artin in his book *Geometric*

---

* Reprinted from *Loo-Keng Hua Selected Papers*. New York: Springer-Verlag, 1983: 281-284. The references in this article are those in this book.

Algebra. From it Hua[88,97] deduced also the fundamental theorem of l-dimensional projective geometry over a skew field. In 1950[97] he extended his theorem to semi-homomorphisms of rings without zero divisors.

Secondly, in 1949 L. K. Hua[91] gave a straightforward proof that "every proper normal subfield of a skew field is contained in its center". This result appears in the literature as the Cartan-Brauer-Hua theorem. Before the work of Hua and Richard Brauer, Henri Cartan's proof had used the complicated device of Galois extensions over subfields. By contrast, Hua's proof requires only the elementary identity: If $ab \neq ba$, then

$$a = \left(b^{-1} - (a-1)^{-1}b^{-1}(a-1)\right)\left(a^{-1}b^{-1}a - (a-1)^{-1}b^{-1}(a-1)\right)^{-1}.$$

In 1950, Hua[96] proved also that "if a skew field is not a field, then its multiplicative group is not meta-abelian".

Classical Groups. Early in 1946, L. K. Hua[73] published his first paper on automorphisms of classical groups, in which he determined the automorphisms of a real symplectic group. Subsequently, in 1948, he[85] determined the automorphisms of a symplectic group over any field of characteristic not 2. The method of Hua for determining the automorphisms of symplectic groups can be applied also to classical groups of other types; but since Dieudonné published his results on the automorphisms of classical groups in 1951, Hua[101] restricted himself to publishing only solutions, by his own method, to a series of problems left open by Dieudonné. The first of these was the determination of automorphisms of $\mathrm{GL}_2(K)$, K being an arbitrary skew field of characteristic $\neq 2$.

Besides $\mathrm{GL}_2(K)$, Hua[101] determined also the automorphisms of $\mathrm{SL}_4(K)$ and $\mathrm{PSL}_4(K)$, where K is a skew field of characteristic not 2, and the automorphism of $\mathrm{O}_4^+(K, f)$, where K is a field of characteristic not 2 and f is a quadratic form of index 2. Afterwards, Hua and Z. X. Wan[105] determined the automorphisms of $\mathrm{SL}_2(K)$ and $\mathrm{PSL}_2(K)$, where K is a skew field of characteristic $\neq 0$, the automorphisms of $\mathrm{SL}_4(K)$ and $\mathrm{PSL}_4(K)$, where K is a skew field of characteristic 2, and they proved also the nonisomorphism of certain linear groups.

Hua's work on the automorphisms of classical groups, shows mastery of the techniques of matrix calculation. The procedure is to start with the low-dimensional cases and to proceed to the higher-dimensional cases by induction, as in [85], for $\mathrm{SP}_{2n}(K)$.

About the structure of classical groups, Hua extended the usual unitary group to the case when the basic field is not necessarily commutative but has an involutive antiautomorphism. He proved that the group $TU_n(K_1S)$ generated by unitary transvections modulo its center is a simple group, if S has index $\geqslant 1$ and that $TU_n(K_1S)$ is the commutator subgroup of $U_n(K_1S)$, if the index of S satisfies $n \geqslant 2V \geqslant 4$.

Hua and I. Reiner[102,106] also determined the automorphisms of $\mathrm{GL}_n(\mathbb{Z})$ and $\mathrm{DGL}_n(\mathbb{Z})$, which was the start of the work on the automorphisms of classical groups over rings. They[92] also proved that $\mathrm{GL}_n(\mathbb{Z})$ is generated by three elements, $\mathrm{SL}_n(\mathbb{Z})$ by two elements, and $\mathrm{Sp}_{2n}(\mathbb{Z})$ by four elements for $n \geqslant 2$. Formerly Poincaré[*] had stated without proof that $\mathrm{Sp}_{2n}(\mathbb{Z})$ is generated by elementary matrices of two simple types, and later Brahana[†] had proved this by showing that every element of $\mathrm{Sp}_{2n}(\mathbb{Z})$ is expressible as a product of matrices taken from some finite set of matrices.

Geometry of Matrices[67,76–78,93,99]. Study of this topic was initiated by Hua and relates to Siegel's work on fractional linear transformations. In it, the points of the space are matrices of a certain kind, for instance, rectangular matrices, symmetric matrices or skew-symmetric matrices of the same size. There is then a group of motions in this space, and the problem is to characterize the group of motions by as few geometric invariants as possible. First, he studied the geometry of matrices of various types over the complex or real fields. Later, he extended his results to the case when the basic field is not necessarily commutative and discovered that the invariant "coherence" is alone sufficient to characterize the group of motions of the space. Take his paper [99] as an example. He proved the fundamental theorem of affine geometry of rectangular matrices: *Let $1 < n \leqslant m$. Then the one-to-one mappings from the set of $n \times m$ matrices over a skew field K onto itself preserving coherence (two matrices M and N are said to be coherent, if the rank of $M-N$ is 1) is necessarily of the form*

$$Z_1 = PZ^\sigma Q + R, \tag{1}$$

where $P = P^{(n)}$ and $Q = Q^{(m)}$ are invertible matrices, R is an $n \times m$ matrix, and σ is an automorphism of K; if $n = m$, then besides (1) we have also

$$Z_1 = PZ'^\tau Q + R,$$

where τ is an anti-automorphism of K. From this theorem he deduced the fundamental theorem of the projective geometry of rectangular matrices (the Grassmann space),

---

[*] Poincaré H. *Rend Circ Mat Palermo*, 1904, **18**: 45–110.
[†] Brahana R R. *Ann of Math*, 1923, **24** (2): 265–270.

and he determined the Jordan isomorphism of total matrix rings over skew fields of characteristic $\neq 2$ and the Lie isomorphism of total matrix rings over skew fields of characteristic $\neq 2, 3$.

Arising from the geometry of matrices and the theory of functions of several complex variables, Hua went on to study the classification problem of matrices; for instance, the classification of complex symmetric and skew-symmetric matrices under the unitary group, of a pair of Hermitian matrices under congruence[66], and of Hermitian matrices under the orthogonal group[76] (*editorial note*: by "elementary divisors of a characteristic matrix" is meant, in current usage, "Jordan blocks" in a Jordan normal form, in the sense that $(X - \alpha)^d$ is an elementary divisor of multiplicity m if and only if the Jordan form has exactly m blocks of $d \times d$ matrices

$$J_d(\alpha) = \begin{pmatrix} \alpha & 1 & & 0 \\ & \alpha & 1 & \\ & & \ddots & 1 \\ 0 & & & \alpha \end{pmatrix}.$$

In [76], on p. 509 four lines from the bottom, read $\overline{\Gamma}'Q(\overline{\Gamma}')^{-1} = T$ for $\overline{\Gamma}Q(\overline{\Gamma}')^{-1} = T$, and on p. 512 line four, read $H = KT_0$ for $K = HT_0$).

苏家驹之代数的五次方程式解法不能成立之理由*

五次方程式经 Abel, Galois 之证明后,一般算学者均认为不可以代数解矣,而《学艺》七卷十号载有苏君之"代数的五次方程式之解法"一文,罗欣读之而研究之,于去年冬亦仿得"代数的六次方程式之解法"矣. 罗对此欣喜异常,意为果能成立,则于算学史中亦可占一席地也,惟自思若不将 Abel 言论驳倒,终不能完全此种理论,故罗沉思于 Abel 之论中,凡一阅月,见其条例精严,无懈可击,后经本社编辑员之暗示,遂从事于苏君解法确否之工作,于六月中遂得其不能成立之理由,罗安敢自秘,特公之于世,尚祈示正焉.

解 法 简 述

用 Sylvester 之分离消去法 (diabylic method of elemination) 将普通形

$$x^5 + p_1 x^4 + p_2 x^3 + p_3 x^2 + p_4 x + p_5 = 0$$

化为可解形 $X^5 + P_1 X^4 + P_2 X^3 + P_3 X^2 + P_4 X + P_5 = 0$ (中有 $P_1 = 0, P_3 = 0, P_2^2 = 5P_4$),而 x, X 有 $X = n_0 + n_1 x + n_2 x^2 + n_3 x^3 + n_4 x^4$ 之关系. P_1, P_2, P_3, P_4, P_5 为 n_0, n_1, n_2, n_3, n_4 之一次、二次、三次、四次、五次齐次函数. $P_1 = 0$,即 n_0 可以 n_1, n_2, n_3, n_4 之一齐次函数表之,以之代入 P_2, P_3, P_4, P_5,则得 n_1, n_2, n_3, n_4 之二、三、四、五次齐次函数,而 P_3 之一般形可写为

$$\begin{aligned}&A_1 n_1^3 + A_2 n_2^3 + A_3 n_3^3 + A_4 n_4^3 + A_5 n_1^2 n_2 + A_6 n_1^2 n_3 + A_7 n_1^2 n_4 + A_8 n_2^2 n_1 + A_9 n_2^2 n_3\\&+ A_{10} n_2^2 n_4 + A_{11} n_3^2 n_5 + A_{12} n_3^2 n_2 + A_{13} n_3^2 n_4 + A_{14} n_4^2 n_1 + A_{15} n_4^2 n_2 + A_{16} n_4^2 n_3\\&+ A_{17} n_1 n_2 n_3 + A_{18} n_1 n_2 n_4 + A_{19} n_1 n_2 n_4 + A_{20} n_2 n_3 n_4,\end{aligned} \quad (\mathrm{I})$$

式中 $A_1, \cdots, A_{20}$ 为 $p_1, \cdots, p_5$ 之函数为已知者.

若令等于下式

$$\begin{aligned}&(a_1 n_1 + a_2 n_2)(a_3 n_1^2 + a_4 n_2^2 + a_5 n_3^2 + a_6 n_4^2 + a_7 n_1 n_2 + a_8 n_1 n_3 + a_9 n_1 n_4 + a_{10} n_2 n_3\\&+ a_{11} n_2 n_4 + a_{12} n_3 n_4) + (a_{13} n_3 + a_{14} n_4)(a_{15} n_1^2 + a_{16} n_2^2 + a_{17} n_3^2 + a_{18} n_4^2 + a_{19} n_1 n_2\\&+ a_{20} n_1 n_3 + a_{21} n_1 n_4 + a_{22} n_2 n_3 + a_{23} n_2 n_4 + a_{24} n_3 n_4),\end{aligned} \quad (\mathrm{II})$$

式中 $a_1, \cdots, a_{24}$ 为未定系数.

---

*科学, 1930, **15**: 307–309.

再设 $a_1n_1 + a_2n_2 = 0$, $a_{13}n_3 + a_{14}n_4 = 0$, 代入 $P_2^2 = 5P_4$ 式中, 则此式为 n_2, n_4 之四次齐次函数, 解之, 则得 n_2, n_4 之比值, 由此可作得 $n_0 : n_1 : n_2 : n_3 : n_4$ 之值, 故普通形可化为上之可解形, 换言之, 即五次方程式可得而解矣.

谬 误 点

罗研究上意知其谬误在 P_3 中, 即 (I) 不能等于 (II) 也. 夫求未定系数 $a_1, \cdots, a_{24}$, 原文亦有求之之二十方程式, 罗为便利讨探计, 特分之为四类, 转录于下:

(一) $a_1a_3 = A_1$, $\qquad\qquad\qquad a_2a_4 = A_2$,

$\quad\ \ a_3a_2 + a_1a_7 = A_5$, $\qquad\qquad a_4a_1 + a_2a_7 = A_8$;

(二) $a_{13}a_{17} = A_3$, $\qquad\qquad\quad\ a_{14}a_{18} = A_4$,

$\quad\ \ a_{17}a_{14} + a_{13}a_{24} = A_{13}$, $\qquad a_{18}a_{13} + a_{14}a_{24} = A_{16}$;

(三) $a_{13}a_{15} + a_1a_3 = A_6$, $\qquad\ \ a_{14}a_{15} + a_1a_9 = A_7$,

$\quad\ \ a_1a_{11} + a_2a_9 = a_{18} - a_{14}a_{19}$, $\quad a_2a_{11} + a_{14}a_{16} = A_{10}$,

$\quad\ \ a_2a_{10} + a_{13}a_{16} = A_9$, $\qquad\quad a_1a_{10} + a_2a_8 = a_{17} - a_{13}a_{19}$;

(四) $a_1a_5 + a_{13}a_{20} = A_{11}$, $\qquad\ \ a_2a_5 + a_{13}a_{22} = A_{12}$,

$\quad\ \ a_2a_{12} + a_{14}a_{22} = A_{19} - a_{13}a_{23}$, $\ a_1a_{12} + a_{13}a_{21} = A_{20} - a_{14}a_{20}$,

$\quad\ \ a_1a_6 + a_{14}a_{21} = A_{14}$, $\qquad\quad a_2a_6 + a_{14}a_{23} = A_{15}$.

依原所谓假 a_7, a_{24} 则由 (一), (二) 得 $a_1, a_2, a_3, a_4, a_{13}, a_{14}, a_{17}, a_{18}$ 之值, 则第二类乃为 $a_8, a_{15}, a_9, a_{11}, a_{16}, a_{10}$ 之联立一次方程式 (设 a_{19} 为已知), 以行列式解之, 知其各分母悉为

$$\Delta = \begin{vmatrix} a_1 & a_{13} & 0 & 0 & 0 & 0 \\ 0 & a_{14} & a_1 & 0 & 0 & 0 \\ 0 & 0 & a_2 & a_1 & 0 & 0 \\ 0 & 0 & 0 & a_2 & a_{14} & 0 \\ 0 & 0 & 0 & 0 & a_{13} & a_2 \\ a_2 & 0 & 0 & 0 & 0 & a_1 \end{vmatrix}.$$

然

$$\begin{vmatrix} a_1 & a_{13} & 0 & 0 & 0 & 0 \\ 0 & a_{14} & a_1 & 0 & 0 & 0 \\ 0 & 0 & a_2 & a_1 & 0 & 0 \\ 0 & 0 & 0 & a_2 & a_{14} & 0 \\ 0 & 0 & 0 & 0 & a_{13} & a_2 \\ a_2 & 0 & 0 & 0 & 0 & a_1 \end{vmatrix} = a_1 \begin{vmatrix} a_{14} & a_1 & 0 & 0 & 0 \\ 0 & a_2 & a_1 & 0 & 0 \\ 0 & 0 & a_2 & a_{14} & 0 \\ 0 & 0 & 0 & a_{13} & a_2 \\ 0 & 0 & 0 & 0 & a_1 \end{vmatrix}$$

$$-a_{13}\begin{vmatrix} 0 & a_1 & 0 & 0 & 0 \\ 0 & a_2 & a_1 & 0 & 0 \\ 0 & 0 & a_2 & a_{14} & 0 \\ 0 & 0 & 0 & a_{13} & a_2 \\ a_2 & 0 & 0 & 0 & a_1 \end{vmatrix}=0,$$

而 $a_8, a_{15}, a_9, a_{11}, a_{16}, a_{10} = \delta/\Delta$.

因 $\Delta = 0$, 故 $a_8, a_{15}, a_9, a_{11}, a_{16}, a_{10}$ 非不定即无限大, 故 (I) 等 (II) 之谬论不攻自破矣. 换言之, 即 P_3 为零不能解得二一次式, 故此法亦不能解五次方程式也.

Geometries of matrices. I. Generalizations of von Staudt's theorem*

It was first shown in the author's recent investigations on the theory of automorphic functions of a matrix-variable that there are three types of geometry playing important roles. Besides their applications, the author obtained a great many results which seem to be interesting in themselves.

The main object of the paper is to generalize a theorem due to von Staudt, which is known as the fundamental theorem of the geometry in the complex domain. The statement of the theorem is:

Every topological transformation of the complex plane into itself, which leaves the relation of harmonic separation invariant, is either a collineation or an anti-collineation.

Since the fields and groups may be varied, several generalizations of von Staudt's theorem will be given. The proofs of the theorems have interesting corollaries.

The paper contains also some fundamental results which will be useful in succeeding papers.

The interest of the paper seems to be not only geometric but also algebraic, for example we shall establish the following purely algebraic theorem:

Let $\mathfrak{M}$ be the module formed by n-rowed symmetric matrices over the complex field. Let Γ be a continuous (additive) automorphism of $\mathfrak{M}$ leaving the rank unaltered and $\Gamma(iX) = i\Gamma(X)$. Then Γ is an inner automorphism of $\mathfrak{M}$, that is, we have a nonsingular matrix T such that

$$\Gamma(X) = TXT'.$$

The author makes the paper self-contained in the sense that no knowledge of the author's contributions to the theory of automorphic functions is assumed.

---

* Presented to the Society, April 28, 1945; received by the editors November 20, 1944. Reprinted from Transactions of the *American Mathematical Society*, 1945, **57**: 441-481.

I. Geometry of symmetric matrices

Let Φ be any field. In I, II, and III, capital Latin letters denote $n \times n$ matrices unless the contrary is stated. But on the contrary, we use $M^{(n,m)}$ to denote an $n \times m$ matrix, and $M^{(n)} = M^{(n,n)}$. I and 0 denote the identity and zero matrices respectively.

Throughout I, we use

$$\mathfrak{F} = \begin{pmatrix} 0 & I \\ -I & 0 \end{pmatrix}, \quad \mathfrak{I} = \begin{pmatrix} I & 0 \\ 0 & I \end{pmatrix},$$

which are $2n$-rowed matrices.

1. Definitions

We make the following definitions.

A pair of matrices (Z_1, Z_2) is said to be *symmetric* if

$$(Z_1, Z_2)\mathfrak{F}(Z_1, Z_2)' = 0,$$

that is, if $Z_1 Z_2' = Z_2 Z_1'$. The pair is said to be *nonsingular* if (Z_1, Z_2) is of rank n.

A $2n \times 2n$ matrix $\mathfrak{T}$ is said to be *symplectic* if

$$\mathfrak{T}\mathfrak{F}\mathfrak{T}' = \mathfrak{F}.$$

Explicitly, let

$$\mathfrak{T} = \begin{pmatrix} A & B \\ C & D \end{pmatrix},$$

then we have

$$AB' = BA', \quad CD' = DC', \quad AD' - BC' = I.$$

Further, it may be easily verified that

$$\mathfrak{T}^{-1} = \begin{pmatrix} D' & -B' \\ -C' & A' \end{pmatrix}$$

is also symplectic.

We define

$$(W_1, W_2) = Q(Z_1, Z_2)\mathfrak{T}$$

to be a *symplectic transformation*, where Q is nonsingular and $\mathfrak{T}$ is symplectic.

Since
$$(W_1, W_2)\mathfrak{F}(W_1, W_2)' = Q(Z_1, Z_2)\mathfrak{T}\mathfrak{F}\mathfrak{T}'(Z_1, Z_2)'Q',$$
a symplectic transformation carries symmetric (nonsingular) pairs into symmetric (nonsingular) pairs.

We identify two nonsingular symmetric pairs of matrices (Z_1, Z_2) and (W_1, W_2) by means of the relation
$$(Z_1, Z_2) = Q(W_1, W_2).$$
It is called a point of the space. The space so defined is unaltered under symplectic transformations, which may be considered as the motions of the space.

If Z_1 and W_1 are both nonsingular and if $(W_1, W_2) = Q(Z_1, Z_2)\mathfrak{T}$, let
$$W = -W_1^{-1}W_2, \quad Z = -Z_1^{-1}Z_2,$$
then W and Z are both symmetric and
$$Z = (AW + B)(CW + D)^{-1}.$$
Thus a symmetric pair of matrices may be considered as homogeneous coordinates of a symmetric matrix. The terminology "geometry of symmetric matrices" is thus justified.

2. Equivalence of points

Theorem 1 *Any two nonsingular symmetric pairs of matrices are equivalent. Or what is the same thing: every nonsingular symmetric pair is equivalent to $(I, 0)$.*

Proof Let (Z_1, Z_2) be a nonsingular symmetric pair.

(1) If Z_1 is nonsingular, we have
$$(Z_1, Z_2) = Z_1(I, Z_1^{-1}Z_2) = Z_1(I, 0)\begin{pmatrix} I & S \\ 0 & I \end{pmatrix},$$
where $S = Z_1^{-1}Z_2$ is symmetric, and then
$$\begin{pmatrix} I & S \\ 0 & I \end{pmatrix}$$
is symplectic.

(2) Suppose Z_1 to be singular. We have nonsingular matrices P and Q such that
$$W_1 = PZ_1Q = \begin{pmatrix} I^{(r)} & 0^{(r,n-r)} \\ 0^{(n-r,r)} & 0^{(n-r)} \end{pmatrix}$$

and
$$(W_1, W_2) = P(Z_1, Z_2) \begin{pmatrix} Q & 0 \\ 0 & Q'^{-1} \end{pmatrix}$$
and
$$W_2 = P Z_2 Q'^{-1} = \begin{pmatrix} s^{(r)} & m^{(r,n-r)} \\ q^{(n-r,r)} & t^{(n-r)} \end{pmatrix}.$$

Since
$$\begin{pmatrix} Q & 0 \\ 0 & Q'^{-1} \end{pmatrix}$$
is symplectic, (W_1, W_2) is nonsingular and symmetric. Consequently s is symmetric and q is a zero matrix.

Let
$$(U_1, U_2) = (W_1, W_2) \begin{pmatrix} I & -S \\ 0 & I \end{pmatrix},$$
where
$$S = \begin{pmatrix} s^{(r)} & 0 \\ 0 & I^{(n-r)} \end{pmatrix}.$$

Then
$$U_1 = W_1, \quad U_2 = -W_1 S + W_2 = \begin{pmatrix} 0 & m \\ 0 & t \end{pmatrix}.$$

Since (U_1, U_2) is nonsingular, $t^{(n-r)}$ is nonsingular. Let
$$(V_1, V_2) = (U_1, U_2) \begin{pmatrix} I & 0 \\ I & I \end{pmatrix},$$
then
$$V_1 = U_1 + U_2 = \begin{pmatrix} I^{(r)} & m \\ 0 & t \end{pmatrix},$$
which is nonsingular. By (1), we have the theorem.

3. Equivalence of point-pairs

Definition Let (Z_1, Z_2) and (W_1, W_2) be two nonsingular symmetric pairs of matrices. We define the rank of
$$(Z_1, Z_2) \mathfrak{F} (W_1, W_2)' = Z_1 W_2' - Z_2 W_1'$$

to be the *arithmetic distance* between the two points represented. Evidently, the notion is independent of the choice of representation. Further, it is invariant under symplectic transformations. In fact, let

$$(Z_1^*, Z_2^*) = Q(Z_1, Z_2)\mathfrak{T}, \quad (W_1^*, W_2^*) = R(W_1, W_2)\mathfrak{T},$$

then

$$(Z_1^*, Z_2^*)\mathfrak{F}(W_1^*, W_2^*)' = Q(Z_1, Z_2)\mathfrak{T}\mathfrak{F}\mathfrak{T}'(W_1, W_2)'R' = Q(Z_1, Z_2)\mathfrak{F}(W_1, W_2)'R'.$$

In nonhomogeneous coordinates, the arithmetic distance between two symmetric matrices W, Z is equal to the rank of $W - Z$.

Theorem 2 *Two point-pairs are equivalent if and only if they have the same arithmetic distance. What is the same thing: every point-pair with arithmetic distance r is equivalent to*

$$(I, 0), \quad (I, I_r),$$

where

$$I_r = \begin{pmatrix} I^{(r)} & 0 \\ 0 & 0 \end{pmatrix}.$$

Proof By Theorem 1, we may assume that the point-pairs are of the form

$$(I, 0), \quad (Z_1, Z_2).$$

The arithmetic distance being r, it follows that Z_2 is of rank r. We have two nonsingular matrices P and Q such that

$$QZ_2P = \begin{pmatrix} I^{(r)} & 0 \\ 0 & 0 \end{pmatrix} = I_r.$$

Then

$$Q(Z_1, Z_2)\begin{pmatrix} P'^{-1} & 0 \\ 0 & P \end{pmatrix} = (T, I_r)$$

and

$$Q(I, 0)\begin{pmatrix} P'^{-1} & 0 \\ 0 & P \end{pmatrix} = QP'^{-1}(I, 0).$$

Since (T, I_r) is a nonsingular symmetric pair, we have, consequently,

$$T = \begin{pmatrix} s^{(r)} & t \\ 0 & p^{(n-r)} \end{pmatrix},$$

where s is symmetric and p is nonsingular. Then

$$\begin{pmatrix} I^{(r)} & -tp^{-1} \\ 0 & p^{-1} \end{pmatrix} (T, I_r) = \left(\begin{pmatrix} s^{(r)} & 0 \\ 0 & I^{(n-r)} \end{pmatrix}, I_r \right).$$

Further

$$\left(\begin{pmatrix} s^{(r)} & 0 \\ 0 & I^{(n-r)} \end{pmatrix}, I_r \right) \begin{pmatrix} I & 0 \\ \begin{pmatrix} I-s & 0 \\ 0 & 0 \end{pmatrix} & I \end{pmatrix} = (I, I_r)$$

and

$$(I, 0) \begin{pmatrix} I & 0 \\ \begin{pmatrix} I-s & 0 \\ 0 & 0 \end{pmatrix} & I \end{pmatrix} = (I, 0).$$

Since

$$\begin{pmatrix} I & 0 \\ \begin{pmatrix} I-s & 0 \\ 0 & 0 \end{pmatrix} & I \end{pmatrix}$$

is symplectic, we have the result.

Definition The points (X_1, X_2) with singular X_1 are called *points at infinity* (or symmetric matrices at infinity). Finite points are those with nonsingular X_1.

Lemma *Any finite number of points may be carried simultaneously into finite points by a symplectic transformation, if Φ is the field of complex numbers.*

Proof (1) Given any symmetric pair of matrices (T_1, T_2), we have a symplectic matrix

$$\begin{pmatrix} P_1 & P_2 \\ T_1 & T_2 \end{pmatrix}.$$

In fact, by Theorem 2, we have a symplectic $\mathfrak{T}$ such that

$$(T_1, T_2) = Q(-I, 0)\mathfrak{T}.$$

Let

$$(P_1, P_2) = Q'^{-1}(0, I)\mathfrak{T}.$$

Then

$$\begin{pmatrix} P_1 & P_2 \\ T_1 & T_2 \end{pmatrix} = \begin{pmatrix} Q'^{-1} & 0 \\ 0 & Q \end{pmatrix} \begin{pmatrix} 0 & I \\ -I & 0 \end{pmatrix} \mathfrak{T},$$

which is evidently symplectic.

(2) For a fixed point (X_1, X_2), the manifold

$$\det((X_1, X_2)\mathfrak{F}(Z_1, Z_2)') = 0$$

is of dimension $n(n+1) - 2$. Let

$$(A_1, A_2), \cdots, (L_1, L_2)$$

be p given points. Then we have p manifolds

$$\det((A_1, A_2)\mathfrak{F}(Z_1, Z_2)') = 0, \cdots, \det((L_1, L_2)\mathfrak{F}(Z_1, Z_2)') = 0.$$

In the space, there is a point (T_1, T_2) which is not on any one of the manifolds. The transformation

$$(Y_1, Y_2) = Q(X_1, X_2) \begin{pmatrix} P_1 & P_2 \\ T_1 & T_2 \end{pmatrix}^{-1} = Q(X_1, X_2) \begin{pmatrix} T_2' & -P_2' \\ -T_1' & P_1' \end{pmatrix}$$

carries evidently the p points into finite points simultaneously.

4. Equivalence of triples of points

Definition 1 A subspace is said to be *normal* if it is equivalent to the subspace formed by symmetric matrices (in nonhomogeneous coordinates) of the form

$$\begin{pmatrix} Z_0^{(r)} & 0 \\ 0 & 0^{(n-r)} \end{pmatrix}.$$

The least possible r is defined to be the *rank of the subspace*.

Definition 2 A triple of points is said to be of *degeneracy* $d = n - r$ if it belongs to a normal subspace of rank r.

Evidently degeneracy is invariant under symplectic transformations.

Theorem 3 *In the complex field, two triples of points are equivalent if and only if they have the same degeneracy and the arithmetic distances between any two corresponding pairs of points are equal.*

Proof Evidently, if two triples are equivalent, they have the same degeneracy and the arithmetic distances between any two corresponding pairs of points are equal.

We prove the converse in six steps.

(1) Every triple with arithmetic distances n, n, r is equivalent to

$$0, \quad I, \quad \begin{pmatrix} -I^{(r)} & 0 \\ 0 & 0 \end{pmatrix} \quad \text{(in nonhomogeneous coordinates)}$$

(notice that now the degeneracy is 0). We use $r(A, B)$ to denote the arithmetic distance between A and B. Let A, B, C be the three points of the triple. Then

$$r(A, B) = r(A, C) = n.$$

By Theorem 2, we may write in homogeneous coordinates

$$A = (I, 0), \quad B = (0, I), \quad C = (Z_1, Z_2).$$

Since $r(A, C) = n$ and Z_2 is nonsingular, we may write C as

$$(S, I),$$

where S is a symmetric matrix of rank r. We have a nonsingular matrix Γ such that

$$\Gamma S \Gamma' = I_r = \begin{pmatrix} I^{(r)} & 0 \\ 0 & 0 \end{pmatrix},$$

then

$$\Gamma \begin{pmatrix} (I, 0) \\ (0, I) \\ (S, I) \end{pmatrix} \begin{pmatrix} \Gamma' & 0 \\ 0 & \Gamma^{-1} \end{pmatrix} = \begin{pmatrix} \Gamma\Gamma'(I, 0) \\ (0, I) \\ (I_r, I) \end{pmatrix}.$$

Thus the triple is equivalent to

$$(I, 0), \quad (0, I), \quad (I_r, I).$$

Since (in the nonhomogeneous coordinate system)

$$0, \ I, \ -I_r$$

is a triple with distances n, n, r, we have the theorem.

(2) Every triple of points with arithmetic distances n, s, t is equivalent to

$$0, \ I, \ \begin{pmatrix} -I^{(p)} & 0 & 0 \\ 0 & 0 & 0 \\ 0 & 0 & I^{(q)} \end{pmatrix},$$

where $p + q = s$, $n - q = t$ (obviously, $s + t \geqslant n$).

In fact, we may assume that

$$A = (I, 0), \quad B = (I, I), \quad C = (Z_1, Z_2).$$

We may determine two nonsingular matrices U, V such that
$$UZ_2V = \begin{pmatrix} I^{(r)} & 0 \\ 0 & 0 \end{pmatrix},$$
where r is the rank of Z_2. If we set
$$G = \begin{pmatrix} V'^{-1} & 0 \\ V - V'^{-1} & V \end{pmatrix},$$
the relations
$$U(I,0)G = UV'^{-1}(I,0),$$
$$U(I,I)G = UV(I,I),$$
$$U(Z_1, Z_2)G = \left(P, \begin{pmatrix} I^{(r)} & 0 \\ 0 & 0 \end{pmatrix}\right)$$
imply that we may assume that
$$Z_1 = P, \quad Z_2 = \begin{pmatrix} I^{(r)} & 0 \\ 0 & 0 \end{pmatrix}.$$
Owing to the symmetry, we have
$$P = \begin{pmatrix} S^{(r)} & W \\ 0 & T \end{pmatrix},$$
where S is symmetric and T is nonsingular. Further, since
$$\begin{pmatrix} I & -WT^{-1} \\ 0 & T^{-1} \end{pmatrix} \left(\begin{pmatrix} S^{(r)} & W \\ 0 & T \end{pmatrix}, \begin{pmatrix} I^{(r)} & 0 \\ 0 & 0 \end{pmatrix} \right) = \left(\begin{pmatrix} S^{(r)} & 0 \\ 0 & I \end{pmatrix}, \begin{pmatrix} I^{(r)} & 0 \\ 0 & 0 \end{pmatrix} \right),$$
we may assume that
$$Z_1 = \begin{pmatrix} S^{(r)} & 0 \\ 0 & I \end{pmatrix}, \quad Z_2 = \begin{pmatrix} I^{(r)} & 0 \\ 0 & 0 \end{pmatrix}.$$
In the normal subspace of rank r, the points $(I^{(r)}, 0^{(r)}), (I^{(r)}, I^{(r)}), (S^{(r)}, I^{(r)})$ are, by (1), equivalent to
$$(I^{(r)}, 0^{(r)}), \quad (I^{(r)}, I^{(r)}), \quad \left(I^{(r)}, \begin{pmatrix} -I^{(p)} & 0 \\ 0 & 0^{(r-p)} \end{pmatrix}\right).$$
Thus, we have, in nonhomogeneous coordinates,
$$\begin{pmatrix} I^{(r)} & 0 \\ 0 & 0^{(n-r)} \end{pmatrix}, \quad \begin{pmatrix} 0^{(r)} & 0 \\ 0 & I^{(n-r)} \end{pmatrix}, \quad \begin{pmatrix} -I^{(p)} & 0 & 0 \\ 0 & 0^{(r-p)} & 0 \\ 0 & 0 & 0^{(n-r)} \end{pmatrix}.$$

The transformation

$$\begin{pmatrix} I^{(r)} & 0 \\ 0 & iI^{(n-r)} \end{pmatrix} \left(Z - \begin{pmatrix} 0^{(r)} & 0 \\ 0 & I^{(n-r)} \end{pmatrix} \right) \begin{pmatrix} I^{(r)} & 0 \\ 0 & iI^{(n-r)} \end{pmatrix} = W$$

carries the three points to the required form.

(3) Now we are going to prove that any three points are equivalent to

$$A = 0, \quad B = b_1 \dotplus \cdots \dotplus b_\lambda, \quad C = c_1 \dotplus \cdots \dotplus c_\lambda{}^{①},$$

where b_ν and c_ν are unit matrices of degree (ν), multiplied with a factor 1, 0, or -1. (1) and (2) are special cases of this. We shall consider another special case with

$$A = 0, \quad B = \begin{pmatrix} 0 & M \\ M' & 0 \end{pmatrix}, \quad C = \begin{pmatrix} 0 & N \\ N' & 0 \end{pmatrix},$$

where

$$M = \begin{pmatrix} 0 & \cdots & 0 \\ & I^{(m)} & \end{pmatrix}, \quad N = \begin{pmatrix} & I^{(m)} & \\ 0 & \cdots & 0 \end{pmatrix}, \quad n = 2m+1.$$

They form a triple with distances $2m, 2m, 2m$.

Now we are going to establish that there exists a symmetric matrix S such that the transformation

$$W = Z(SZ + I)^{-1}$$

will carry the three points to

$$A = 0, \quad B = \begin{pmatrix} 0 & 0 \\ 0 & B_1^{(n-1)} \end{pmatrix}, \quad C = \begin{pmatrix} 1 & 0 \\ 0 & C_1^{(n-1)} \end{pmatrix},$$

where B_1 is nonsingular. In fact S is given by

$$\begin{pmatrix} 1 & 0 & -1 \\ 0 & 0 & 0 \\ -1 & 0 & 1 \end{pmatrix}, \quad \begin{pmatrix} 1 & 0 & 0 & -1 & 0 \\ 0 & 0 & 0 & 0 & 0 \\ 0 & 0 & 0 & 0 & 0 \\ -1 & 0 & 0 & 1 & 0 \\ 0 & 0 & 0 & 0 & 0 \end{pmatrix},$$

and so on. The general form may be obtained easily. Applying the results obtained in (2) to

$$0^{(n-1)}, \quad B_1^{(n-1)}, \quad C_1^{(n-1)},$$

---

① $\dotplus$ and $\sum'$ denote direct sums.

we have the conclusion.

(4) Let B, C be a nonsingular pair of symmetric matrices (in the ordinary sense), that is, we have λ and μ such that

$$\det(\lambda B + \mu C) \neq 0.$$

Suppose C is nonsingular; the conclusion announced in (3) is true by (2). Otherwise ($\lambda \neq 0$) we have Γ such that

$$\Gamma'(\lambda B + \mu C)\Gamma = I,$$

$$\Gamma' C \Gamma = \begin{pmatrix} C_1^{(r)} & 0 \\ 0 & 0 \end{pmatrix}, \quad C_1^{(r)} \text{ nonsingular}.$$

Then

$$\lambda \Gamma' B \Gamma = \begin{pmatrix} I^{(r)} - \mu C_1^{(r)} & 0 \\ 0 & I^{(n-r)} \end{pmatrix}.$$

Applying the results of (2) to

$$0, \quad \frac{1}{\lambda}(I^{(r)} - \mu C_1^{(r)}), \quad C_1^{(r)}$$

and

$$0, \quad \frac{1}{\lambda} I^{(n-r)}, \quad 0,$$

we have the result announced in (3).

(5) Finally, for any pair of symmetric matrices (cf. the lemma of §3)

$$B, \quad C,$$

we have a nonsingular matrix Γ such that

$$\Gamma B \Gamma' = b_1 \dotplus \cdots \dotplus b_\lambda$$

and

$$\Gamma C \Gamma' = c_1 \dotplus \cdots \dotplus c_\lambda,$$

where

$$(b_\nu, c_\nu)$$

is either the pair discussed in (4) or the pair discussed in (3), hence the results in (3).

(6) By a rearrangement and some evident modifications, for a triple of points with degeneracy t, we have

$$A = 0^{(p)} \dotplus 0^{(q)} \dotplus 0^{(r)} \dotplus 0^{(s)} \dotplus 0^{(t)},$$
$$B = I^{(p)} \dotplus 0^{(q)} \dotplus I^{(r)} \dotplus I^{(s)} \dotplus 0^{(t)},$$
$$C = -I^{(p)} \dotplus I^{(q)} \dotplus 0^{(r)} \dotplus I^{(s)} \dotplus 0^{(t)},$$

which is the only possible form. The arithmetic distances between two points are given by
$$a = r(B,C) = p+q+r,$$
$$b = r(C,A) = p+q+s,$$
$$c = r(A,B) = p+r+s.$$

Thus, for given t, a, b, c, if the equations are soluble, the solution is unique. We have therefore the theorem.

The conditions for solubility are
$$n - t \geqslant a, b, c,$$
$$a + b + c \geqslant 2(n-t). \tag{1}$$

In terms of a "triangle" we have the following theorem.

Theorem 4 *A triangle of degeneracy t with sides a, b, c exists if and only if (1) holds. If it exists, it is unique apart from equivalence.*

Incidentally, we have
$$a + b \geqslant 2(n-t) - c \geqslant c,$$
equality holds if and only if $c = a+b = n-t$.

The "triangle-relation"
$$a+b \geqslant c, \quad b+c \geqslant a, \quad c+a \geqslant b$$

does not guarantee the existence of triangles with a given degeneracy, for example, $n=2$, $t=0$, $a=b=c=1$. But we have the following theorem.

Theorem 5 *Given the lengths of three sides a, b, c ($\leqslant n$), where the sum of every two is greater than the third one, there are λ non-equivalent triangles, where*
$$\lambda = \begin{cases} [(a+b+c)/2] - \max(a,b,c) + 1, & \text{for } n \geqslant [(a+b+c)/2]\text{\textcircled{1}}, \\ n - \max(a,b,c) + 1, & \text{for } n < [(a+b+c)/2]. \end{cases}$$

Proof From $a+b \geqslant c$, $b+c \geqslant a$, $c+a \geqslant b$, we have

---

① $[x]$ denotes the integral part of x.

$$a+b+c \geqslant 2\max(a,b,c).$$

There always exists a t such that

$$a+b+c \geqslant 2(n-t) \geqslant 2\max(a,b,c).$$

Then

$$\max(0, n-[(a+b+c)/2]) \leqslant t \leqslant n-\max(a,b,c).$$

Thus, the number of t's is equal to

$$n-\max(a,b,c)-\max(0, n-[(a+b+c)/2])+1$$
$$=\min(n,[(a+b+c)/2])-\max(a,b,c)+1.$$

Corollary 1 If one of the sides is of length n, the triangle is unique.

Corollary 2 If the sum of two sides is equal to the third, then the triangle is unique.

5. Equivalence of quadruples of points

Definition Let Z_1, Z_2, Z_3, Z_4 be four points in the nonhomogeneous coordinate-system. The matrix

$$(Z_1-Z_3)(Z_1-Z_4)^{-1}(Z_2-Z_4)(Z_2-Z_3)^{-1}$$

is defined to be the cross-ratio-matrix of the four points, and it is denoted by

$$(Z_1, Z_2; Z_3, Z_4).$$

It is defined only when $Z_1 - Z_4$ and $Z_2 - Z_3$ are nonsingular.

In the homogeneous coordinate-system, we let P_1, P_2, P_3, P_4 be four points with coordinates

$$(X_1, Y_1), \quad (X_2, Y_2), \quad (X_3, Y_3), \quad (X_4, Y_4).$$

In terms of

$$\langle P_i, P_j \rangle = (X_j, Y_j)\mathfrak{F}(X_i, Y_i)',$$

the cross-ratio-matrix is defined by

$$(P_1, P_2; P_3, P_4) = \langle P_1, P_3 \rangle \langle P_1, P_4 \rangle^{-1} \langle P_2, P_4 \rangle \langle P_2, P_3 \rangle^{-1},$$

provided that it is not meaningless.

Let P_i^* be the point with coordinates
$$(X_i^*, Y_i^*) = Q_i(X_i, Y_i)\mathfrak{T},$$
where $\mathfrak{T}$ is symplectic; then
$$\begin{aligned}\langle P_i^*, P_j^*\rangle &= (X_j^*, Y_j^*)\mathfrak{F}(X_i^*, Y_i^*)' \\ &= Q_j(X_j, Y_j)\mathfrak{F}(X_i, Y_i)'Q_i' = Q_j\langle P_i, P_j\rangle Q_i'.\end{aligned}$$
Therefore
$$\begin{aligned}(P_1^*, P_2^*; P_3^*, P_4^*) &= \langle P_1^*, P_3^*\rangle\langle P_1^*, P_4^*\rangle^{-1}\langle P_2^*, P_4^*\rangle\langle P_2^*, P_3^*\rangle^{-1} \\ &= Q_3\langle P_1, P_3\rangle Q_1' Q_1'^{-1}\langle P_1, P_4\rangle Q_4 Q_4^{-1}\langle P_2, P_4\rangle Q_2' Q_2'^{-1}\langle P_2, P_3\rangle^{-1}Q_3^{-1} \\ &= Q_3(P_1, P_2; P_3, P_4)Q_3^{-1},\end{aligned}$$
and we now state the following theorem.

Theorem 6 *In an algebraically closed field, two quadruples of points, no two of the points having arithmetic distance less than n, are equivalent if and only if their cross-ratio-matrices are equivalent.*

In order to prove Theorem 6, we need to establish the following theorem.

Theorem 7 *In the algebraically closed field, any quadruple of points, no two of which have arithmetic distance less than n, is equivalent to*

$$0, \quad \infty, \quad \sum_{1 \leq i \leq \nu}{}' a_i, \quad \sum_{1 \leq i \leq \nu}{}' b_i,$$

where

$$a_i = \begin{pmatrix} 0 & \cdots & 0 & 1 \\ 0 & \cdots & 1 & 0 \\ \vdots & & \vdots & \vdots \\ 1 & \cdots & 0 & 0 \end{pmatrix}, \quad b_i = \begin{pmatrix} 0 & 0 & \cdots & 0 & \lambda_i \\ 0 & 0 & \cdots & \lambda_i & 1 \\ \vdots & \vdots & & \vdots & \vdots \\ \lambda_i & 1 & \cdots & 0 & 0 \end{pmatrix}, \quad \lambda_i \neq 0 \text{ or } 1.$$

Proof In homogeneous coordinates, we may write the four points as
$$(0, I), \quad (I, 0), \quad (Z_1, Z_2), \quad (W_1, W_2).$$
Since no two of the arithmetic distances are less than n_1, Z_1, Z_2, W_1, W_2 are all nonsingular. We may write them in the nonhomogeneous coordinates as
$$0, \quad \infty, \quad S_1, \quad S_2.$$
We have a nonsingular matrix T such that
$$TS_1 T' = \sum{}' a_i, \quad TS_2 T' = \sum{}' b_i.$$

The theorem follows.

The proof of Theorem 6 is now evident.

Remark The equivalence of quadruples in any field seems to be more difficult. The condition in Theorem 6 is insufficient for the real case (a signature system is required).

Definition We define a quadruple of points satisfying
$$(P_1, P_2; P_3, P_4) = -I$$
to be a *harmonic range*.

Evidently a harmonic range is invariant under a symplectic transformation.

6. Von Staudt's theorem in the complex number field

Now we let Φ be the field formed by complex numbers.

We use $\overline{Z}$ to denote the conjugate complex matrix of Z. The transformation
$$(W_1, W_2) = Q(\overline{Z}_1, \overline{Z}_2)\mathfrak{T}$$
carrying a symmetric pair (W_1, W_2) into a symmetric pair (Z_1, Z_2) is called *anti-symplectic* if Q is nonsingular and $\mathfrak{T}$ symplectic.

Theorem 8 *A transformation satisfying the following conditions:*

(1) *one-to-one and continuous;*

(2) *carrying symmetric matrices into symmetric matrices;*

(3) *keeping arithmetic distance invariant;*

(4) *keeping the harmonic relation invariant*

is either a symplectic or an anti-symplectic transformation.

Proof Let Γ be the transformation considered. Taking three points A, B, C (symmetric matrices), no two of which have arithmetic distance less than n, let A_1, B_1, C_1 be their images. By (3), the arithmetic distance between any two of A_1, B_1, C_1 is n. Let $\mathfrak{T}_1$ and $\mathfrak{T}_2$ be two symplectic transformations carrying respectively A, B, C and A_1, B_1, C_1 into $0, I, \infty$, in accordance with Theorem 3. Then, without loss of generality, we may assume that
$$0 = \Gamma(0), \quad I = \Gamma(I), \quad \infty = \Gamma(\infty).$$

Since
$$Z, \quad Z_1, \quad (Z+Z_1)/2, \quad \infty$$
form a harmonic range, we have
$$\Gamma(Z) + \Gamma(Z_1) = \Gamma(Z+Z_1).$$

Consequently,
$$\Gamma(rZ) = r\Gamma(Z)$$
for all rational r. By continuity, this holds for all real r.

Now we introduce the following notations:
$$E_{ii} = (p_{st}), \quad p_{st} = \begin{cases} 1, & \text{if } s = t = i, \\ 0, & \text{otherwise} \end{cases}$$
and
$$E_{ij} = (q_{st}), \quad q_{st} = \begin{cases} 1, & \text{if } s = i, t = j \text{ or } s = j, t = i, \\ 0, & \text{otherwise.} \end{cases}$$
Let
$$\Gamma(E_{ii}) = M_i.$$
Since M_i is of rank 1 and symmetric, we have
$$M_i = (\lambda_{i1}, \cdots, \lambda_{in})'(\lambda_{i1}, \cdots, \lambda_{in}).$$
Let
$$\Lambda = (\lambda_{ij}).$$
Then
$$I = \Gamma(I) = \sum_{i=1}^{n} \Gamma(E_{ii}) = \sum_{i=1}^{n} M_i$$
$$= \sum_{i=1}^{n} (\lambda_{i1}, \cdots, \lambda_{in})'(\lambda_{i1}, \cdots, \lambda_{in})$$
$$= \sum_{i=1}^{n} (\lambda_{ij}\lambda_{ik}) = \left(\sum_{i=1}^{n} \lambda_{ij}\lambda_{ik} \right)$$
$$= \Lambda'\Lambda.$$

That is, Λ is an orthogonal matrix
$$(\lambda_{i1}, \cdots, \lambda_{in})\Lambda' = (\delta_{i1}, \cdots, \delta_{in}),$$
where δ_{ij} is Kronecker's delta. Thus
$$\Lambda\Gamma(E_{ii})\Lambda' = E_{ii}.$$
Let
$$\Delta(Z) = \Lambda\Gamma(Z)\Lambda',$$
then Δ has the same property as Γ, that is,

$$\Delta(Z+Z_1) = \Delta(Z) + \Delta(Z_1),$$

$$0 = \Delta(0), \quad I = \Delta(I), \quad \infty = \Delta(\infty)$$

and
$$E_{ii} = \Delta(E_{ii}).$$

Let
$$\Delta(E_{ij}) = M = (m_{st}), \quad i \neq j,$$

M is of rank 2. Since
$$E_{ij} + \lambda E_{ii} + E_{jj}/\lambda$$

is of rank 1, owing to the invariance of arithmetic distance, the matrix

$$M + \lambda E_{ii} + E_{jj}/\lambda \tag{1}$$

is also of rank 1 for all λ. We are going to prove that $M = \pm E_{ij}$. In fact, we may assume that $i = 1, j = 2$. The two-rowed minor of (1)

$$\begin{vmatrix} m_{11} + \lambda & m_{12} \\ m_{12} & m_{22} + 1/\lambda \end{vmatrix} = 0$$

for any λ, that is
$$m_{11}m_{22} - m_{12}^2 + m_{11}/\lambda + m_{22}\lambda + 1 = 0,$$

that is
$$m_{11} = m_{22} = 0, \quad m_{12} = \pm 1.$$

Further
$$\begin{vmatrix} m_{11} + \lambda & m_{1t} \\ m_{1t} & m_{tt} \end{vmatrix} = 0, \quad t \geqslant 3,$$

for all λ, then $m_{tt} = 0$, $m_{1t} = 0$ for all $t \geqslant 3$. Finally

$$\begin{vmatrix} 0 & m_{st} \\ m_{st} & 0 \end{vmatrix} = 0 \quad \text{if} \quad (s,t) \neq (1,2),$$

then $m_{st} = 0$ for $(s,t) \neq (1,2)$. Hence, we have
$$M = \pm E_{12}.$$

Thus
$$\Delta(E_{ij}) = \pm E_{ij}.$$

Let
$$D = [\varepsilon_1, \cdots, \varepsilon_n], \quad \varepsilon_p = \pm 1.$$
Then
$$D\Delta(E_{ij})D' = \pm\varepsilon_i\varepsilon_j E_{ij}.$$
Thus we may choose ε properly so that
$$D\Delta(E_{1i})D' = E_{1i}.$$
Let $D\Delta D' = \Pi$. Then Π has all properties of Δ and further
$$\Pi(E_{1i}) = E_{1i}.$$

Now we consider E_{ij}. Without loss of generality we take $(i, j) = (2, 3)$. Then, if $\Pi(E_{23}) = -E_{23}$, we have
$$\Pi(E_{11} + E_{33} + E_{12} + E_{13} + E_{23}) = E_{11} + E_{33} + E_{12} + E_{13} - E_{23},$$
since
$$\begin{vmatrix} 1 & 1 & 1 \\ 1 & 0 & \varepsilon \\ 1 & \varepsilon & 1 \end{vmatrix} = -(\varepsilon - 1)^2,$$
which is equal to zero for $\varepsilon = 1$ and not zero for $\varepsilon = -1$. Consequently the ranks of $E_{11} + E_{33} + E_{12} + E_{13} + E_{23}$ and $E_{11} + E_{33} + E_{12} + E_{13} - E_{23}$ are not equal. This is impossible.

Thus, we have
$$\Pi(X) = X$$
for all real X (if we do not use continuity, it holds for all rational X). We may assume Γ to be Π.

Further for real Y, the four points $Yi, -Yi, Y, -Y$ form a harmonic range, while
$$\Gamma(Y) = Y, \quad \Gamma(-Y) = -Y,$$
thus we have
$$\Gamma(iY)Y^{-1} = -Y\Gamma(iY)^{-1}.$$
In particular,
$$(\Gamma(iI))^2 = -I.$$

Then
$$\Gamma(iI) = iJ,$$
where J is an involutory symmetric matrix, that is $J^2 = I$ and $J = J'$. We have a matrix T (not necessarily orthogonal) such that
$$J = T'T.$$
Let
$$T'^{-1}\Gamma(Z)T^{-1} = \Phi(Z),$$
we have then
$$\Phi(iI) = iI.$$
Let
$$\nabla(Z) = -i\Phi(iZ).$$
Then
$$\nabla(0) = 0, \quad \nabla(I) = I, \quad \nabla(\infty) = \infty,$$
so the ranks of Z and $\nabla(Z)$ are equal. By the method used before, we have
$$\nabla_1 = B'\nabla B,$$
such that
$$\nabla_1(X) = X,$$
for all real X. Thus, we have finally that
$$\Gamma(X + iY) = \Gamma(X) + \Gamma(iY) = X + iA'YA,$$
where A is independent of X and Y.

Now we have
$$A'YA \cdot Y^{-1} = Y(A'YA)^{-1},$$
that is,
$$(A'YAY^{-1})^2 = I,$$
for all real Y. Here we introduce a lemma.

Lemma *Let A be a nonsingular matrix. If*
$$(A'YAY^{-1})^2 = I$$
for all symmetric Y then $A = \rho I$, where $\rho = \pm 1$ or $\pm i$.

If the lemma is true, then
$$\Gamma(X+iY)=X+iY \text{ or } X-iY.$$
The theorem is proved.

Proof of the lemma (1) We have a nonsingular matrix Γ such that
$$\Gamma^{-1}A\Gamma = B$$
and
$$B = J_1 \dotplus \cdots \dotplus J_\nu,$$
where J_i is a Jordan matrix of degree n_i. Evidently
$$(B'(\Gamma'Y\Gamma)B(\Gamma'Y\Gamma)^{-1})^2 = \Gamma'(A'YAY^{-1})^2\Gamma'^{-1} = I.$$
Thus it is sufficient to prove the theorem for B instead of A.

(2) We shall prove $n_i = 1$. In fact, if
$$J_i = \begin{pmatrix} \lambda & 1 & \cdots & 0 \\ 0 & \lambda & \cdots & 0 \\ \vdots & \vdots & & \vdots \\ 0 & 0 & \cdots & \lambda \end{pmatrix},$$
then
$$(J_i'IJ_iI)^2 = \begin{pmatrix} \lambda^2 & \lambda & 0 & \cdots & 0 \\ \lambda & 1+\lambda^2 & \lambda & \cdots & 0 \\ \vdots & \vdots & \vdots & & \vdots \\ 0 & 0 & 0 & \cdots & 1+\lambda^2 \end{pmatrix}^2 = \begin{pmatrix} \lambda^2(\lambda^2+1) & 2\lambda^3+\lambda & \lambda^2 & \cdots \\ \cdot & \cdot & \cdot & \cdots \\ \cdot & \cdot & \cdot & \cdots \\ \cdot & \cdot & \cdot & \cdots \end{pmatrix} = I,$$
which is impossible. Thus $n_i = 1$, that is,
$$B = [\varepsilon_1, \cdots, \varepsilon_n], \quad \varepsilon_\nu \neq 0,$$
which is a diagonal matrix.

(3) Putting
$$Y = \begin{pmatrix} 1 & 1 & 0 & 0 & \cdots & 0 \\ 1 & 0 & 0 & 0 & \cdots & 0 \\ 0 & 0 & 1 & 0 & \cdots & 0 \\ 0 & 0 & 0 & 1 & \cdots & 0 \\ \vdots & \vdots & \vdots & \vdots & & \vdots \\ 0 & 0 & 0 & 0 & \cdots & 1 \end{pmatrix},$$

we have

$$\left(\begin{pmatrix}\varepsilon_1 & 0 \\ 0 & \varepsilon_2\end{pmatrix}\begin{pmatrix}1 & 1 \\ 1 & 0\end{pmatrix}\begin{pmatrix}\varepsilon_1 & 0 \\ 0 & \varepsilon_2\end{pmatrix}\begin{pmatrix}0 & 1 \\ 1 & -1\end{pmatrix}\right)^2 = \begin{pmatrix}\varepsilon_1^2\varepsilon_2^2 & 2\varepsilon_1^2\varepsilon_2(\varepsilon_1-\varepsilon_2) \\ 0 & \varepsilon_1^2\varepsilon_2^2\end{pmatrix} = \begin{pmatrix}1 & 0 \\ 0 & 1\end{pmatrix},$$

which implies

$$\varepsilon_1 = \varepsilon_2.$$

Similarly

$$B = \varepsilon I, \quad \varepsilon \neq 0.$$

Then $A = \varepsilon I$. For $Y = I$, we have $\varepsilon^4 = 1$, that is $\varepsilon = \pm 1, \pm i$. The lemma is thus completely proved.

7. Remarks

The following results are contained in the proof of Theorem 8:

Theorem 9 *Let Φ be the complex field, and let $\mathfrak{M}$ be a module formed by symmetric matrices over Φ. Let Γ be an additive continuous automorphism of $\mathfrak{M}$ leaving the rank invariant, so that Γ satisfies:*

(i) $\Gamma(X) \in \mathfrak{M}$, *if* $X \in \mathfrak{M}$;
(ii) $\Gamma(X+Y) = \Gamma(X) + \Gamma(Y)$, *if* $X \in \mathfrak{M}$ *and* $Y \in \mathfrak{M}$;
(iii) $\Gamma(iX) = i\Gamma(X)$; *and*
(iv) $\Gamma(X)$ *has the same rank as* X.

Then $\Gamma(X)$ is an inner automorphism, that is

$$\Gamma(X) = TXT'$$

for certain T.

In the case of the real field the situation is more complicated. In Theorem 9, we require an additional condition that the signature of $\Gamma(X)$ is the same as that of X.

The analogue of Theorem 8 in the real field is more complicated. Since Theorem 7 is not true in case of the real field, degeneracy and lengths of sides do not characterize the equivalence of triples of points, for example, there does not exist a real symplectic transformation Γ satisfying

$$\Gamma(0) = 0, \quad \Gamma(\infty) = \infty, \quad \Gamma\begin{pmatrix}2 & 0 \\ 0 & -2\end{pmatrix} = \begin{pmatrix}-2 & 0 \\ 0 & -2\end{pmatrix}.$$

In fact the transformation satisfying the first two of these relations is of the form

$$\Gamma(Z) = CZC',$$

where C is nonsingular and real. It keeps the signature invariant. By means of the signature of a triple, we may obtain an analogue of Theorem 9 in the real field.

II. Geometry of skew-symmetric matrices

Throughout II, we use
$$\mathfrak{F}_1 = \begin{pmatrix} 0 & I \\ I & 0 \end{pmatrix}, \quad \mathfrak{I} = \begin{pmatrix} I & 0 \\ 0 & I \end{pmatrix}.$$

We let $n = 2m$.

8. Notions

A pair of matrices (Z_1, Z_2) is said to be *skew-symmetric* if
$$(Z_1, Z_2)\mathfrak{F}_1(Z_1, Z_2)' = 0,$$
that is,
$$Z_1 Z_2' = -Z_2 Z_1'.$$

A $2n \times 2n$ matrix $\mathfrak{T}$ is said to be $\mathfrak{F}_1$-orthogonal, if
$$\mathfrak{T}\mathfrak{F}_1\mathfrak{T}' = \mathfrak{F}_1.$$

We define
$$(W_1, W_2) = Q(Z_1, Z_2)\mathfrak{T}$$
to be an $\mathfrak{F}_1$-orthogonal transformation if Q is nonsingular.

The transformation carries nonsingular skew-symmetric pairs into nonsingular skew-symmetric pairs.

The nonsingular skew-symmetric pair of matrices may be considered as the homogeneous coordinates of a skew-symmetric matrix.

It is easy to verify that the geometry so obtained (analogous to I) is *transitive*, that is, any two points of the space are equivalent.

We define the rank of
$$(Z_1, Z_2)\mathfrak{F}_1(W_1, W_2)'$$
to be the *arithmetic distance* between the two points represented by (Z_1, Z_2) and (W_1, W_2).

We have also:

Two point-pairs are equivalent if and only if they have the same arithmetic distance.

We may also define the cross-ratio-matrix of four points P_1, P_2, P_3, P_4, $P_i = (X_i, Y_i)$ $(i = 1, 2, 3, 4)$ by

$$\langle P_1, P_3 \rangle \langle P_1, P_4 \rangle^{-1} \langle P_2, P_4 \rangle \langle P_2, P_3 \rangle^{-1},$$

where

$$\langle P_i, P_j \rangle = (X_j, Y_j) \mathfrak{F}_1 (X_i, Y_i)'.$$

The analogue of Theorem 6 is also true.

If the cross-ratio-matrix is equal to $-I$, we define P_1, P_2, P_3, P_4 to be a harmonic range.

9. An algebraic theorem

On the ground of similarity, the following statement seems to be true.

Let Γ be a continuous (additive) automorphism of the module formed by all skew-symmetric matrices, such that $\Gamma(iX) = i\Gamma(X)$, and that the rank is left invariant. Then $\Gamma(X)$ is an inner automorphism.

Unfortunately, this statement is false and so the situation becomes more complicated. For $n = 2$,

$$\Gamma \begin{pmatrix} 0 & a & b & c \\ -a & 0 & d & e \\ -b & -d & 0 & f \\ -c & -e & -f & 0 \end{pmatrix} = \begin{pmatrix} 0 & a & b & d \\ -a & 0 & c & e \\ -b & -c & 0 & f \\ -d & -e & -f & 0 \end{pmatrix}$$

is an automorphism but not an inner automorphism.

It is an automorphism of the required kind, since the principal minors form equal sets, say

$$(af - be + cd)^2;\ a^2,\ b^2,\ c^2,\ d^2,\ e^2,\ f^2.$$

It is not an inner automorphism. In fact, we write

$$\Gamma \begin{pmatrix} P & Q \\ -Q' & R \end{pmatrix} = \begin{pmatrix} P & Q' \\ -Q & R \end{pmatrix},$$

where P and R are two-rowed skew-symmetric matrices. Suppose it is an inner automorphism, that is, that there exists a nonsingular matrix

$$\begin{pmatrix} A & B \\ C & D \end{pmatrix},$$

such that

$$\begin{pmatrix} A & B \\ C & D \end{pmatrix} \begin{pmatrix} P & Q \\ -Q' & R \end{pmatrix} \begin{pmatrix} A & B \\ C & D \end{pmatrix}' = \begin{pmatrix} P & Q' \\ -Q & R \end{pmatrix}, \qquad (1)$$

for all P, Q, R.

In particular, if $P = R = 0$, $Q = I$, we have

$$\begin{pmatrix} A & B \\ C & D \end{pmatrix} \begin{pmatrix} 0 & I \\ -I & 0 \end{pmatrix} \begin{pmatrix} A & B \\ C & D \end{pmatrix}' = \begin{pmatrix} 0 & I \\ -I & 0 \end{pmatrix}. \qquad (2)$$

Combining (1) and (2), we have

$$\begin{pmatrix} A & B \\ C & D \end{pmatrix} \begin{pmatrix} Q & -P \\ R & Q' \end{pmatrix} = \begin{pmatrix} Q' & -P \\ R & Q \end{pmatrix} \begin{pmatrix} A & B \\ C & D \end{pmatrix}.$$

Putting $P = R = 0$, we have

$$AQ = Q'A, \quad BQ' = Q'B, \quad CQ = QC, \quad DQ' = QD$$

for any Q. Consequently, we obtain

$$B = \beta I, \quad C = \gamma I, \quad A = D = 0.$$

But

$$\begin{pmatrix} 0 & \beta I \\ \gamma I & 0 \end{pmatrix} \begin{pmatrix} P & Q \\ -Q' & R \end{pmatrix} \begin{pmatrix} 0 & \gamma I \\ \beta I & 0 \end{pmatrix} = \begin{pmatrix} \beta^2 R & -\beta\gamma Q' \\ \beta\gamma Q & \gamma^2 P \end{pmatrix},$$

which, in general, is not equal to

$$\begin{pmatrix} P & Q' \\ -Q & R \end{pmatrix}.$$

Thus the automorphism is *not* an inner automorphism.

The above argument suggests that in general we might have $m - 1$ basic automorphisms:

(i) $a_{14} \rightleftarrows a_{23}$, other elements invariant;
(ii) $a_{14} \rightleftarrows a_{23}$, $a_{16} \rightleftarrows a_{25}$, other elements invariant;
(iii) $a_{14} \rightleftarrows a_{23}$, $a_{16} \rightleftarrows a_{25}$, $a_{18} \rightleftarrows a_{27}$, other elements invariant;

$\cdots\cdots\cdots$

$(m-1)$ $a_{14} \rightleftarrows a_{23}, \cdots, a_{1,2m} \rightleftarrows a_{2,2m-1}$, other elements invariant.

Such a reasonable suggestion is a false one, since for $m \geqslant 3$, "$a_{14} \rightleftarrows a_{23}$" does not keep the rank invariant, for example,

$$\Gamma \begin{pmatrix} 0 & 0 & 0 & 0 & 1 & 0 \\ 0 & 0 & 1 & 0 & 0 & 1 \\ 0 & -1 & 0 & 0 & -1 & 0 \\ 0 & 0 & 0 & 0 & 0 & 0 \\ -1 & 0 & 1 & 0 & 0 & 1 \\ 0 & -1 & 0 & 0 & -1 & 0 \end{pmatrix} = \begin{pmatrix} 0 & 0 & 0 & 1 & 1 & 0 \\ 0 & 0 & 0 & 0 & 0 & 1 \\ 0 & 0 & 0 & 0 & -1 & 0 \\ -1 & 0 & 0 & 0 & 0 & 0 \\ -1 & 0 & 1 & 0 & 0 & 1 \\ 0 & -1 & 0 & 0 & -1 & 0 \end{pmatrix}.$$

Here one matrix is singular while the other is nonsingular.

Theorem 10 Let Φ be the field of complex numbers. Let $\mathfrak{M}$ be the module formed by all skew-symmetric matrices over Φ. Let Γ be a continuous (additive) automorphism of Φ leaving the rank invariant and $\Gamma(iX) = i\Gamma(X)$. Then, for $m \neq 2$, Γ is an inner automorphism. For $m = 2$ there exists a nonsingular matrix T such that

$$\Gamma(X) = TX_\nu T',$$

where X_ν is either X or

$$\begin{pmatrix} 0 & a_{12} & a_{13} & a_{23} \\ -a_{12} & 0 & a_{14} & a_{24} \\ -a_{13} & -a_{14} & 0 & a_{34} \\ -a_{23} & -a_{24} & -a_{34} & 0 \end{pmatrix}.$$

Proof (i) Evidently, the automorphisms

$$Y = TX_\nu T'$$

satisfy the requirement, where T is nonsingular.

(ii) The additive property may be stated as

$$\Gamma(X+Y) = \Gamma(X) + \Gamma(Y), \tag{1}$$

for any two X and Y belonging to $\mathfrak{M}$. Putting $X = Y = 0$, we have

$$\Gamma(0) = 0. \tag{2}$$

It is also very easy to deduce that

$$\Gamma(rX) = r\Gamma(X) \tag{3}$$

for any rational r. By continuity, it holds for any real r. Since $\Gamma(iX) = i\Gamma(X)$, the relation holds for all complex r.

Let
$$A = \Gamma\left(\begin{pmatrix} 0 & 1 \\ -1 & 0 \end{pmatrix} \dotplus \cdots \dotplus \begin{pmatrix} 0 & 1 \\ -1 & 0 \end{pmatrix}\right),$$
where A is a nonsingular skew-symmetric matrix. There exists a matrix Q such that
$$QAQ' = \begin{pmatrix} 0 & 1 \\ -1 & 0 \end{pmatrix} \dotplus \cdots \dotplus \begin{pmatrix} 0 & 1 \\ -1 & 0 \end{pmatrix}.$$

Let $\Gamma_1(X) = Q\Gamma(X)Q'$, then Γ_1 is an automorphism satisfying the properties given in the theorem and
$$\Gamma_1(J) = J, \tag{4}$$
where
$$J = \begin{pmatrix} 0 & 1 \\ -1 & 0 \end{pmatrix} \dotplus \cdots \dotplus \begin{pmatrix} 0 & 1 \\ -1 & 0 \end{pmatrix}.$$

Write Γ instead of Γ_1 (in the following we shall repeat this procedure by the simple statement "we may let Γ satisfy (4)").

(iii) Let $\lambda_1, \cdots, \lambda_m$ be m distinct numbers, and let
$$A = \begin{pmatrix} 0 & \lambda_1 \\ -\lambda_1 & 0 \end{pmatrix} \dotplus \cdots \dotplus \begin{pmatrix} 0 & \lambda_m \\ -\lambda_m & 0 \end{pmatrix}.$$

Consider the two pairs of matrices A, J and $\Gamma(A)$, J. Since $\Gamma(A - \lambda J) = \Gamma(A) - \lambda J$, the characteristic roots of $\Gamma(A - \lambda J)$ are also $\lambda_1, \cdots, \lambda_m$ (each is a double root). We have a nonsingular matrix M such that
$$M\Gamma(A)M' = A, \quad MJM' = J. \tag{5}$$

Now we are going to prove that M can be chosen independent of the λ's. Write $M = M_{\lambda_1, \cdots, \lambda_m}$. We have
$$\Gamma\left(\begin{pmatrix} 0 & 0 \\ 0 & 0 \end{pmatrix} \dotplus \cdots \dotplus \begin{pmatrix} 0 & 1 \\ -1 & 0 \end{pmatrix} \dotplus \cdots \dotplus \begin{pmatrix} 0 & 0 \\ 0 & 0 \end{pmatrix}\right)$$
$$= M_i \left(\begin{pmatrix} 0 & 0 \\ 0 & 0 \end{pmatrix} \dotplus \cdots \dotplus \begin{pmatrix} 0 & 1 \\ -1 & 0 \end{pmatrix} \dotplus \cdots \dotplus \begin{pmatrix} 0 & 0 \\ 0 & 0 \end{pmatrix}\right) M_i'^{①},$$

---

① The term different from the zero-matrix is the ith term of the sum.

where $M_i = M_{\lambda_1,\cdots,\lambda_m}$ with $\lambda_i = 1$ and $\lambda_j = 0$ for $j \neq i$. In this expression, only the $(2i-1)$th and $2i$th columns are significant. Let P be a matrix having $(2i-1)$th and $2i$th columns in common with M_i for $i = 1, 2, \cdots, n$. Then

$$\Gamma(A) = PAP'.$$

Putting $\lambda_1 = \cdots = \lambda_m = 1$, we find that P is nonsingular.

Now we may let

$$\Gamma(A) = A, \tag{6}$$

where

$$A = \begin{pmatrix} 0 & \lambda_1 \\ -\lambda_1 & 0 \end{pmatrix} \dotplus \cdots \dotplus \begin{pmatrix} 0 & \lambda_m \\ -\lambda_m & 0 \end{pmatrix}.$$

(iv) The theorem is evident for $m = 1$.

Now we take $m = 2$. Let

$$\Gamma \begin{pmatrix} 0 & a & b & c \\ -a & 0 & d & e \\ -b & -d & 0 & f \\ -c & -e & -f & 0 \end{pmatrix} = \begin{pmatrix} 0 & a' & b' & c' \\ -a' & 0 & d' & e' \\ -b' & -d' & 0 & f' \\ -c' & -e' & -f' & 0 \end{pmatrix}.$$

Since

$$\begin{vmatrix} 0 & a-\lambda & b & c \\ -a+\lambda & 0 & d & e \\ -b & -d & 0 & f-\mu \\ -c & -e & -f+\mu & 0 \end{vmatrix} = 0,$$

that is, $(a-\lambda)(f-\mu) - be + dc = 0$, if and only if

$$\begin{vmatrix} 0 & a'-\lambda & b' & c' \\ -a'+\lambda & 0 & d' & e' \\ -b' & -d' & 0 & f'-\mu \\ -c' & -e' & -f'+\mu & 0 \end{vmatrix} = 0,$$

we have

$$a = a', \quad f = f', \tag{7}$$

$$be - cd = b'e' - c'd'. \tag{8}$$

Now we consider

$$\Gamma \begin{pmatrix} \cdot & 1 & 0 \\ & 0 & 1 \\ * & & \cdot \end{pmatrix} = \begin{pmatrix} \cdot & & M \\ * & & \cdot \end{pmatrix},$$

$$\Gamma\begin{pmatrix} \cdot & 1 & 0 \\ & 0 & 0 \\ * & \cdot & \end{pmatrix} = \begin{pmatrix} \cdot & M_1 \\ * & \cdot \end{pmatrix},$$

where "$\cdot$" stands for zero-matrix and "$*$" stands for a matrix which either is evident or has no essential significance in the consideration (this convention will be retained for the rest of the paper). Then M is of rank 2, M_1 and $M - M_1$ are of ranks less than or equal to 1, that is

$$|M_1 - \lambda M| = 0$$

has two characteristic roots 0 and 1. We can find two matrices P and Q of determinant 1, such that

$$PMQ = \begin{pmatrix} 1 & 0 \\ 0 & 1 \end{pmatrix}, \quad PM_1Q = \begin{pmatrix} 1 & 0 \\ 0 & 0 \end{pmatrix}.$$

Therefore

$$\begin{pmatrix} P & 0 \\ 0 & Q' \end{pmatrix}\begin{pmatrix} \cdot & M \\ * & \cdot \end{pmatrix}\begin{pmatrix} P' & 0 \\ 0 & Q \end{pmatrix} = \begin{pmatrix} \cdot & 1 & 0 \\ & 0 & 1 \\ * & \cdot & \end{pmatrix}$$

and

$$\begin{pmatrix} P & 0 \\ 0 & Q' \end{pmatrix}\begin{pmatrix} \cdot & M_1 \\ * & \cdot \end{pmatrix}\begin{pmatrix} P' & 0 \\ 0 & Q \end{pmatrix} = \begin{pmatrix} \cdot & 1 & 0 \\ & 0 & 0 \\ * & \cdot & \end{pmatrix}.$$

Since

$$\begin{pmatrix} P & 0 \\ 0 & Q' \end{pmatrix}\begin{pmatrix} 0 & \lambda & & \cdot \\ -\lambda & 0 & & \\ & & 0 & \mu \\ \cdot & & -\mu & 0 \end{pmatrix}\begin{pmatrix} P' & 0 \\ 0 & Q \end{pmatrix} = \begin{pmatrix} 0 & \lambda & & \cdot \\ -\lambda & 0 & & \\ & & 0 & \mu \\ \cdot & & -\mu & 0 \end{pmatrix},$$

we may let

$$\Gamma\begin{pmatrix} \cdot & b & 0 \\ & 0 & e \\ * & \cdot & \end{pmatrix} = \begin{pmatrix} \cdot & b & 0 \\ & 0 & e \\ * & \cdot & \end{pmatrix}.$$

We deduce easily that

$$b = b', \quad e = e'. \tag{9}$$

From (8) we have

$$cd = c'd'.$$

In particular, we have

$$\Gamma \begin{pmatrix} \cdot & 0 & 1 \\ \cdot & 1 & 0 \\ * & \cdot & \end{pmatrix} = \begin{pmatrix} \cdot & 0 & c' \\ \cdot & d' & 0 \\ * & \cdot & \end{pmatrix}, \quad c'd' = 1.$$

Since

$$\begin{pmatrix} (d')^{1/2} & 0 & 0 & 0 \\ 0 & (d')^{-1/2} & 0 & 0 \\ 0 & 0 & (d')^{-1/2} & 0 \\ 0 & 0 & 0 & (d')^{1/2} \end{pmatrix} \begin{pmatrix} 0 & a & b & c' \\ -a & 0 & d' & e \\ -b & -d' & 0 & f \\ -c' & -e & -f & 0 \end{pmatrix}$$

$$\times \begin{pmatrix} (d')^{1/2} & 0 & 0 & 0 \\ 0 & (d')^{-1/2} & 0 & 0 \\ 0 & 0 & (d')^{-1/2} & 0 \\ 0 & 0 & 0 & (d')^{1/2} \end{pmatrix} = \begin{pmatrix} 0 & a & b & 1 \\ -a & 0 & 1 & e \\ -b & -1 & 0 & f \\ -1 & -e & -f & 0 \end{pmatrix},$$

we may let

$$\Gamma \begin{pmatrix} 0 & 0 & 0 & 1 \\ 0 & 0 & 1 & 0 \\ 0 & -1 & 0 & 0 \\ -1 & 0 & 0 & 0 \end{pmatrix} = \begin{pmatrix} 0 & 0 & 0 & 1 \\ 0 & 0 & 1 & 0 \\ 0 & -1 & 0 & 0 \\ -1 & 0 & 0 & 0 \end{pmatrix}.$$

Finally, we have

$$\Gamma \begin{pmatrix} 0 & 0 & 0 & 1 \\ 0 & 0 & 0 & 0 \\ 0 & 0 & 0 & 0 \\ -1 & 0 & 0 & 0 \end{pmatrix} = \begin{pmatrix} 0 & 0 & 0 & \lambda \\ 0 & 0 & \mu & 0 \\ 0 & -\mu & 0 & 0 \\ -\lambda & 0 & 0 & 0 \end{pmatrix}.$$

Then

$$\left| \begin{pmatrix} 0 & \lambda \\ \mu & 0 \end{pmatrix} - k \begin{pmatrix} 0 & 1 \\ 1 & 0 \end{pmatrix} \right| = 0$$

for $k = 1$ and 0, so we have either

$$\Gamma(X) = X$$

or

$$\Gamma(X) = X_1.$$

(v) For the sake of simplicity, we give the proof for $m = 3$. The method is valid for any m.

Let $M = (a_{ij})$, $M_1 = (a'_{ij})$ and
$$\Gamma(M) = M_1.$$

Since, by (iii),

$$\Gamma \begin{pmatrix} \cdot & \cdot & & \cdot & \\ & 0 & \lambda & & \cdot \\ \cdot & -\lambda & 0 & & \\ & & & 0 & \mu \\ \cdot & \cdot & & -\mu & 0 \end{pmatrix} = \begin{pmatrix} \cdot & \cdot & & \cdot & \\ & 0 & \lambda & & \cdot \\ \cdot & -\lambda & 0 & & \\ & & & 0 & \mu \\ \cdot & \cdot & & -\mu & 0 \end{pmatrix},$$

the determinants of

$$M - \begin{pmatrix} \cdot & \cdot & & \cdot & \\ & 0 & \lambda & & \cdot \\ \cdot & -\lambda & 0 & & \\ & & & 0 & \mu \\ \cdot & \cdot & & -\mu & 0 \end{pmatrix} \quad \text{and} \quad M_1 - \begin{pmatrix} \cdot & \cdot & & \cdot & \\ & 0 & \lambda & & \cdot \\ \cdot & -\lambda & 0 & & \\ & & & 0 & \mu \\ \cdot & \cdot & & -\mu & 0 \end{pmatrix}$$

are identically equal. Comparing the coefficients of $\lambda^2 \mu^2$, we find $a_{12} = a'_{12}$. Similarly, we deduce
$$a_{34} = a'_{34}, \quad a_{56} = a'_{56}.$$

Now we let
$$\Gamma \begin{pmatrix} \cdot & 1 & 0 & \cdot \\ & 0 & 1 & \\ * & \cdot & \cdot & \\ \cdot & \cdot & \cdot & \end{pmatrix} = M_1.$$

Since for $\lambda \mu = 1$,

$$\left| \begin{pmatrix} \cdot & \begin{pmatrix} 1 & 0 \\ 0 & 1 \end{pmatrix} & \cdot \\ * & \cdot & \end{pmatrix} - \begin{pmatrix} & 0 & \lambda & & \cdot \\ & -\lambda & 0 & & \\ & & & 0 & \mu \\ \cdot & & & -\mu & 0 \end{pmatrix} \right| = 0,$$

it follows that

$$M_1 - \begin{pmatrix} & 0 & \lambda & & \cdot \\ & -\lambda & 0 & & \\ & & & 0 & \mu & \\ \cdot & & & -\mu & 0 & \\ & & & & & \cdot \end{pmatrix}$$

is of rank not greater than 2 for all λ, μ satisfying $\lambda\mu = 1$. Thus we have

$$\begin{vmatrix} 0 & \lambda & a'_{13} & a'_{14} \\ -\lambda & 0 & a'_{23} & a'_{24} \\ & & 0 & \mu \\ * & & -\mu & 0 \end{vmatrix} = \begin{vmatrix} 0 & \lambda & a'_{13} & a'_{15} \\ -\lambda & 0 & a'_{23} & a'_{25} \\ & & 0 & a'_{35} \\ * & & -a'_{35} & 0 \end{vmatrix} = \begin{vmatrix} 0 & \lambda & a'_{13} & a'_{16} \\ -\lambda & 0 & a'_{23} & a'_{26} \\ & & 0 & a'_{36} \\ * & & -a'_{36} & 0 \end{vmatrix}$$

$$= \begin{vmatrix} 0 & \lambda & a'_{14} & a'_{15} \\ -\lambda & 0 & a'_{24} & a'_{25} \\ & & 0 & a'_{45} \\ * & & -a'_{45} & 0 \end{vmatrix} = \begin{vmatrix} 0 & \lambda & a'_{14} & a'_{16} \\ -\lambda & 0 & a'_{24} & a'_{26} \\ & & 0 & a'_{46} \\ * & & -a'_{46} & 0 \end{vmatrix} = \begin{vmatrix} 0 & \lambda & a'_{15} & a'_{16} \\ -\lambda & 0 & a'_{25} & a'_{26} \\ & & \cdot & \cdot \\ * & & \cdot & \cdot \end{vmatrix} = 0.$$

The first equation gives

$$\begin{vmatrix} a'_{13} & a'_{14} \\ a'_{23} & a'_{24} \end{vmatrix} = 1,$$

so we may take

$$\begin{pmatrix} a'_{13} & a'_{14} \\ a'_{23} & a'_{24} \end{pmatrix} = \begin{pmatrix} 1 & 0 \\ 0 & 1 \end{pmatrix}$$

(in fact, we can choose a suitable Q such that

$$\begin{pmatrix} I & 0 & 0 \\ 0 & Q & 0 \\ 0 & 0 & I \end{pmatrix} M \begin{pmatrix} I & 0 & 0 \\ 0 & Q' & 0 \\ 0 & 0 & I \end{pmatrix}$$

has the required form).

Then, from the system of equations, we have

$$a'_{35} = a'_{36} = a'_{45} = a'_{46} = a'_{15} = a'_{16} = a'_{25} = a'_{26} = 0.$$

Thus, we may let

$$\Gamma \begin{pmatrix} \cdot & 1 & 0 & \cdot \\ \cdot & 0 & 1 & \cdot \\ * & \cdot & \cdot & \cdot \\ \cdot & \cdot & \cdot & \cdot \end{pmatrix} = \begin{pmatrix} \cdot & 1 & 0 & \cdot \\ \cdot & 0 & 1 & \cdot \\ * & \cdot & \cdot & \cdot \\ \cdot & \cdot & \cdot & \cdot \end{pmatrix}.$$

Let

$$\Gamma \begin{pmatrix} \cdot & 1 & 0 & \cdot \\ \cdot & 0 & 0 & \cdot \\ * & \cdot & \cdot & \cdot \\ \cdot & \cdot & \cdot & \cdot \end{pmatrix} = \begin{pmatrix} \cdot & P & * \\ * & \cdot & * \\ * & * & \cdot \end{pmatrix} = M_1.$$

Since M_1 is of rank 2, we have $|P| = 0$. Since

$$\begin{pmatrix} \cdot & 1 & 0 & \cdot \\ \cdot & 0 & 1 & \cdot \\ * & \cdot & \cdot & \cdot \\ \cdot & \cdot & \cdot & \cdot \end{pmatrix} - M_1 \tag{10}$$

is of rank 2, we have $|P - I| = 0$. There is also a matrix Q such that

$$QPQ^{-1} = \begin{pmatrix} 1 & 0 \\ 0 & 0 \end{pmatrix}.$$

Thus we may assume that

$$\Gamma \begin{pmatrix} \cdot & 1 & 0 & \cdot \\ \cdot & 0 & 0 & \cdot \\ * & \cdot & \cdot & \cdot \\ \cdot & \cdot & \cdot & \cdot \end{pmatrix} = \begin{pmatrix} \cdot & 1 & 0 & a'_{15} & a'_{16} \\ \cdot & 0 & 0 & a'_{25} & a'_{26} \\ * & \cdot & \cdot & a'_{35} & a'_{36} \\ \cdot & \cdot & \cdot & a'_{45} & a'_{46} \\ * & * & \cdot & \cdot & \cdot \end{pmatrix}.$$

By consideration of the ranks of the previous matrix and the matrix given in (10) we find

$$a'_{15} = a'_{16} = a'_{25} = a'_{26} = a'_{35} = a'_{36} = a'_{45} = a'_{46} = 0.$$

Consider again a general skew-symmetric matrix M and its image $M_1 = \Gamma(M)$. By the same method used for $m = 2$, we have either

$$M_1 = \begin{pmatrix} 0 & a_{12} & a_{13} & a_{14} & a'_{15} & a'_{16} \\ & 0 & a_{23} & a_{24} & a'_{25} & a'_{26} \\ & & 0 & a_{34} & a'_{35} & a'_{36} \\ & & & 0 & a'_{45} & a'_{46} \\ & * & & & 0 & a'_{56} \\ & & & & & 0 \end{pmatrix}$$

or

$$M_1 = \begin{pmatrix} 0 & a_{12} & a_{13} & a_{23} & a'_{15} & a'_{16} \\ & 0 & a_{14} & a_{24} & a'_{25} & a'_{26} \\ & & 0 & a_{34} & a'_{35} & a'_{36} \\ & & & 0 & a'_{45} & a'_{46} \\ & * & & & 0 & a'_{56} \\ & & & & & 0 \end{pmatrix}.$$

Repeating the process for
$$\begin{pmatrix} a_{15} & a_{16} \\ a_{25} & a_{26} \end{pmatrix},$$
we obtain either
$$a_{25} \rightleftarrows a_{25}, \quad a_{16} \rightleftarrows a_{16},$$
or
$$a_{25} \rightleftarrows a_{16}, \quad a_{16} \rightleftarrows a_{25}.$$

For the equivalence of "$a_{14} \rightleftarrows a_{23}$" and "$a_{16} \rightleftarrows a_{25}$", we have three cases: (α) $\Gamma(M) = M_1$, M_1 is obtained by replacing $a_{35}, a_{36}, a_{45}, a_{46}$ by $a'_{35}, a'_{36}, a'_{45}, a'_{46}$ in M,

$$(\beta) \qquad \Gamma(M) = M_2 = \begin{pmatrix} 0 & a_{12} & a_{13} & a_{23} & a_{15} & a_{16} \\ & 0 & a_{14} & a_{24} & a_{25} & a_{26} \\ & & 0 & a_{34} & a'_{35} & a'_{36} \\ & & & 0 & a'_{45} & a'_{46} \\ & * & & & 0 & a_{56} \\ & & & & & 0 \end{pmatrix},$$

$$(\gamma) \qquad \Gamma(M) = M_3 = \begin{pmatrix} 0 & a_{12} & a_{13} & a_{23} & a_{15} & a_{25} \\ & 0 & a_{14} & a_{24} & a_{16} & a_{26} \\ & & 0 & a_{34} & a'_{35} & a'_{36} \\ & & & 0 & a'_{45} & a'_{46} \\ & * & & & 0 & a_{56} \\ & & & & & 0 \end{pmatrix}.$$

(α) We leave $a_{35}, a_{36}, a_{45}, a_{46}$ arbitrary. Putting $a_{16} = a_{24} = a_{34} = 1$, $a_{25} = x$, and the others equal to 0, we have $\det(M) = (x - a_{35})^2$, $\det(M_1) = (x - a'_{35})^2$. We have $\det(M) = 0$ if and only if $\det(M_1) = 0$, that is $a_{35} = a'_{35}$. Next putting $a_{16} = a_{23} = a_{34} = 1$, $a_{25} = x$, and the others equal to 0, we obtain $a_{45} = a'_{45}$. Putting $a_{15} = a_{24} = a_{14} = 1$, $a_{26} = x$, and the others equal to 0 and putting $a_{15} = a_{23} = a_{34} = 1$, $a_{26} = x$ and the others equal to 0, we have respectively
$$a_{36} = a'_{36}, \quad a_{46} = a'_{46}.$$
Thus we have
$$\Gamma(M) = M.$$

(β) and (γ) Putting $a_{24} = a_{34} = 1$, $a_{16} = a_{25} = x$ and others equal to 0, we have $a_{35} = a'_{35}$. Further, if we put

$$a_{12} = a_{13} = 0, \ a_{14} = x, \ a_{15} = 1, \ a_{16} = 0, \ a_{23} = y, \ a_{24} = a_{25} = 0, \ a_{26} = 1,$$
$$a_{34} = 0, \ a_{35} = -1, \ a_{36} = a_{45} = a_{46} = 0, \ a_{56} = 1,$$

then
$$d(M) = (x(y-1))^2, \quad d(M_1) = (y(x-1))^2.$$

By putting $x = -1$, $y = +1$, we see that this is impossible.

The general proof may be arranged in the following steps:

(a) Dividing the matrix into m^2 2-rowed matrices;

(b) Choosing the first row of the small matrices as in the case $m = 3$ and applying the analogous method as above to the image;

(c) Determining the other small matrices by the method given for $m = 3$ (from (10) et seq.);

(d) Considering the 6-rowed minors we find that the exceptional case appearing for $m = 2$ cannot exist for $m \geqslant 3$.

10. Another generalization of von Staudt's theorem

The transformation
$$(W_1, W_2) = Q(\overline{Z}_1, \overline{Z}_2)\mathfrak{T}$$
is called *anti-orthogonal* if Q is nonsingular and $\mathfrak{T}$ is $\mathfrak{F}_1$-orthogonal.

Lemma Let A be a nonsingular matrix. Suppose that
$$(A'YAY^{-1})^2 = I$$
holds for all skew-symmetric Y. For $m = 1$, A is a matrix of determinant ± 1. For $m > 1$, then
$$A = \rho I, \quad \text{or} \quad A = \rho T[1, \cdots, 1, -1]T^{-1},$$
where $\rho = \pm 1, \pm i$, and conversely.

Proof (i) The result is evident for $m = 1$.

For $m > 1$, $A = \rho I$ evidently satisfies the equation. Now we prove that $A = \rho T[1, \cdots, 1, -1]T^{-1}$ satisfies the equation.

We write
$$T'YT = \begin{pmatrix} Y_1^{(n-1)} & v \\ -v' & 0 \end{pmatrix}, \quad (T'YT)^{-1} = \begin{pmatrix} Y_1^* & v^* \\ -v^{*\prime} & 0 \end{pmatrix}.$$

Then
$$\begin{pmatrix} I & 0 \\ 0 & I \end{pmatrix} = \begin{pmatrix} Y_1 & v \\ -v' & 0 \end{pmatrix} \begin{pmatrix} Y_1^* & v^* \\ -v^{*\prime} & 0 \end{pmatrix} = \begin{pmatrix} Y_1 Y_1^* - vv^{*\prime} & Y_1 v^* \\ -v' Y_1^* & -v' v^* \end{pmatrix}.$$

Further
$$T'A'YAY^{-1}T'^{-1} = \rho^2 [1, \cdots, 1, -1] \begin{pmatrix} Y_1 & v \\ -v' & 0 \end{pmatrix} [1, \cdots, 1, -1] \begin{pmatrix} Y_1^* & v^* \\ -v^{*\prime} & 0 \end{pmatrix}$$
$$= \rho^2 \begin{pmatrix} Y_1 Y_1^* + vv^{*\prime} & Y_1 v^* \\ v' Y_1^* & v' v^* \end{pmatrix} = \rho^2 \begin{pmatrix} I + 2vv^{*\prime} & 0 \\ 0 & -1 \end{pmatrix}.$$

Since $v^{*\prime} v = -1$, we have
$$(I^{(n-1)} + 2vv^{*\prime})^2 = I + 4vv^{*\prime} + 4vv^{*\prime} vv^{*\prime} = I^{(n-1)},$$
hence the result.

(ii) As in the proof of the lemma of §6, we may assume that
$$A = J_1 \dotplus J_2 \dotplus \cdots,$$
where J_ν is again of degree n_ν. The number of odd n_ν's is always even.

(iii) We consider the case
$$A = J^{(n)},$$
where n is even. Write
$$J = \begin{pmatrix} p & q \\ 0 & p \end{pmatrix}, \quad p = p^{(n/2)}, \quad q = q^{(n/2)},$$
where
$$p = \begin{pmatrix} \lambda & 1 & 0 & \cdots & 0 \\ 0 & \lambda & 1 & \cdots & 0 \\ \vdots & \vdots & \vdots & & \vdots \\ 0 & 0 & 0 & \cdots & \lambda \end{pmatrix}, \quad q = \begin{pmatrix} 0 & 0 & \cdots & 0 \\ 0 & 0 & \cdots & 0 \\ \vdots & \vdots & & \vdots \\ 1 & 0 & \cdots & 0 \end{pmatrix}.$$

Take
$$Y = \begin{pmatrix} 0 & I^{(n/2)} \\ -I^{(n/2)} & 0 \end{pmatrix},$$
then
$$I = (A'YAY^{-1})^2 = \left(\begin{pmatrix} p' & 0 \\ q' & p' \end{pmatrix} \begin{pmatrix} 0 & I \\ -I & 0 \end{pmatrix} \begin{pmatrix} p & q \\ 0 & p \end{pmatrix} \begin{pmatrix} 0 & -I \\ I & 0 \end{pmatrix} \right)^2$$
$$= \begin{pmatrix} (p'p)^2 & * \\ * & * \end{pmatrix}.$$

implies $(p'p)^2 = I^{(n/2)}$, which is impossible for $n > 2$ (that is $m > 1$).

(iv) Let
$$A = J_1 \dotplus J_2, \quad n_1 + n_2 = n,$$
where n_1 and n_2 are both odd. Let $n_1 = n_2$. Write
$$A = \begin{pmatrix} p & 0 \\ 0 & q \end{pmatrix}, \quad Y = \begin{pmatrix} 0 & I^{(n/2)} \\ -I & 0 \end{pmatrix}.$$
Then, we have
$$(A'YAY^{-1})^2 = \begin{pmatrix} (p'q)^2 & * \\ * & * \end{pmatrix} = I,$$
which is possible only for $n_1 = n_2 = 1$. Further let $n_1 > n_2$. We write
$$A = \begin{pmatrix} p & q \\ 0 & r \end{pmatrix}, \quad Y = \begin{pmatrix} 0 & I^{(n/2)} \\ -I & 0 \end{pmatrix}.$$
Then
$$(A'YAY^{-1})^2 = \begin{pmatrix} (p'r)^2 & * \\ * & * \end{pmatrix} = I,$$
which is impossible for $n_1 > 3$. For $n_1 = 3$ and $n_2 = 1$, we have consequently
$$A = \begin{pmatrix} \lambda & 1 & 0 & 0 \\ 0 & \lambda & 1 & 0 \\ 0 & 0 & \lambda & 0 \\ 0 & 0 & 0 & \mu \end{pmatrix}, \quad \mu = -\lambda, \quad \lambda^4 = 1.$$
Taking
$$Y = \begin{pmatrix} \cdot & 2 & 1 \\ & 1 & 1 \\ * & \cdot & \end{pmatrix},$$
we find this also to be impossible.

Thus each of the numbers
$$n_1, n_2, \cdots$$
must be either 1 or 2.

(v) It is easily seen that no two of the n_ν's can be 2. In fact
$$A = \begin{pmatrix} \lambda_1 & 1 & 0 & 0 \\ 0 & \lambda_1 & 0 & 0 \\ 0 & 0 & \lambda_2 & 1 \\ 0 & 0 & 0 & \lambda_2 \end{pmatrix} = \begin{pmatrix} p & \cdot \\ \cdot & q \end{pmatrix},$$

say. Taking
$$Y = \begin{pmatrix} 0 & I \\ -I & 0 \end{pmatrix},$$
we find this to be impossible.

(vi) Further, if one of the n_ν's is 2, then n is equal to 2. In fact, suppose that
$$A = \begin{pmatrix} \lambda_1 & 1 & & \\ 0 & \lambda_1 & & \cdot \\ & & \lambda_2 & 0 \\ & \cdot & 0 & \lambda_3 \end{pmatrix}.$$

Taking
$$Y = \begin{pmatrix} 0 & I \\ -I & 0 \end{pmatrix},$$
we have
$$\lambda_2 = -\lambda_3, \quad \lambda_1^2 \lambda_2^2 = 1.$$

Taking further
$$Y = \begin{pmatrix} 0 & 0 & 1 & 1 \\ 0 & 0 & 1 & 0 \\ -1 & -1 & 0 & 0 \\ -1 & 0 & 0 & 0 \end{pmatrix},$$
we have $\lambda_2 = \lambda_3$. Both results cannot hold simultaneously.

(vii) Suppose $n \geqslant 4$. Let
$$A = [\lambda_1, \lambda_2, \lambda_3, \lambda_4].$$

Taking
$$Y = \begin{pmatrix} 0 & 0 & 1 & 1 \\ 0 & 0 & 1 & 0 \\ -1 & -1 & 0 & 0 \\ -1 & 0 & 0 & 0 \end{pmatrix},$$
we have
$$(A'YAY^{-1})^2 = \begin{pmatrix} \lambda_1^2 \lambda_4^2 & \lambda_1(\lambda_1\lambda_4+\lambda_2\lambda_3)(\lambda_3-\lambda_4) & \cdot & \cdot \\ \cdot & \lambda_2^2 \lambda_3^2 & \cdot & \cdot \\ \cdot & \cdot & \lambda_2^2 \lambda_3^2 & \lambda_3(\lambda_1\lambda_4+\lambda_2\lambda_3)(\lambda_1-\lambda_2) \\ \cdot & \cdot & \cdot & \lambda_1^2 \lambda_4^2 \end{pmatrix} = I,$$
(1)

which implies
$$\lambda_1^2\lambda_4^2 = \lambda_2^2\lambda_3^2 = 1.$$

Since $\lambda_1, \lambda_2, \lambda_3, \lambda_4$ can be permuted, we have
$$\lambda_i^2\lambda_j^2 = 1 \quad \text{for all } i,j \ (i \neq j).$$

Thus
$$\lambda_1^4 = 1, \quad \lambda_1^2 = \lambda_2^2 = \lambda_3^2 = \lambda_4^2 = \pm 1.$$

In general, $A = [\lambda_1, \cdots, \lambda_n]$, where $\lambda_1^2 = \cdots = \lambda_n^2 = \pm 1$. By choosing a suitable ρ in the lemma, we may consider the case with
$$\lambda_1^2 = \lambda_2^2 = \cdots = \lambda_n^2 = 1.$$

If among the λ's there occur two positive and two negative numbers, we take $\lambda_1 = \lambda_4 = 1, \lambda_2 = \lambda_3 = -1$. Then (1) is impossible. Thus we have the lemma.

Theorem 11 *A transformation satisfying the following conditions:*

(1) one-to-one and continuous;

(2) carrying skew-symmetric matrices into skew-symmetric matrices;

(3) keeping arithmetic distance invariant,

and

(4) keeping the harmonic relation invariant

is for $n \neq 4$ either $\mathfrak{F}_1$-orthogonal or anti-orthogonal. In the case $n = 4$, the transformation is either $\mathfrak{F}_1$-orthogonal or anti-orthogonal, or is equivalent to

$$\Gamma(Z) = Z_1, \quad \text{where} \quad Z = \begin{pmatrix} p & q \\ -q' & r \end{pmatrix}, \quad Z_1 = \begin{pmatrix} p & q' \\ -q & r \end{pmatrix},$$

or equivalent to
$$\Gamma(Z) = \overline{Z}_1.$$

Proof (i) A triple of points, no two of which have arithmetic distance less than n, is equivalent to
$$0, \quad \begin{pmatrix} 0 & I^{(m)} \\ -I^{(m)} & 0 \end{pmatrix}, \quad \infty.$$

We may let the transformation satisfy
$$\Gamma(0) = 0, \quad \Gamma(\mathfrak{F}) = \mathfrak{F}, \quad \Gamma(\infty) = \infty.$$

(ii) As in the symmetric case, we have

$$\Gamma(Z) + \Gamma(Z_1) = \Gamma(Z + Z_1).$$

Again, for $n \neq 4$, we have, analogous to the symmetric case,

$$\Gamma(X + iY) = X + iA'YA.$$

Since Y, $-Y$ are separated harmonically by iY and $-iY$ for real Y, we have

$$\Gamma(Yi)Y^{-1} = -Y(\Gamma(Yi))^{-1},$$

that is,

$$(A'YAY^{-1})^2 = I$$

for all real Y. Now we suppose $m \geqslant 3^{①}$, then we have

$$\Gamma(X + iY) = X \pm iY$$

or

$$\Gamma(X + iY) = X \pm iT'^{-1}[1, \cdots, 1, -1]T'YT[1, \cdots, 1, -1]T^{-1}.$$

The first case is what we require. Changing variables in the second case we may let

$$\Gamma(X + iY) = X \pm i[1, \cdots, 1, -1]Y[1, \cdots, 1, -1].$$

Since

$$\begin{vmatrix} 0 & 0 & 1 & 1 \\ 0 & 0 & i & i \\ * & \cdot & & \end{vmatrix} = 0 \quad \text{and} \quad \begin{vmatrix} 0 & 0 & 1 & 1 \\ 0 & 0 & i & -i \\ * & \cdot & & \end{vmatrix} \neq 0,$$

the rank is not invariant. The last case does not satisfy our requirement.

In case $m = 2$, a great deal of special consideration is needed. Apart from the lemma, we require the solutions of

$$(A'Y_1AY^{-1})^2 = I,$$

where

$$Y = \begin{pmatrix} P & Q \\ -Q' & R \end{pmatrix}, \quad Y_1 = \begin{pmatrix} P & Q' \\ -Q & R \end{pmatrix}.$$

---

① For $m = 1$, the result is almost evident.

The proof of the lemma establishes that either

$$A = \rho I$$

or

$$A = \rho T[1,\ 1,\ 1,\ -1]T^{-1}$$

(the Q used here is always symmetric).

As in the preceding proof we have four cases,

$$\Gamma(X+iY) = X \pm iA'YA, \quad \text{or} \quad X_1 \pm iA'Y_1A',$$

$$\text{or} \quad X_1 \pm iA'YA, \quad \text{or} \quad X \pm iA'Y_1A.$$

By the previous argument, for $m \geqslant 3$, we have, for the first two cases,

$$\Gamma(Z) = Z,\ \overline{Z},\ Z_1,\ \overline{Z}_1,$$

where $Z = X + iY$, and Z_1 is obtained from Z by the process yielding Y_1 from Y.

Next, if $\Gamma(X+iY) = X_1 \pm iA'YA$, we have either

$$\Gamma(X+iY) = X_1 \pm iY \quad \text{or} \quad X_1 \pm i[1,\ 1,\ 1,\ -1]Y[1,\ 1,\ 1,\ -1].$$

Putting

$$Z = \begin{pmatrix} 0 & 0 & i & i \\ 0 & 0 & 1 & 1 \\ -i & -1 & 0 & 0 \\ -i & -1 & 0 & 0 \end{pmatrix},$$

$$\Gamma(Z) = \begin{pmatrix} 0 & 0 & i & i+1 \\ 0 & 0 & 0 & 1 \\ -i & 0 & 0 & 0 \\ -i-1 & -1 & 0 & 0 \end{pmatrix} \quad \text{or} \quad \begin{pmatrix} 0 & 0 & i & -i+1 \\ 0 & 0 & 0 & 1 \\ -i & 0 & 0 & 0 \\ i-1 & -1 & 0 & 0 \end{pmatrix},$$

we see that both these automorphisms could render a singular matrix nonsingular. Thus both these cases are ruled out.

Finally, the possibility of

$$\Gamma(X+iY) = X \pm iAY_1A'$$

may be treated in a similar way.

III. Geometry of Hermitian matrices

The geometry of symmetric matrices in the real domain is closely analogous to the present geometry.

11. Notions

We define an *Hermitian pair* (Z_1, Z_2) of matrices by

$$(\overline{Z}_1, \overline{Z}_2)\mathfrak{F}(Z_1, Z_2)' = 0.$$

A conjunctively symplectic matrix $\mathfrak{T}$ is defined by [1]

$$\mathfrak{T}^*\mathfrak{F}\mathfrak{T}' = \mathfrak{F}.$$

We define a conjunctively symplectic transformation by

$$(W_1, W_2) = Q(Z_1, Z_2)\mathfrak{T}.$$

We identify two nonsingular Hermitian pairs of matrices (Z_1, Z_2) and (W_1, W_2) by means of the relation

$$(Z_1, Z_2) = Q(W_1, W_2).$$

It is called a point of the space. Evidently, the space so defined is transitive. The rank of

$$(\overline{W}_1, \overline{W}_2)\mathfrak{F}(Z_1, Z_2)'$$

is defined to be the arithmetic distance of the points (W_1, W_2) and (Z_1, Z_2). Two pairs of points are equivalent if and only if they have the same arithmetic distance.

Let P_1, P_2, P_3 be three points no two of which have arithmetic distance less than n. Then they are equivalent to the three points

$$0, \quad \infty, \quad K(= [1, \cdots, 1, -1, \cdots, -1]).$$

The signature of K is defined to be the signature of the range P_1, P_2, P_3 (the order of points is significant).

Evidently the signature is invariant under the group. We also may say that two triples of points are in the same sense if they have the same signature. We may

---

[1] $\mathfrak{T}^*$ denotes the conjugate complex matrix of $\mathfrak{T}$.

prove that if two ranges are in the same sense, there is a conjunctively symplectic transformation carrying one into the other.

As to the equivalence of quadruples of points, a great deal of difficulty arises from the fact that the existing treatments of the theory of Hermitian forms are incomplete. We shall give elsewhere a complete classification and then its application to the present problem will be immediate.

12. A further generalization of von Staudt's theorem

Theorem 12 *Let Γ be an additive continuous automorphism of the module formed by all Hermitian matrices keeping rank and signature invariant. Then Γ is either an inner automorphism or an anti-automorphism $(Z \to \overline{PZP'})$.*

Proof (Cf. the results of §6)   We have $\Gamma(0) = 0$. Let

$$\Gamma(I) = H_0.$$

Since H_0 is positive definite, we may let

$$\Gamma(I) = I.$$

As in the proof of Theorem 8 (in the real field), we may let

$$\Gamma(X) = X$$

for all real symmetric X.

Let Y be any real skew-symmetric matrix, and let

$$\Gamma(iY) = H.$$

Since

$$\det(X + iY) = 0$$

if and only if

$$\det(X + H) = 0,$$

by Hilbert's theorem on polynomial ideals, we have an integer ρ such that

$$(\det(X + iY))^\rho \equiv 0 \,(\mathrm{mod}\, \det(X + H))$$

and

$$(\det(X + H))^\rho \equiv 0 \,(\mathrm{mod}\, \det(X + iY)),$$

in the polynomial ring formed from the real field by adjunction of the elements of X. Let $X = X' = (x_{ij})$. We write

$$\det(X + iY) = f_1 x_{11} + g_1,$$

$$\det(X+H) = f_2 x_{11} + g_2,$$

where f_1, f_2, g_1, g_2 are elements in the ring $\Re$ (generated by the elements of X omitting x_{11}). Since

$$(f_2(f_1 x_{11} + g_1) - f_1(f_2 x_{11} + g_2))^\rho \equiv 0 \,(\mathrm{mod}\, \det(X+H)),$$

and since $f_2 g_1 - f_1 g_2$ is independent of x_{11}, we have

$$f_2 g_1 - f_1 g_2 = 0.$$

Since the determinant of an Hermitian matrix is an irreducible polynomial in its elements, f_1 and f_2 are irreducible and f_1 and g_1 have no common divisor. Consequently we have

$$f_1 = f_2, \quad g_1 = g_2.$$

In this procedure, we have to compare one of the coefficients.

Thus
$$\det(X + iY) = \det(X + H).$$

Consequently each principal minor of iY equals the corresponding principal minor of H. We complete the proof with the aid of the following lemma.

Lemma *If two Hermitian matrices H and K have the same principal minors of orders* 1, 2, 3, *then (for $\exp(x) = e^x$)*

$$H = [\exp(i\theta_1), \cdots, \exp(i\theta_n)] K^* [\exp(-i\theta_1), \cdots, \exp(-i\theta_n)],$$

where K^* is obtained from K by replacing k_{rs} by either k_{rs} or $\bar{k}_{rs}$.

From the lemma, we may let

$$h_{rs} = \pm i y_{rs}.$$

Since one of

$$\begin{pmatrix} 2 & i & -i \\ -i & 0 & 1 \\ i & 1 & 0 \end{pmatrix} \text{ and } \begin{pmatrix} 2 & i & -i \\ -i & 0 & 1 \\ i & 1 & 0 \end{pmatrix}$$

is singular and the other is not, we have

$$H = \pm i Y.$$

The proof of the lemma is straightforward. Considering the 1-rowed principal minors of $H = (h_{rs})$, $K = (k_{rs})$, we have

Since
$$h_{rr} = k_{rr}.$$

$$\begin{vmatrix} h_{rr} & h_{rs} \\ \bar{h}_{rs} & h_{ss} \end{vmatrix} = \begin{vmatrix} h_{rr} & k_{rs} \\ \bar{k}_{rs} & h_{ss} \end{vmatrix},$$

we have
$$|h_{rs}|^2 = |k_{rs}|^2.$$

We may choose $\theta_1, \cdots, \theta_n$ such that the matrix

$$[\exp(i\theta_1), \cdots, \exp(i\theta_n)] H [\exp(-i\theta_1), \cdots, \exp(-i\theta_n)]$$

has real $h_{12}, h_{13}, \cdots, h_{1n}$. We may let

$$h_{1i} = k_{1i}$$

be real and positive. Consider

$$\begin{vmatrix} h_{11} & h_{1i} & h_{ij} \\ \bar{h}_{1i} & h_{ii} & h_{ij} \\ \bar{h}_{1j} & \bar{h}_{ij} & h_{jj} \end{vmatrix} = \begin{vmatrix} k_{11} & k_{1i} & k_{1j} \\ \bar{k}_{1i} & k_{ii} & k_{ij} \\ \bar{k}_{1j} & \bar{k}_{ij} & k_{jj} \end{vmatrix}.$$

We have
$$h_{ij} + \bar{h}_{ij} = k_{ij} + \bar{k}_{ij},$$

then letting $h_{ij} = \alpha + \beta i$, $k_{ij} = \gamma + \delta i$, we have

$$\alpha^2 + \beta^2 = \gamma^2 + \delta^2, \quad \alpha = \gamma.$$

Thus $\beta = \pm \delta$ and we have

$$h_{ij} = k_{ij} \text{ or } \bar{k}_{ij}.$$

Theorem 13 *A transformation satisfying the following conditions:*

(1) *one-to-one and continuous;*

(2) *carrying Hermitian matrices into Hermitian matrices;*

(3) *keeping arithmetic distance invariant;*

(4) *keeping sense of triples of points invariant;*

(5) *keeping the harmonic relation invariant*

is either a conjunctively symplectic or a conjunctively anti-symplectic transformation.

The proof is omitted because of the similarity to the real analogue of Theorem 8 (cf. §7).

IV. Geometry of rectangular matrices

13. Subgeometries of the geometry of unitary matrices

The geometry studied in III may also be interpreted as the geometry of *unitary matrices*. Since the matrix
$$\mathfrak{F} = i \begin{pmatrix} 0 & I \\ -I & 0 \end{pmatrix}$$
is of signature 0, we may use
$$\mathfrak{F} = \begin{pmatrix} I & 0 \\ 0 & -I \end{pmatrix}$$
instead of it. Then the pair of matrices (Z_1, Z_2), satisfying
$$(\overline{Z}_1, \overline{Z}_2) \begin{pmatrix} I & 0 \\ 0 & -I \end{pmatrix} (Z_1, Z_2)' = 0,$$
that is
$$\overline{Z}_1 Z_1' - \overline{Z}_2 Z_2' = 0, \quad I - (\overline{Z}_1^{-1}\overline{Z}_2)(Z_1^{-1}Z_2)' = 0,$$
is the homogeneous representation of a unitary matrix. We can generalize the idea a little further. Let
$$\mathfrak{F}_2 = \begin{pmatrix} I^{(n)} & 0 \\ 0 & -I^{(m)} \end{pmatrix}, \quad n \geqslant m.$$
The matrix $\Gamma^{(m+n)}$ satisfying
$$\overline{\Gamma} \mathfrak{F}_2 \Gamma' = \mathfrak{F}_2$$
is called conjunctive with signature (n, m). The pair $(Z_1^{(n)}, Z_2^{(n,m)})$ of matrices satisfying
$$(\overline{Z}_1, \overline{Z}_2) \mathfrak{F}_2 (Z_1, Z_2)' = 0$$
is called an (n, m)-unitary pair.

Instead of going into the details of this geometry we shall be content to make the following remark.

Let
$$W_1^{(n,m)} = Z_1^{-1} Z_2.$$
Then we have
$$\overline{W}_1 W_1' = I.$$

W_1 is formed by m columns of a unitary matrix. Thus the geometry may be considered as a subgeometry of the geometry of unitary matrices by identifying the elements with the same m columns as an element of the subgeometry. This may be described in short as "the process of projection".

14. Remarks

(i) The condition "one-to-one" is redundant, since the invariance of arithmetic distance implies it.

(ii) The continuity for the real case is also very probably redundant (cf. Sierpinski's contribution to the solution of the functional equation $f(x+y) = f(x) + f(y)$).

(iii) The geometry of pairs of matrices $(Z_1^{(n)}, Z_2^{(n,m)})$ with the group given in IV has interesting applications to the study of automorphic functions. It is *not* an analogue of projective geometry but of non-Euclidean geometry.

(iv) Analogous to IV, we may establish a geometry of real rectangular matrices.

Finally the author should like to express his warmest thanks to the referee for his help with the manuscript.

Geometries of matrices. I_1. Arithmetical construction*

A discussion of the redundancy of the conditions involved in the generalizations of von Staudt's theorem should be preceded by a study of the involutions. As an illustration and a supplement to part I, we give here a discussion of the geometry of 2-rowed symmetric matrices; we intend to treat the general case at a later occasion. More definitely, the condition concerning the harmonic separation is a consequence of the invariance of arithmetic distance. For the real case, even continuity (as well as the condition concerning the signature) is redundant (a proof for even order symmetric matrices has been obtained). As to the general discussion, some knowledge concerning involutions seems to be indispensable; the author will come back to it later.

Throughout the paper, the notations in I are taken over and we assume that $n = 2$.

1. Normal subspaces

Theorem 1   *Given two matrices Z_1 and Z_2 with arithmetic distance $r(Z_1, Z_2) = 1$, the points Z satisfying*

$$r(Z, Z_1) \leqslant 1, \quad r(Z, Z_2) \leqslant 1$$

form a normal subspace.

Proof   Without loss of generality, we may take

$$Z_1 = \begin{pmatrix} 1 & 0 \\ 0 & 0 \end{pmatrix}, \quad Z_2 = \begin{pmatrix} -1 & 0 \\ 0 & 0 \end{pmatrix}.$$

Let

$$Z = \begin{pmatrix} x & y \\ y & z \end{pmatrix}.$$

Then we have

$$(x \pm 1)z - y^2 = 0,$$

---

* Presented to the Society, April 28, 1945; received by the editors January 2, 1945. Reprinted from *Transactions of the American Mathematical Society*, 1945, **57**: 482–490.

Geometries of matrices. I_1. Arithmetical construction

that is, $z = y = 0$. The theorem is now evident.

Definition The normal subspace obtained in Theorem 1 is said to be spanned by Z_1 and Z_2.

Definition Two normal subspaces are said to be *complementary* if there is one and only one pair of matrices, one from each subspace, with arithmetic distance less than 2.

Theorem 2 *Two complementary subspaces may be carried simultaneously to*

$$\begin{pmatrix} x & 0 \\ 0 & 0 \end{pmatrix}, \quad \begin{pmatrix} 0 & 0 \\ 0 & y \end{pmatrix}.$$

Proof We may let

$$\begin{pmatrix} 1 & 0 \\ 0 & 0 \end{pmatrix}, \quad \begin{pmatrix} -1 & 0 \\ 0 & 0 \end{pmatrix}$$

be the matrices to span the first subspace, and

$$\begin{pmatrix} 0 & 0 \\ 0 & 1 \end{pmatrix}, \quad X$$

be the matrices to span the second subspace. In fact, by a theorem of I, we may carry any three points A, B, C with

$$r(A, B) = 1, \quad r(A, C) = r(B, C) = 2$$

into

$$A_1 = \begin{pmatrix} 1 & 0 \\ 0 & 0 \end{pmatrix}, \quad B_1 = \begin{pmatrix} -1 & 0 \\ 0 & 0 \end{pmatrix}, \quad C_1 = \begin{pmatrix} 0 & 0 \\ 0 & 1 \end{pmatrix}.$$

Since

$$r\left(\begin{pmatrix} 0 & 0 \\ 0 & 1 \end{pmatrix}, X\right) = 1,$$

we have

$$X = \begin{pmatrix} a^2 & ab \\ ab & b^2 + 1 \end{pmatrix}.$$

If $a \neq 0$, the matrix

$$\begin{pmatrix} x & 0 \\ 0 & 0 \end{pmatrix}, \quad x = a^2/(b^2 + 1)$$

of the first subspace is at distance 1 from X. This contradicts our hypothesis. Thus the second is spanned by

$$\begin{pmatrix} 0 & 0 \\ 0 & 1 \end{pmatrix}, \quad \begin{pmatrix} 0 & 0 \\ 0 & b^2+1 \end{pmatrix}.$$

The theorem follows.

Consequently two complementary subspaces have a unique matrix in common.

Theorem 3 *Let $\mathfrak{S}$ denote the set of matrices S such that*

$$r(S, P) = 2,$$

where P is the common matrix of two complementary normal subspaces. Then $\mathfrak{S}$ is transitive under the subgroup leaving the two subspaces invariant.

Proof Let the two complementary subspaces be

$$\begin{pmatrix} x & 0 \\ 0 & 0 \end{pmatrix}, \quad \begin{pmatrix} 0 & 0 \\ 0 & y \end{pmatrix}$$

and S be the points in $\mathfrak{S}$. By the hypothesis with $P = 0$, we know that S is nonsingular. The transformation

$$Z_1 = Z(-S^{-1}Z + I)^{-1}$$

carries S into ∞, and it leaves the subspaces

$$\begin{pmatrix} x & 0 \\ 0 & 0 \end{pmatrix}, \quad \begin{pmatrix} 0 & 0 \\ 0 & y \end{pmatrix}$$

invariant.

2. Direct sum of complementary subspaces

Definition The subspace formed by points of the form

$$\begin{pmatrix} x & 0 \\ 0 & y \end{pmatrix}$$

is called a (completely) *reducible subspace* which is the direct sum of the subspaces

$$\begin{pmatrix} x & 0 \\ 0 & 0 \end{pmatrix}, \quad \begin{pmatrix} 0 & 0 \\ 0 & y \end{pmatrix}.$$

Now we shall give its arithmetic construction. We take, by Theorem 2, I as any point of the set $\mathfrak{S}$. Find P satisfying

$$r\left(P,\begin{pmatrix} x & 0 \\ 0 & 0 \end{pmatrix}\right) = r\left(P,\begin{pmatrix} 0 & 0 \\ 0 & y \end{pmatrix}\right) = r(P, I) = 1.$$

Let
$$P = \begin{pmatrix} a^2+1 & ab \\ ab & b^2+1 \end{pmatrix},$$
we have
$$(a^2+1-x)(b^2+1) = a^2 b^2, \quad (a^2+1)(b^2+1-y) = a^2 b^2.$$

Then, $a^2 = y(1-x)/(xy-x-y)$, $b^2 = x(1-y)/(xy-x-y)$.

Thus we have two solutions:
$$P_{\pm} = \begin{pmatrix} 1 & 0 \\ 0 & 1 \end{pmatrix} + \frac{1}{xy-x-y}\begin{pmatrix} y(1-x) & \pm(xy(1-x)(1-y))^{1/2} \\ \pm(xy(1-x)(1-y))^{1/2} & x(1-y) \end{pmatrix}.$$

Finally, we find all K satisfying
$$r(K, P_+) = r(K, P_-) = 1.$$

Putting
$$K = I + \frac{1}{(xy-x-y)}\begin{pmatrix} k_1 & k_2 \\ k_2 & k_3 \end{pmatrix},$$
we have
$$(k_1 - y(1-x))(k_3 - x(1-y)) - (k_2 \pm (xy(1-x)(1-y))^{1/2})^2 = 0.$$

Then $k_2 = 0$, $k_1 = (1+\rho)y(1-x)$, $k_3 = (1+1/\rho)x(1-y)$.

The matrices of the form
$$K = \begin{pmatrix} 1+(1+\rho)y(1-x) & 0 \\ 0 & 1+(1+1/\rho)x(1-y) \end{pmatrix}$$
run over all matrices of the reducible space.

Since we use the arithmetical notion only, we have the following general definition.

Definition Given two complementary subspaces $\mathfrak{X}$ and $\mathfrak{Y}$, let Q be a point of the set $\mathfrak{S}$. The reducible space (or direct sum of both subspaces) is defined by the aggregate of points K such that
$$r(P_+, K) = r(P_-, K) = 1,$$

where P_+ and P_- are both solutions of
$$r(Q, P) = r(P, X) = r(P, Y) = 1$$
and X and Y belong to $\mathfrak{X}$ and $\mathfrak{Y}$ respectively.

As a consequence of Theorems 2 and 3 we have the following theorem.

Theorem 4 *The aggregate of reducible spaces is transitive.*

3. Involutions

Given
$$Z = \begin{pmatrix} z_1 & z_2 \\ z_2 & z_3 \end{pmatrix},$$
we wish to find all matrices
$$\begin{pmatrix} x & 0 \\ 0 & y \end{pmatrix}$$
of a reducible space such that
$$r\left(\begin{pmatrix} x & 0 \\ 0 & y \end{pmatrix}, \begin{pmatrix} z_1 & z_2 \\ z_2 & z_3 \end{pmatrix}\right) = 1,$$
consequently
$$(x - z_1)(y - z_3) - z_2^2 = 0.$$
This set is denoted by Σ. To each matrix Z we have a set Σ. Conversely, to each Σ, we have two matrices
$$Z \text{ and } \begin{pmatrix} z_1 & -z_2 \\ -z_2 & z_3 \end{pmatrix} = \begin{pmatrix} 1 & 0 \\ 0 & -1 \end{pmatrix} Z \begin{pmatrix} 1 & 0 \\ 0 & -1 \end{pmatrix}.$$

Thus we obtain a transformation
$$Z_1 = \begin{pmatrix} 1 & 0 \\ 0 & -1 \end{pmatrix} Z \begin{pmatrix} 1 & 0 \\ 0 & -1 \end{pmatrix}, \tag{1}$$
which is called an *involution of the first kind*. Further, each point of the reducible space is a fixed point of (1); there are no other fixed points.

Since
$$\frac{1}{2} \begin{pmatrix} 1 & -1 \\ 1 & 1 \end{pmatrix} \begin{pmatrix} 1 & 0 \\ 0 & -1 \end{pmatrix} \begin{pmatrix} 1 & 1 \\ -1 & 1 \end{pmatrix} = \begin{pmatrix} 0 & 1 \\ 1 & 0 \end{pmatrix},$$
we have an equivalent involution

$$Z_1 = \begin{pmatrix} 0 & 1 \\ 1 & 0 \end{pmatrix} Z \begin{pmatrix} 0 & 1 \\ 1 & 0 \end{pmatrix} \tag{2}$$

and, since

$$\begin{pmatrix} 1 & i \\ i & 1 \end{pmatrix}^{-1} \begin{pmatrix} 1 & 0 \\ 0 & -1 \end{pmatrix} \begin{pmatrix} 1 & i \\ i & 1 \end{pmatrix} = \begin{pmatrix} 0 & i \\ -i & 0 \end{pmatrix},$$

we have another equivalent involution

$$Z_1 = -\begin{pmatrix} 0 & 1 \\ -1 & 0 \end{pmatrix} Z \begin{pmatrix} 0 & -1 \\ 1 & 0 \end{pmatrix}. \tag{3}$$

It may be shown that the most general form of involutions of the first kind[①] is

$$Z_1 = (PZ - K_1)(K_2 Z + P')^{-1}, \tag{4}$$

where K_1 and K_2 are skew symmetric and

$$\mathfrak{F} = \begin{pmatrix} P & -K_1 \\ K_2 & P' \end{pmatrix}$$

is symplectic. We use $\mathfrak{F}_1, \mathfrak{F}_2, \mathfrak{F}_3$ to denote the matrices of (1), (2) and (3) respectively. We may easily verify that

$$\mathfrak{F}_i \mathfrak{F}_j = -\mathfrak{F}_j \mathfrak{F}_i$$

and

$$\mathfrak{F}_1 \mathfrak{F}_2 \mathfrak{F}_3 = \begin{pmatrix} iI & 0 \\ 0 & -iI \end{pmatrix},$$

that is

$$Z_1 = -Z. \tag{5}$$

This is called an involution of the second kind.

4. Commutative involutions

Let $\mathfrak{S}$ and $\mathfrak{S}_0$ be two reducible subspaces associated with two involutions of the first kind, $\mathfrak{F}$ and $\mathfrak{F}_0$ respectively (it may be shown easily that, if $\mathfrak{F}\mathfrak{F}_0 = \mathfrak{F}_0\mathfrak{F}$, then $\mathfrak{F}_0$ carries $\mathfrak{S}$ into itself pointwise).

[①] The general definition of an involution of the first or second kind is that $\mathfrak{F}^2 = \mathfrak{T}$ or $-\mathfrak{T}$. It may be verified easily that they are equivalent symplectically to (1) or (5) respectively. A detailed study of involutions forms the subject of II, which will appear soon. For this reason, the author omits some details of the discussion. Certainly, for $n = 2$, the properties used can all be verified directly and easily.

If $\mathfrak{F}$ carries $\mathfrak{S}_0$ into itself, but not pointwise, then

$$\mathfrak{F}_0\mathfrak{F} = -\mathfrak{F}\mathfrak{F}_0.$$

In fact, $\mathfrak{F}^{-1}\mathfrak{F}_0\mathfrak{F}$ leaves $\mathfrak{S}_0$ fixed pointwise. Thus

$$\mathfrak{F}^{-1}\mathfrak{F}_0\mathfrak{F}_1 = \pm\mathfrak{F}_0,$$

the upper sign is ruled out, since $\mathfrak{F}$ does not leave $\mathfrak{S}_0$ pointwise fixed.

Theorem 5 *Any pair of commutative involutions of the first kind may be carried into $\mathfrak{F}_1$ and $\mathfrak{F}_2$ simultaneously.*

Proof The first one may be assumed to be $\mathfrak{F}_1$. Let the second be given by (4). Then

$$\begin{pmatrix} P & -K_1 \\ K_2 & P' \end{pmatrix} \begin{pmatrix} 1 & 0 & 0 & 0 \\ 0 & -1 & 0 & 0 \\ 0 & 0 & 1 & 0 \\ 0 & 0 & 0 & -1 \end{pmatrix} = -\begin{pmatrix} 1 & 0 & 0 & 0 \\ 0 & -1 & 0 & 0 \\ 0 & 0 & 1 & 0 \\ 0 & 0 & 0 & -1 \end{pmatrix} \begin{pmatrix} P & -K_1 \\ K_2 & P' \end{pmatrix}$$

and we have

$$P = \begin{pmatrix} 0 & p_1 \\ p_2 & 0 \end{pmatrix}, \quad K_1 = \begin{pmatrix} 0 & k_1 \\ -k_1 & 0 \end{pmatrix}, \quad K_2 = \begin{pmatrix} 0 & k_2 \\ -k_2 & 0 \end{pmatrix},$$

where

$$p_1 p_2 + k_1 k_2 = 1.$$

If

$$\begin{pmatrix} p_1 - \lambda & -k_1 \\ k_2 & p_2 - \lambda \end{pmatrix}$$

has only simple elementary divisors, we have a, b, c, d such that

$$\begin{pmatrix} a & b \\ c & d \end{pmatrix} \begin{pmatrix} p_1 & -k_1 \\ k_2 & p_2 \end{pmatrix} \begin{pmatrix} a & b \\ c & d \end{pmatrix}^{-1} = \begin{pmatrix} \lambda & 0 \\ 0 & 1/\lambda \end{pmatrix}, \quad ad - bc = 1.$$

Then

$$\begin{pmatrix} 0 & a & 0 & b \\ a & 0 & b & 0 \\ 0 & c & 0 & d \\ c & 0 & d & 0 \end{pmatrix} \begin{pmatrix} P & -K_1 \\ K_2 & P' \end{pmatrix} \begin{pmatrix} 0 & d & 0 & -b \\ d & 0 & -b & 0 \\ 0 & -c & 0 & a \\ -c & 0 & a & 0 \end{pmatrix}$$

Geometries of matrices. I₁. Arithmetical construction

$$= \begin{pmatrix} 0 & \lambda & 0 & 0 \\ 1/\lambda & 0 & 0 & 0 \\ 0 & 0 & 0 & 1/\lambda \\ 0 & 0 & \lambda & 0 \end{pmatrix}$$

(notice that the transformation carries $\mathfrak{F}_1$ into $-\mathfrak{F}_1$, but they denote the same transformation). Further

$$\begin{pmatrix} 1 & 0 & 0 & 0 \\ 0 & \lambda & 0 & 0 \\ 0 & 0 & 1 & 0 \\ 0 & 0 & 0 & 1/\lambda \end{pmatrix}$$

carries $\mathfrak{F}$ into $\mathfrak{F}_2$. If

$$\begin{pmatrix} p_1 - \lambda & -k_1 \\ k_2 & p_2 - \lambda \end{pmatrix}$$

has a double elementary divisor, we may take

$$\begin{pmatrix} p_1 & -k_1 \\ k_2 & p_2 \end{pmatrix} = \pm \begin{pmatrix} 1 & 0 \\ 1 & 1 \end{pmatrix}.$$

Now we have to consider the case

$$\begin{pmatrix} 0 & 1 & 0 & 0 \\ 1 & 0 & 0 & 0 \\ 0 & 1 & 0 & 1 \\ -1 & 0 & 1 & 0 \end{pmatrix}.$$

The transformation

$$\begin{pmatrix} 1 & 0 & 0 & 0 \\ 0 & 1 & 0 & 0 \\ -1 & 0 & 1 & 0 \\ 0 & 0 & 0 & 1 \end{pmatrix}$$

carries $\mathfrak{F}$ into $\mathfrak{F}_2$. Thus, we have the theorem.

Theorem 6 Any triple of commutative involutions of the first kind may be carried into $\mathfrak{F}_1$, $\mathfrak{F}_2$ and $\mathfrak{F}_3$ simultaneously.

Proof We may assume that the first two are $\mathfrak{F}_1$ and $\mathfrak{F}_2$. Let the third one be $\mathfrak{F}$. Since $\mathfrak{F}_1\mathfrak{F} = -\mathfrak{F}\mathfrak{F}_1$, we have

$$\mathfrak{F} = \begin{pmatrix} 0 & p_1 & 0 & -k_1 \\ p_2 & 0 & k_1 & 0 \\ 0 & k_2 & 0 & p_2 \\ -k_2 & 0 & p_1 & 0 \end{pmatrix}, \quad p_1p_2 + k_1k_2 = 1.$$

Since $\mathfrak{F}_2\mathfrak{F} = -\mathfrak{F}\mathfrak{F}_2$, we have $p_1 = -p_2$. We may change it into $\mathfrak{F}_3$, since

$$\begin{vmatrix} p_1 - \lambda & -k_1 \\ k_2 & -p_1 - \lambda \end{vmatrix} = \lambda^2 + 1.$$

5. Involution of the second kind

An involution of the second kind has two isolated fixed points with arithmetic distance 2. Conversely, any two given points, with arithmetic distance 2, will serve as the isolated fixed points of an involution of the second kind, which is uniquely determined by them. In fact, let 0 and ∞ be fixed points, then the transformation takes the form

$$Z_1 = -AZA', \quad A^2 = I.$$

We have P such that

$$PAP' = \begin{cases} \pm I, \\ \pm \begin{pmatrix} 1 & 0 \\ 0 & -1 \end{pmatrix}. \end{cases}$$

The latter case cannot happen, since then 0 would not be an isolated fixed point. Thus we have

$$Z_1 = -Z.$$

Now we may define harmonic ranges. Four points Z_1, Z_2, Z_3, Z_4, no two of them with arithmetic distance less than 2, are said to form a harmonic range, if the involution determined by Z_1, Z_2 permutes Z_3 and Z_4.

Analytically, we let

$$(Z_1, Z_2, Z_3, Z_4) = ((Z_1 - Z_3)(Z_1 - Z_4)^{-1})((Z_2 - Z_3)(Z_2 - Z_4)^{-1})^{-1}.$$

The involution

$$(Z - Z_1)(Z - Z_2)^{-1} = -(Z^* - Z_1)(Z^* - Z_2)^{-1}$$

carries Z_3 into Z_4. Evidently

$$(Z_1, Z_2, Z_3, Z_4) = -I.$$

This condition is sufficient as well as necessary.

Thus the "invariance of arithmetic distances" implies the invariance of "harmonic range".

Orthogonal classification of Hermitian matrices*

1. Introduction

Elsewhere[①] the author established the symplectic classification of Hermitian matrices, which has applications to the geometry of symmetric matrices. It is the purpose of the present paper to treat the analogous problem: the orthogonal classification of Hermitian matrices. In other words, it may also be described as the quasi-unitary classification of symmetric matrices. Besides their own interest, the results of the present paper have applications in the geometries of skew-symmetric and symmetric matrices, as well as in the theory of automorphic functions of a matrix variable.

It should be noted that the method previously used is only applicable to the present problem for matrices of even order. In order to establish the general solution, we introduce here a different method.

Unless the contrary is stated, throughout the paper, capital latin letters denote n-rowed square matrices. Further let M' and $\overline{M}$ respectively denote the transposed matrix and conjugate imaginary matrix of M. I denotes the identity matrix.

2. Statement of the problems

Two Hermitian matrices H and K are said to be *conjunctive orthogonally*, if there is an orthogonal matrix P such that

$$PHP' = K. \qquad (1)$$

From (1) and the orthogonality of P we deduce immediately that

$$P\overline{H}HP' = \overline{K}K. \qquad (2)$$

Therefore, if two Hermitian matrices H and K are conjunctive orthogonally, the elementary divisors of the characteristic matrices of $\overline{H}H$ and $\overline{K}K$ are the same .

In the converse, two problems arise:

1. Does $\overline{H}H$ take any prescribed elementary divisors? The answer is in the negative. Then we ask further:

---

* Presented to the Society, February 23, 1946; received by the editors, October 18, 1945. Reprinted from *Transactions of the American Mathematical Society*, 1946, **59**(3): 508-523.

① *Amer. J. Math.*, 1944, **66**: 531-563

2. For a given admissible set of elementary divisors, is there a unique H, apart from orthogonal conjunctiveness, such that $\overline{H}H$ takes the given set as its elementary divisors?

The answer is also in the negative. More definitely, the elementary divisors of $\overline{H}H$ characterize the orthogonal conjunctiveness of H with a reservation about the uncertainty of signs related to "signature".

More generally, apparently, we have the following problem. Two pairs (H,S) and (H_1,S_1) of matrices, where H and H_1 are Hermitian and S and S_1 are symmetric, are said to be equivalent if there exists a nonsingular matrix P such that

$$\overline{P}HP' = H_1 \tag{3}$$

and

$$PSP' = S_1. \tag{4}$$

Consequently, we have

$$\overline{P}HS^{-1}\overline{HS}^{-1}\overline{P}^{-1} = H_1 S_1^{-1} \overline{H_1}\, \overline{S_1}^{-1}, \tag{5}$$

if S is nonsingular. Hereafter we discuss only the case with nonsingular S. Analogously, we have also two problems.

3. Solution of problem 1

Theorem 1 *Let T be a symmetric matrix. The equation*

$$\overline{H}H = T \tag{6}$$

is soluble in an Hermitian H, if and only if the elementary divisors of T have the following two properties:

(i) Each elementary divisor corresponding to a negative characteristic root must occur an even number of times;

(ii) Complex elementary divisors must occur only in complex conjugate pairs.

Analogously, we have

Theorem 1' *The equation*

$$HS^{-1}\overline{HS}^{-1} = Q \tag{7}$$

is soluble in symmetric S and Hermitian H if and only if the elementary divisors of Q satisfy (i) and (ii) of Theorem 1.

Equivalence of both theorems. If $S^{-1} = \Gamma\Gamma'$ in (7), we have

$$(\overline{\Gamma}'H\Gamma)(\Gamma'\overline{H}\,\overline{\Gamma}) = \overline{\Gamma}'Q(\overline{\Gamma}')^{-1}. \tag{8}$$

Thus, Theorem 1′ implies Theorem 1. Conversely, since we have Γ such that $\overline{\Gamma}Q(\overline{\Gamma}')^{-1} = T$, Theorem 1 also implies Theorem 1′.

The sole difficulty lies in proving the property (i).

Proof of the theorems (1) Suppose that

$$T = \begin{pmatrix} q_1 & 0 \\ 0 & q_2 \end{pmatrix}, \tag{9}$$

where q_1 and $\bar{q}_2$ have no characteristic root in common. Let

$$H = \begin{pmatrix} h_1 & k \\ \bar{k}' & h_2 \end{pmatrix}. \tag{10}$$

From (6), (9) and (10), we have

$$\bar{h}_1 h_1 + \bar{k}k' = q_1, \tag{11}$$

$$\bar{h}_1 k + \bar{k}h_2 = 0 \tag{12}$$

and

$$k'k + \bar{h}_2 h_2 = q_2. \tag{13}$$

From (11), (12) and (13), we deduce immediately

$$q_1\bar{k} = \bar{h}_1 h_1 \bar{k} + \bar{k}k'\bar{k} = -\bar{h}_1 k\bar{h}_2 + \bar{k}k'\bar{k} = \bar{k}(h_2\bar{h}_2 + \bar{k}'k) = \bar{k}\bar{q}_2.$$

Since q_1 and $\bar{q}_2$ have no characteristic root in common, we have $k = 0$. Thus (6) is soluble if and only if both equations

$$h_1\bar{h}_1 = q_1, \quad h_2\bar{h}_2 = q_2$$

are soluble in Hermitian h_1 and h_2. Thus we have to consider only either that T has a real root or that T has a pair of conjugate imaginary roots.

(2) Now we introduce the notations:

$$j^{(p)} = \begin{pmatrix} 0 & 0 & \cdots & 0 & 1 \\ 0 & 0 & \cdots & 1 & 0 \\ \vdots & \vdots & & \vdots & \vdots \\ 1 & 0 & \cdots & 0 & 0 \end{pmatrix} = (a_{ij}),$$

where $a_{ij} = 1$ if $i+j = p+1$ and 0 otherwise. Let $j_\alpha^{(p)}$ be the ordinary Jordan's canonical form[1] of a matrix with elementary divisor $(x-\alpha)^p$. Then jj_α is symmetric.

It is well known that there exists a real polynomial $f(x)$ such that

$$f(j_1)^2 = j_1.$$

Let $s = \alpha^{1/2} j f(j_1)$ which is symmetric, since

$$jj_1^\sigma = j_1' j j_1^{\sigma-1} = \cdots = j_1'^\sigma j.$$

It can be verified easily that s satisfies

$$sjs = \alpha jj_1. \tag{14}$$

(3) Now we consider the case thet T has only a pair of conjugate imaginary characteristic roots. Evidently the elementary divisors of $\overline{H}H - \lambda I$ and $H\overline{H} - \lambda I$ are the same; the solvability, therefore, implies the property (ii) of the theorem.

Conversely, the pair of symmetric matrices (I,T) is congruent to a pair of matrices (T_1, T_2) which is a direct sum of the following pairs:

$$\begin{pmatrix} j & 0 \\ 0 & j \end{pmatrix}, \quad \begin{pmatrix} \alpha jj_1 & 0 \\ 0 & \bar\alpha jj_1 \end{pmatrix}.$$

Let

$$\Gamma\Gamma' = T_1, \quad \Gamma T \Gamma' = T_2.$$

Then, (6) implies

$$(\Gamma \overline{H} \overline{\Gamma}') \overline{T}_1^{-1} (\overline{\Gamma} H \Gamma') = T_2 \tag{15}$$

and conversely.

Since T_1 is real and $T_1^2 = I$, the solvability of (6) is equivalent to that of

$$\overline{H} T_1 H = T_2.$$

Since the equation

$$\begin{pmatrix} 0 & s \\ \bar s & 0 \end{pmatrix} \begin{pmatrix} j & 0 \\ 0 & j \end{pmatrix} \begin{pmatrix} 0 & \bar s \\ s & 0 \end{pmatrix} = \begin{pmatrix} \alpha jj_1 & 0 \\ 0 & \bar\alpha jj_1 \end{pmatrix}$$

---

[1] For example, $j_\alpha^{(2)} = \begin{pmatrix} \alpha & 1 \\ 0 & \alpha \end{pmatrix}$.

is equivalent to $sjs = \alpha jj_1$, which is soluble in s by (14), the theorem is proved for the case with complex characteristic roots.

(4) The easiest case is that in which T has a positive root. Since we may assume that (T_1, T_2) of (3) is a direct sum of

$$(j, \alpha jj_1), \quad \alpha > 0,$$

by (14), there exists a real s such that

$$sjs = \alpha jj_1$$

and the theorem is now proved.

(5) Finally, there comes the difficult case that α is negative. Without loss of generality, we may assume that $\alpha = -1$.

Since, by (14), we have a symmetric matrix s such that

$$\begin{pmatrix} 0 & s \\ \bar{s} & 0 \end{pmatrix} \begin{pmatrix} j & 0 \\ 0 & j \end{pmatrix} \begin{pmatrix} 0 & \bar{s} \\ s & 0 \end{pmatrix} = \begin{pmatrix} -jj_1 & 0 \\ 0 & -jj_1 \end{pmatrix},$$

we see that if (i) is satisfied, (7) is soluble. The sole difficulty lies in the converse.

(6) Now we write

$$T_0 = j^{(s_1)} \dotplus \cdots \dotplus j^{(s_e)}, \tag{16}$$

$$T_1 = j_{-1}^{(s_1)} \dotplus \cdots \dotplus j_{-1}^{(s_e)}. \tag{17}$$

Let $K = HT_0$, then the equation

$$\overline{H}T_1 H = T_0 \tag{18}$$

takes the form

$$\overline{K}T_0 T_1 K = I. \tag{19}$$

From (19), we deduce immediately that

$$KT_0 T_1 = KT_0 T_1 \overline{K} T_0 T_1 K = T_0 T_1 K. \tag{20}$$

Let

$$K = (k_{ij}), \quad k_{ij} = k_{ij}^{(s_i, s_j)}.$$

From (20), we deduce

$$k_{ij} j_{-1}^{(s_j)} = j_{-1}^{(s_i)} k_{ij}. \tag{21}$$

It is known that

$$k_{ij} = \begin{pmatrix} a_{ij} & b_{ij} & c_{ij} & \cdots \\ 0 & a_{ij} & b_{ij} & \cdots \\ \cdots & \cdots & \cdots & \cdots \\ \cdots & \cdots & \cdots & a_{ij} \end{pmatrix}, \quad \text{if } s_i = s_j \tag{22}$$

and in case $s_j > s_k$, we add $s_j - s_k$ rows of zeros below the matrix of the form (21) and in case $s_j < s_k$, we add $s_k - s_j$ columns of zeros before the matrix of the form (21).

From (19), we find

$$\sum_j \bar{k}_{i_j} j_{-1}^{(s_j)} k_{jk} = \delta_{ik} I^{(s_k)}.$$

Now we consider the element at the (1,1) position of the case $s_i = s_k$. By (21),

$$\sum_j \bar{k}_{ij} k_{jk} j_{-1}^{(s_j)} = \delta_{ik} I^{(s_k)}. \tag{23}$$

By the constitution of the first column of j_{-1}, we need only consider the element at the (1,1) position of $k_{ij} k_{jk}$. It equals zero, if $s_j < s_k$ and if $s_j > s_k$. Thus the sum (23) runs only over all those j's with $s_j = s_k$.

Let s be any integer occurring in the set $s_1, \cdots, s_e$. Without loss of generality, we may assume that

$$s_i = s \text{ for } 1 \leqslant i \leqslant m, \quad s_i \neq s \text{ for } m < i \leqslant e.$$

Then, from (23), we deduce

$$-\sum_{j=1}^m \bar{a}_{ij} a_{jk} = \delta_{ik}, \quad 1 \leqslant i, k \leqslant m,$$

that is $\bar{A}A = -I^{(m)}$ if we let $A = (a_{ij})_{1 \leqslant i,j \leqslant m}$. Taking determinants of both sides, we have $|d(A)|^2 = (-1)^m$, which can not hold for odd m.

The theorem is now proved.

4. Square root of an orthogonal matrix

Now we require a result very likely due to Hilton, but his paper is not available here. Accordingly we state the rediscovered result in the following without proof.

We say a matrix P is *orthogonal with respect to a nonsingular symmetric matrix S*, if

$$PSP' = S.$$

In particular, for $S = I$, we omit the phrase "with respect to I".

Two matrices A and B, orthogonal with respect to S, are said to be similar orthogonally, if there exists a matrix P orthogonal with respect to S such that $A = PBP^{-1}$.

Theorem 2 (Hilton) *Every orthogonal matrix is similar orthogonally to a direct sum of orthogonal matrices with elementary divisors either of the form*

(i) $\qquad\qquad\qquad (x-\alpha)^r, \quad (x-1/\alpha)^r,$

or of the form

(ii) $\qquad\qquad\qquad (x \pm 1)^r, \quad for\ odd\ r.$

For the first case, the matrix orthogonal with respect to

$$S = \begin{pmatrix} 0 & I^{(r)} \\ I^{(r)} & 0 \end{pmatrix}$$

is similar to the matrix

$$M = M^{(2r)} = \begin{pmatrix} (j_\alpha^{(r)})' & 0 \\ 0 & (j_\alpha^{(r)})^{-1} \end{pmatrix} \qquad (24)$$

orthogonally, and for the second case, the matrix orthogonal with respect to

$$S_1 = \begin{pmatrix} 0 & 0 & I^{(p)} \\ 0 & 1 & 0 \\ I^{(p)} & 0 & 0 \end{pmatrix}$$

is similar to

$$N = N^{(r)} = \begin{pmatrix} (j_{\pm 1}^{(p)})' & 0 & 0 \\ v & 1 & 0 \\ -(j_{\pm 1}^{(p)})^{-1}v'v/2 & -(j_{\pm 1}^{(p)})^{-1}v' & (j_{\pm 1}^{(p)})^{-1} \end{pmatrix} \qquad (25)$$

orthogonally, where $v=(0,0,\cdots,0,1)$ and $2p+1=r$.

Evidently, the squares of the matrices M and N are matrices M_1 and N_1, orthogonal with respect to S and S_1, with elementary divisors

(i′) $\qquad\qquad\qquad (x-\alpha^2)^r, \quad (x-\alpha^{-2})^r$

and

(ii′) $\qquad\qquad\qquad (x+1)^r,$

respectively.

With a slight modification we have the following assertion: for given orthogonal M_1 and N_1 with elementary divisors (i′) and (ii′) respectively, we can find orthogonal matrices M and N with the elementary divisors (i) and (ii) respectively and with α in the right half-plane in the plane of complex numbers. Thus, except when α is negative, an orthogonal matrix has a square root matrix.

Theorem 3 *Let R be an orthogonal matrix without a negative root and suppose that we have an Hermitian matrix H such that*

$$RH = H\overline{R}'. \qquad (26)$$

Then there exists an orthogonal matrix Q with characteristic roots on the right half-plane such that $Q^2 = R$ and

$$QH = H\overline{Q}'. \qquad (27)$$

Proof We may assume that R is a direct sum of the matrices

$$R = r_1 \dotplus r_2 \dotplus \cdots \dotplus r_e,$$

where r_i has elementary divisors either of the form (i') or of the form (ii'). They are denoted typically by M_1 and N_1 respectively. We construct M and N accordingly; they are denoted by $q_1, \cdots, q_e$. Let

$$Q = q_1 \dotplus q_2 \dotplus \cdots \dotplus q_e.$$

Then, evidently, we have $Q^2 = R$.

Further, from

$$RH = H\overline{R}',$$

we have

$$r_i h_{ij} = h_{ij} \overline{r}'_j, \qquad (28)$$

and if this implies

$$q_i h_{ij} = h_{ij} \overline{q}'_j, \qquad (29)$$

we have $QH = H\bar{Q}'$ and consequently we have the theorem.

In order to verify that (28) implies (29), we need only verify that

$$M^2 X = X M^2 \quad \text{and} \quad N^2 X = X N^2$$

implies

$$MX = XM \quad \text{and} \quad NX = XN,$$

respectively, where M and N are defined exactly as in (24) and (25). Straightforward calculation establishes both implications.

5. Solution of the second problem

Theorem 4 If $H\overline{H} = K\overline{K}$, where H and K are both Hermitian, then there exist two matrices P and Q such that

$$\overline{P}HP' = \begin{pmatrix} h_1 & 0 \\ 0 & h_2 \end{pmatrix}$$

and

$$\overline{Q}KQ' = \begin{pmatrix} -h_1 & 0 \\ 0 & h_2 \end{pmatrix}.$$

Proof Since $K^{-1}H = R$ is orthogonal, we have an orthogonal matrix T such that

$$T^{-1}RT = \begin{pmatrix} r_1 & 0 \\ 0 & r_2 \end{pmatrix} = R_1,$$

where r_1 contains all negative characteristic roots of R_1. Then, we have

$$H_1 = K_1 R_1 = \overline{R}_1' K_1,$$

where $H_1 = \overline{T}'HT$, $K_1 = \overline{T}'KT$ are both Hermitian. Since r_1 and $\bar{r}_2$ have no characteristic root in common, we have

$$K_1 = \begin{pmatrix} k_1 & 0 \\ 0 & k_2 \end{pmatrix}$$

and

$$H_1 = \begin{pmatrix} h_1 & 0 \\ 0 & h_2 \end{pmatrix},$$

where

$$k_1 r_1 = \bar{r}_1' k_1$$

and

$$k_2 r_2 = \bar{r}_2' k_2.$$

By Theorem 3, we have an orthogonal q_2 such that $q_2^2 = r_2$ and

$$h_2 = k_2 q_2^2 = \bar{q}_2' k_2 q_2.$$

Further we have an orthogonal matrix q_1 such that $q_1^2 = -r_1$ and

$$h_1 = -k_1 q_1^2 = -\bar{q}_1' k_1 q_1.$$

The theorem follows.

6. Explicit result

The result previously obtained may be concluded in the following theorem.

Theorem 5 Let H be a nonsingular Hermitian matrix and S be a nonsingular symmetrix matrix. The elementary divisors of the characteristic matrix of

$$HS^{-1}\overline{H}\,\overline{S}^{-1} \tag{30}$$

are of the following three types:

(i) $(x-\alpha)^\lambda$, for real $\alpha > 0$;

(ii) $(x-\beta)^\lambda$, $(x-\beta)^\lambda$, for real $\beta < 0$

and

(iii) $(x-\gamma)^\lambda$, $(x-\bar\gamma)^\lambda$, for complex γ.

Let $\alpha_1 = \alpha^{1/2}$, $\beta_1 = i(-\beta)^{1/2}$ and $\gamma_1 = \gamma^{1/2}$; the determination is taken on the right half-plane. Then the pair of matrices (H,S) is equivalent to[①]

$$\begin{cases} H_1 = \sum_\alpha{'}(\pm jj\alpha_1) \dotplus \sum_\beta{'}\begin{pmatrix} 0 & jj\beta_1 \\ \bar j\bar j\beta_1 & 0 \end{pmatrix} \dotplus \sum_\gamma{'}\pm\begin{pmatrix} 0 & jj\gamma_1 \\ \bar j\bar j\gamma_1 & 0 \end{pmatrix}, \\ S_1 = \sum{'}(\pm j) \dotplus \sum{'}\begin{pmatrix} j & 0 \\ 0 & j \end{pmatrix} \dotplus \sum{'}\pm\begin{pmatrix} j & 0 \\ 0 & j \end{pmatrix}. \end{cases} \tag{31}$$

In the theorem we have to justify only a point that the pair of matrices

$$\left(\begin{pmatrix} 0 & jj\beta_1 \\ \bar j\bar j\beta_1 & 0 \end{pmatrix}, \begin{pmatrix} j & 0 \\ 0 & j \end{pmatrix}\right)$$

is equivalent to

$$\left(-\begin{pmatrix} 0 & jj\beta_1 \\ \bar j\bar j\beta_1 & 0 \end{pmatrix}, -\begin{pmatrix} j & 0 \\ 0 & j \end{pmatrix}\right).$$

In fact, we have a nonsingular matrix γ such that $\gamma j\bar j\beta_1 \gamma' = -jj\beta_1$ and $\gamma j\gamma' = j$. Let

$$P = \begin{pmatrix} 0 & i\bar\gamma \\ i\gamma & 0 \end{pmatrix}.$$

Then

$$\bar P \begin{pmatrix} 0 & jj\beta_1 \\ \bar j\bar j\beta_1 & 0 \end{pmatrix} P' = -\begin{pmatrix} 0 & jj\beta_1 \\ \bar j\bar j\beta_1 & 0 \end{pmatrix}$$

---

① $\sum{'}$ denotes direct sum.

and
$$P\begin{pmatrix} j & 0 \\ 0 & j \end{pmatrix} P' = -\begin{pmatrix} j & 0 \\ 0 & j \end{pmatrix}.$$

The situation of the indeterminate signs corresponding to a positive root or to a pair of conjugate complex roots will be clarified by introducing the concept of signature system.

Suppose that $HS^{-1}\overline{H}\,\overline{S}^{-1}$ has an elementary divisor $(x-\alpha)^\lambda$ ($\alpha > 0$) repeated m times. Let p and q be the number of positive signs and negative signs appearing in the expression (31) corresponding to the elementary divisor $(x-\alpha)^\lambda$. Then (p,q), $p+q=m$, is called the signature corresponding to the elementary divisor $(x-\alpha)^\lambda$. Similarly we define the signature corresponding to

$$(x-\gamma)^\lambda(x-\bar{\gamma})^\lambda$$

for complex γ. The totality of elementary divisors and their corresponding signatures is called the elementary divisors with signature system of the pair of matrices (H,S). Evidently, (H,S) and (H_1,S_1) are equivalent, if they have the same elementary divisors with the same signature system. The converse of this statement is also true. The proof of this fact can be constructed by adapting the results to be obtained in §§8-10 and the method used in (6) of §3. Owing to the similarity, we give here no details of the proof.

7. Anti-involutions

An orthogonal matrix T satisfying

$$T\overline{T} = I \tag{32}$$

is called an *anti-involution of the first kind,* and that satisfying

$$T\overline{T} = -I \tag{33}$$

is called an *anti-involution of the second kind.*

Since $TT' = I$, it is evident that the involution T of the first kind is an Hermitian matrix. By Theorem 5, we have an orthogonal matrix P such that

$$PT\bar{P}' = [1,\cdots,1,-1,\cdots,-1]$$

which is a diagonal matrix.

In case of the second kind, iT is Hermitian, the roots of

$$|(iT)(\overline{iT}) - \lambda I| = 0$$

are all equal to -1. The case can only happen for n even. There exists an orthogonal matrix P such that

$$P(iT)\overline{P}' = \begin{pmatrix} 0 & iI^{(p)} \\ -iI^{(p)} & 0 \end{pmatrix},$$

where $n = 2p$.

8. Automorphs

Now we are going to study the group formed by all the matrices P satisfying

$$PSP' = S \tag{34}$$

and

$$\overline{P}HP' = H. \tag{35}$$

By Theorem 5, we may write

$$S = S_1^{(p)} \dotplus S_2^{(q)}, \qquad p + q = n,$$
$$H = H_1^{(p)} \dotplus H_2^{(q)},$$

where the two matrices

$$H_1 S_1^{-1} \overline{H}_1 \overline{S}_1^{-1} \quad \text{and} \quad H_2 S_2^{-1} \overline{H}_2 \overline{S}_2^{-1}$$

have no characteristic root in common. From (34) and (35), we have

$$\overline{P} H S^{-1} = H S^{-1} P. \tag{36}$$

It follows immediately that P is decomposable as

$$P = \begin{pmatrix} P_1^{(p)} & 0 \\ 0 & P_2^{(q)} \end{pmatrix}.$$

Therefore, without loss of generality, we treat the problem with (i) $HS^{-1}\overline{H}\,\overline{S}^{-1}$ having only a pair of conjugate roots, (ii) $HS^{-1}\overline{H}\,\overline{S}^{-1}$ having only a positive root, and (iii) $HS^{-1}\overline{H}\,\overline{S}^{-1}$ having only a negative root.

We shall discuss the cases separately.

9. The case with a pair of complex roots

Now, without loss of generality, we take

$$S = \sum_{1\leqslant i\leqslant e}{}' \pm \begin{pmatrix} j^{(s_i)} & 0 \\ 0 & j^{(s_i)} \end{pmatrix}$$

and

$$H = \sum_{1\leqslant i\leqslant e}{}' \pm \begin{pmatrix} 0 & jj_\gamma^{(s_i)} \\ j\bar{j}_\gamma^{(s_i)} & 0 \end{pmatrix},$$

where $s_1 \geqslant s_2 \geqslant \cdots \geqslant s_e$. Putting

$$J^{(p)} = \sum_{1\leqslant i\leqslant e}{}'(\pm j^{(s_i)})$$

and

$$J_\gamma^{(p)} = \sum_{1\leqslant i\leqslant e}{}'(\pm jj_\gamma^{(s_i)}),$$

we treat, without loss of generality, the case that

$$S = \begin{pmatrix} J & 0 \\ 0 & J \end{pmatrix}, \quad H = \begin{pmatrix} 0 & J_\gamma \\ \bar{J}_\gamma & 0 \end{pmatrix}.$$

Let

$$P = \begin{pmatrix} A & B \\ C & D \end{pmatrix}, \quad A = A^{(p)},$$

and so on. Then, from (36), we have

$$\begin{pmatrix} \bar{A} & \bar{B} \\ \bar{C} & \bar{D} \end{pmatrix} \begin{pmatrix} 0 & J_\gamma J \\ \bar{J}_\gamma J & 0 \end{pmatrix} = \begin{pmatrix} 0 & J_\gamma J \\ \bar{J}_\gamma J & 0 \end{pmatrix} \begin{pmatrix} A & B \\ C & D \end{pmatrix},$$

that is

$$\bar{B}\bar{J}_\gamma J = J_\gamma J C, \quad \bar{A} J_\gamma J = J_\gamma J D,$$
$$\bar{D}\bar{J}_\gamma J = \bar{J}_\gamma J A, \quad \bar{C} J_\gamma J = \bar{J}_\gamma J B. \tag{37}$$

Then, we have

$$B(J_\gamma J)^2 = (\bar{J}_\gamma J)\bar{C}(J_\gamma J) = (\bar{J}_\gamma J)^2 B.$$

It follows that $B = 0$ since the characteristic root of $(J_\gamma J)^2$ is equal to γ^2 and $\gamma^2 \neq \bar{\gamma}^2$. Consequently $C = 0$.

Further, from (37), we have

$$A(\bar{J}_\gamma J)^2 = (\bar{J}_\gamma J)^2 A \tag{38}$$

and
$$D = (J_\gamma J)^{-1} \overline{A} (J_\gamma J). \tag{39}$$

From (34) we deduce
$$AJA' = J. \tag{40}$$

Conversely, for any A satisfying (38) and (40), the matrix
$$P = \begin{pmatrix} A & 0 \\ 0 & (J_\gamma J)^{-1} \overline{A}(J_\gamma J) \end{pmatrix} \tag{41}$$
satisfies our requirement.

Then we need to find those A satisfying both (38) and (40).

Let
$$A = (I+Q)(I-Q)^{-1}$$
(the exceptional case $|I+A| = 0$ is negligible in counting of parameters).

From (38) and (40), we have
$$Q(\overline{J}_\gamma J)^2 = (\overline{J}_\gamma J)^2 Q$$
and
$$QJ + JQ' = 0.$$

Let $QJ = K$, which is a skew symmetric matrix and which satisfies
$$K(J\overline{J}_\gamma)^2 = (J\overline{J}_\gamma)^2 K \tag{42}$$
(notice that $JJ_\gamma J = J'_{\gamma'}$).

Putting
$$K = (k_{ij})_{1 \leqslant i,j \leqslant e}, \quad k_{ij} = k_{ij}^{(s_i, s_j)}$$
and
$$k_{ij} = -k'_{ji},$$
we have
$$k_{ij}(\overline{j}_\gamma^{(s_j)})^2 = (\overline{j}_\gamma^{(s_j)})'^2 k_{ij}.$$

It follows immediately that

$$k_{ij} = \begin{pmatrix} x_1 & x_2 & \cdots & x_{s_{j-1}} & x_{s_j} \\ x_2 & x_3 & \cdots & x_{s_j} & 0 \\ \vdots & \vdots & & \vdots & \vdots \\ x_{s_j} & 0 & \cdots & 0 & 0 \\ 0 & 0 & \cdots & 0 & 0 \\ \vdots & \vdots & & \vdots & \vdots \end{pmatrix},$$

which contains s_i rows, if $s_j < s_i$ with addition of a sufficient number of zero rows, and which contains $2s_j$ parameters (a complex number is counted as two parameters). Further in case $i = j$ evidently k_{ii} equals zero.

Therefore K depends on

$$2s_2 + 2s_3 + \cdots + 2s_e$$
$$+2s_3 + \cdots + 2s_e$$
$$\cdots \cdots \cdots$$
$$+2s_e$$
$$= 2s_2 + 4s_3 + \cdots + 2(e-1)s_e$$

parameters.

10. The case with a positive root

Now we take

$$S = \sum_{1 \leq i \leq e}{}'(\pm j^{(s_i)}),$$
$$H = \sum_{1 \leq i \leq e}{}'(\pm jj_\alpha^{(s_i)}), \quad \alpha > 0,$$

where $s_1 \geq s_2 \geq \cdots \geq s_e$. Let

$$P = (I + Q)(I - Q)^{-1}.$$

We find, from (34) and (35), that

$$QS + SQ' = 0$$

and

$$\overline{Q}H + HQ' = 0.$$

Putting $QS = K$, which is skew symmetric, we have

$$\overline{K}S^{-1}H - HS^{-1}K = 0. \tag{43}$$

Since $\alpha \neq -\alpha$, we find that K is real. Let

$$K = (k_{ij}), \quad k_{ij} = k_{ij}^{(s_i, s_j)}.$$

Then

$$\bar{k}_{ij} j_\alpha^{(s_j)} = (j^{(s_j)} j_\alpha^{(s_i)} j^{(s_i)}) k_{ij} = j^{(s_i)'} k_{ij}.$$

In a manner similar to that of §9, it can be shown that the number of parameters is equal to

$$2s_2 + 4s_3 + \cdots + 2(e-1)s_e.$$

11. The case with a negative root

Now we may let

$$S = \sum_{1 \leq i \leq e}{}' \begin{pmatrix} j^{(s_i)} & 0 \\ 0 & j^{(s_i)} \end{pmatrix},$$

$$H = \sum_{1 \leq i \leq e}{}' i \begin{pmatrix} 0 & jj_\beta^{(s_i)} \\ -jj_\beta^{(s_i)} & 0 \end{pmatrix},$$

where $\beta > 0$. Let

$$J = \sum_{1 \leq i \leq e}{}' j^{(s_i)}, \quad J_\beta = \sum_{1 \leq i \leq e}{}' j^{(s_i)} j_\beta^{(s_i)}.$$

We may consider, without loss of generality, the case

$$S = \begin{pmatrix} J & 0 \\ 0 & J \end{pmatrix}, \quad H = i \begin{pmatrix} 0 & J_\beta \\ -J_\beta & 0 \end{pmatrix}.$$

Letting

$$P = (I + Q)(I - Q)^{-1}$$

and

$$Q \begin{pmatrix} J & 0 \\ 0 & J \end{pmatrix} = K,$$

which is skew symmetric, we find, from (36), that

$$\overline{K} \begin{pmatrix} 0 & JJ_\beta \\ -JJ_\beta & 0 \end{pmatrix} = \begin{pmatrix} 0 & J_\beta J \\ -J_\beta J & 0 \end{pmatrix} K. \tag{44}$$

Let
$$K = \begin{pmatrix} K_1 & L \\ -L' & K_2 \end{pmatrix}. \tag{45}$$

From (44), we have

$$\overline{L}JJ_\beta = J_\beta JL', \quad \overline{L}'JJ_\beta = J_\beta JL, \tag{46}$$

$$\overline{K}_1 JJ_\beta = J_\beta JK_2, \quad \overline{K}_2 JJ_\beta = J_\beta JK_1. \tag{47}$$

Putting $\overline{L}JJ_\beta = T$, which is Hermitian by (46), we have

$$T(JJ_\beta)^2 = (J_\beta J)^2 T, \tag{48}$$

$$K_1(JJ_\beta)^2 = (J_\beta J)^2 K_1 \tag{49}$$

and

$$K_2 = (J_\beta J)^{-1} \overline{K}_1 (JJ_\beta). \tag{50}$$

Consequently, for T and K_1 satisfying (48) and (49) and K_2 defined by (50), we have K defined by (45) to meet our requirement.

The number of parameters of K_1 is equal to

$$2s_2 + 4s_3 + \cdots + 2(e-1)s_e$$

and the number of T is equal to

$$2s_2 + 4s_3 + \cdots + 2(e-1)s_e + s_1 + s_2 + \cdots + s_e.$$

The total number of parameters of K is, therefore, equal to

$$s_1 + 5s_2 + 9s_3 + \cdots + (4e-3)s_e.$$

12. Automorphs (continuation)

As a consequence of the previous results we have the following statement: let the roots of $HS^{-1}\overline{H}\,\overline{S}^{-1}$ be

$$\alpha_1, \cdots, \alpha_\lambda \quad (\alpha > 0)$$

with the multiplicities

$$p_1, \cdots, p_\lambda;$$

$$\beta_1, \cdots, \beta_\mu; \quad \bar{\beta}_1, \cdots, \bar{\beta}_\mu \quad (\beta \text{ complex})$$

with the multiplicities
$$q_1, \cdots, q_\mu; \quad q_1, \cdots, q_\mu$$
and
$$\gamma_1, \cdots, \gamma_\nu \quad (\gamma < 0)$$
with the multiplicities
$$2r_1, \cdots, 2r_\nu$$
respectively. Then *the group of automorphs of (H,S) depends on at least*
$$r_1 + r_2 + \cdots + r_\nu \tag{51}$$
parameters (evidently, this is a best possible constant).

Theorem 6 *Given λ and an Hermitian matrix H, the symmetric matrices S satisfying*
$$d(HS^{-1}\overline{H}\,\overline{S}^{-1} - \lambda I) = 0 \tag{52}$$
depends on $n(n+1) - 1, n(n+1) - 2$, and $n(n+1) - 3$ parameters according as λ is positive, complex, and negative respectively.

Proof The group of conjunctive automorphs of H, that is, the group formed by Γ satisfying
$$\overline{\Gamma} H \Gamma' = H,$$
depends on n^2 parameters. Now we ask what is the number of parameters of distinct
$$\Gamma S \Gamma'.$$
Since the matrix Γ satisfying
$$\overline{\Gamma} H \Gamma' = H, \quad \Gamma S \Gamma' = S$$
depends on not more than $r_1 + \cdots + r_\nu$ parameters, if the roots of $HS^{-1}\overline{H}\,\overline{S}^{-1}$ are given at the beginning of the section, the totality of different symmetric matrices $\Gamma S \Gamma'$ depends on $n^2 - r_1 - \cdots - r_\nu$ parameters.

For a given positive λ, the manifold formed by S (varying all the other roots) depends on $n^2 + (n-1)$ parameters. Similarly we have the result for a complex root.

For a given negative root, the number of parameters of other roots is equal to $n-2$. Further the different symmetric $\Gamma S \Gamma'$, for all roots being given, depends on at most $n^2 - 1$ parameters (in case all the others are non-negative $n^2 - 1$ is the exact number). Thus the total number of parameters is equal to $n^2 - n - 3$.

Finally, the author wishes to express his warmest thanks to the referee for his help with the manuscript.

Geometries of matrices. II. Study of involutions in the geometry of symmetric matrices*

1. Introduction

The paper contains a detailed study of the involutions in the geometry of symmetric matrices over the complex field. It is one of the aims of the paper to establish the following theorem:

A topological automorphism of the group formed by the symplectic transformations is either an inner automorphism or an anti-symplectic transformation.

More precisely, we identify two symplectic matrices $\mathfrak{T}$ and $-\mathfrak{T}$ as a symplectic transformation $\mathfrak{T}_0$. A continuous automorphism of the group formed by $\mathfrak{T}_0$ is either of the form $\mathfrak{P}_0\mathfrak{T}_0\mathfrak{P}_0^{-1}$ or $\mathfrak{P}_0\mathfrak{T}_0^*\mathfrak{P}_0^{-1}$, where $\mathfrak{P}_0$ denotes a symplectic transformation and $\mathfrak{T}_0^*$ is the conjugate complex of $\mathfrak{T}_0$.

The following result, which can also be derived from Mohr's results[1] on the representations of the symplectic group, can be obtained as an immediate consequence of our present theorem: Every topological automorphism of the group formed by all symplectic matrices (that is, we do not identify $\mathfrak{T}$ and $-\mathfrak{T}$) is either an inner automorphism or the conjugate complex of an inner automorphism. Actually, by means of the method used in the paper, an independent proof of this result can be obtained which is much simpler than that of the first theorem, since the distinction between involutions of the first and the second kind now is apparent.

In the course of our discussion, we find the explicit normal forms of involutions and anti-involutions. The manifold of the fixed points of all sorts of involutions has also been determined completely.

As an introduction, several types of geometries keeping an involution or an anti-involution as absolute have been enumerated. Those obtained from anti-involutions are generalizations of non-Euclidean geometries and those obtained from involutions give us several new types of geometries, whose real analogy (which will be given

---

* Presented to the Society, December 29, 1946; received by the editors February 27, 1946. Reprinted from *Transactions of the American Mathematical Society*, 1947, **61**: 193–228.

[1] Göttingen Dissertation, 1933. The author is indebted to the referee for this reference, but unfortunately it is not available in China.

elsewhere later) is a generalization of Möbius geometry of circles.

Furthermore, the author shows that every symplectic transformation is a product of *two* involutions and *four* anti-involutions, and that for $n = 2^\sigma \tau$, τ odd, in the space of symmetric matrices of order n, we have at most $\sigma + 3$ pairs of points of which any two pairs separate each other harmonically.

Algebraically speaking, the last result is equivalent to the following one: let $\mathfrak{T}_1, \cdots, \mathfrak{T}_s$ be symplectic matrices of order $2n$ satisfying

$$\mathfrak{T}_i^2 = -\mathfrak{J}, \qquad \mathfrak{T}_i \mathfrak{T}_j = -\mathfrak{T}_j \mathfrak{T}_i,$$

where $\mathfrak{J}$ denotes the $2n$-rowed identity. Then $s \leqslant \sigma + 3$ and this maximum is attained. As a by-product the author establishes also that if $S_1, \cdots, S_s$ are n-rowed symmetric matrices satisfying $S_i^2 = -I, S_i S_j = -S_j S_i$, then $s \leqslant \sigma + 1$. In case of skew symmetric matrices, we have $s \leqslant \sigma$. These maximums are all attained.

As in I[①], capital latin letters denote $n \times n$ matrices unless the contrary is stated. On the other hand, we use $M^{(l,m)}$ to denote an $l \times m$ matrix and $M^{(m)} = M^{(m,m)}$. I and 0 denote the identity and zero matrices respectively. We use also

$$\mathfrak{F} = \begin{pmatrix} 0 & I \\ -I & 0 \end{pmatrix}, \qquad \mathfrak{F} = \begin{pmatrix} I & 0 \\ 0 & I \end{pmatrix},$$

which are $2n$-rowed matrices. p and q denote two integers satisfying $p + q = n$.

2. Classification of involutions

First of all we identify the transformations which have the same effect in the space of symmetric matrices.

Theorem 1 *In the space of symmetric matrices (in homogeneous coordinates), two substitutions*

$$Q(W_1, W_2) = (Z_1, Z_2)\mathfrak{T} \tag{1}$$

and

$$Q_0(W_1, W_2) = (Z_1, Z_2)\mathfrak{F}_0 \tag{2}$$

induce the same mapping of the space, if and only if

$$\mathfrak{T} = \pm \mathfrak{F}_0. \tag{3}$$

[①] The first paper of the series will be referred to as I (*Trans. Amer. Math. Soc.*, 1945, **57**: 441-481).

In particular, (1) carries every point of the space into itself, if and only if

$$\mathfrak{T} = \pm \mathfrak{J}.$$

Proof It is sufficient to establish the second part of the theorem. Putting $Z_1 = 0$, we have, by the supposition, $W_1 = 0$. Hence $C = 0$, if we put

$$\mathfrak{T} = \begin{pmatrix} A & B \\ C & D \end{pmatrix}. \tag{4}$$

Similarly, putting $Z_2 = 0$, we find $B = 0$. Now the transformation becomes

$$W = AZA'$$

in the nonhomogeneous coordinate system. By the assumption,

$$Z = AZA' \tag{5}$$

holds for all Z, consequently $A = \pm I$.

From Theorem 1, we deduce at once the following theorem.

Theorem 2 *A symplectic transformation*

$$Q(W_1, W_2) = (Z_1, Z_2)\mathfrak{T}, \quad \mathfrak{T} = \begin{pmatrix} A & B \\ C & D \end{pmatrix}$$

is an involution if and only if

$$\begin{pmatrix} A & B \\ C & D \end{pmatrix} = \pm \begin{pmatrix} D' & -B' \\ -C' & A' \end{pmatrix}, \tag{6}$$

that is, $\mathfrak{T}\mathfrak{F}$ *is either skew symmetric or symmetric according as* $\mathfrak{T}^2 = \mathfrak{J}$ *or* $-\mathfrak{J}$.

By an involution, we understand a symplectic transformation whose square induces the identity mapping. Similarly, we define an anti-involution as an anti-symplectic mapping whose square induces the identity mapping.

Theorem 3 *An anti-symplectic transformation*

$$Q(W_1, W_2) = (\bar{Z}_1, \bar{Z}_2)\mathfrak{T}, \quad \mathfrak{T} = \begin{pmatrix} A & B \\ C & D \end{pmatrix} \tag{7}$$

is an anti-involution, if and only if

$$\begin{pmatrix} A & B \\ C & D \end{pmatrix}^* = \pm \begin{pmatrix} D' & -B' \\ -C' & A' \end{pmatrix}, \tag{8}$$

that is, $\mathfrak{TF}$ is either a skew Hermitian or an Hermitian matrix according as $\mathfrak{TT}^* = \mathfrak{J}$ or $-\mathfrak{J}$.

Proof From (7) and
$$Q^*(Z_1, Z_2) = (\bar{P}_1, \bar{P}_2)\mathfrak{T}, \tag{9}$$

we have
$$\bar{Q}^*Q(W_1, W_2) = (P_1, P_2)\mathfrak{T}^*\mathfrak{T}.$$

This represents the identical mapping, if and only if $\mathfrak{T}^*\mathfrak{T} = \pm\mathfrak{J}$. Further, since $\mathfrak{T}$ is symplectic, that is $\mathfrak{TFT}' = \mathfrak{F}$, we have
$$\mathfrak{T}^*\mathfrak{F} = \mathfrak{T}^*(\mathfrak{TFT}') = \pm\mathfrak{FT}' = \mp(\mathfrak{T}^*\mathfrak{F})^{*\prime}.$$

The theorem is now evident.

Definition 1 The involutions satisfying $\mathfrak{T}^2 = \mathfrak{J}$ are called *involutions of the first kind* and those satisfying $\mathfrak{T}^2 = -\mathfrak{J}$ are called *involutions of the second kind*.

Definition 2 The anti-involutions satisfying $\mathfrak{TT}^* = \mathfrak{J}$ are called *anti-involutions of the first kind* and those satisfying $\mathfrak{TT}^* = -\mathfrak{J}$ are called *anti-involutions of the second kind*.

3. Normal form of involutions

Suppose that $\mathfrak{T}$ is an involution of the first kind. By Theorem 2, $\mathfrak{TF}$ is skew symmetric. Now we consider the pair of skew symmetric matrices $(\mathfrak{TF}, \mathfrak{F})$. The elementary divisors of the matrix $\mathfrak{TF} - \lambda\mathfrak{F} = (\mathfrak{T} - \lambda\mathfrak{J})\mathfrak{F}$ are those of $\mathfrak{T} - \lambda\mathfrak{J}$. Since $\mathfrak{T}^2 = \mathfrak{J}$, the elementary divisors are all simple and the characteristic roots are ± 1. Since the determinant of $\mathfrak{T}$ is equal to 1, the multiplicity of the root -1 is even. The pair of skew symmetric matrices
$$\left(\begin{pmatrix} 0 & H \\ -H & 0 \end{pmatrix}, \mathfrak{F} \right), \quad H = [1, \cdots, 1, -1, \cdots, -1]$$

has the same elementary divisors as the pair of matrices $(\mathfrak{TF}, \mathfrak{F})$. Here H is a diagonal matrix with p terms 1 and q terms -1. Hence we have a matrix $\mathfrak{P}$ such that
$$\mathfrak{P}(\mathfrak{TF}, \mathfrak{F})\mathfrak{P}' = \left(\begin{pmatrix} 0 & H \\ -H & 0 \end{pmatrix}, \mathfrak{F} \right).$$

Thus $\mathfrak{P}$ is symplectic, and
$$\mathfrak{P}\mathfrak{T}\mathfrak{P}^{-1} = -\mathfrak{P}\mathfrak{T}\mathfrak{F}\mathfrak{P}'\mathfrak{F} = -\begin{pmatrix} 0 & H \\ -H & 0 \end{pmatrix} \begin{pmatrix} 0 & I \\ -I & 0 \end{pmatrix} = \begin{pmatrix} H & 0 \\ 0 & H \end{pmatrix},$$

which gives the transformation
$$W = HZH. \tag{10}$$

Therefore we have the following theorem.

Theorem 4 *Every involution of the first kind is equivalent to* (10) *symplectically, where H is a diagonal matrix with p positive 1s' and q negative 1s' and $p \leqslant q$. Further, no two of these involutions are equivalent.*

The last sentence can be justified by considering the multiplicity of the characteristic root 1 of the symplectic matrix.

Definition 3 This involution is said to be *of signature (p, q)*.

Now we consider an involution $\mathfrak{T}$ of the second kind. Then we have a pair of matrices $(\mathfrak{TF}, \mathfrak{F})$; the first matrix is symmetric and the second skew symmetric. The characteristic roots of $\mathfrak{T}$ are $\pm i$ and the elementary divisors are all simple. Let[①] $d(\mathfrak{T} - \lambda \mathfrak{J}) = f(\lambda)$. Since $\mathfrak{TF} = \mathfrak{F}\mathfrak{T}'^{-1}$, we have

$$f(\lambda) = d(\mathfrak{T} - \lambda \mathfrak{J})d(\mathfrak{F}) = d(\mathfrak{F}\mathfrak{T}'^{-1} - \lambda \mathfrak{F})$$
$$= \lambda^{2n} d\left(\mathfrak{T}' - \frac{1}{\lambda}\mathfrak{J}\right) = \lambda^{2n} f\left(\frac{1}{\lambda}\right).$$

Hence the multiplicities of the characteristic roots i and $-i$ are equal.

Since

$$\begin{pmatrix} I & 0 \\ 0 & I \end{pmatrix} - \lambda \begin{pmatrix} 0 & I \\ -I & 0 \end{pmatrix}$$

has the same elementary divisors as $\mathfrak{TF} - \lambda \mathfrak{F}$, we have a symplectic matrix $\mathfrak{P}$ carrying $\mathfrak{T}$ into

$$\begin{pmatrix} 0 & I \\ -I & 0 \end{pmatrix},$$

which corresponds to the transformation

$$W = -Z^{-1}. \tag{11}$$

Hence, we have the following theorem.

Theorem 5 *Every involution of the second kind is equivalent to* (11) *symplectically.*

Remarks (1) For $n = 1$, no involution of the first kind exists.

---

① $d(X)$ denotes the determinant of the matrix X.

(2) The following normal form of an involution of the second kind is sometimes useful:
$$W = -Z. \tag{12}$$

(3) In the case $n = 2p$, we sometimes use
$$W = \begin{pmatrix} 0 & I^{(p)} \\ I^{(p)} & 0 \end{pmatrix} Z \begin{pmatrix} 0 & I^{(p)} \\ I^{(p)} & 0 \end{pmatrix} \tag{13}$$

as the normal form of an involution of the signature (p, p).

Evidently, two simplectic transformations with matrices $\mathfrak{T}_1$ and $\mathfrak{T}_2$ are commutative if and only if
$$\mathfrak{T}_1 \mathfrak{T}_2 = \pm \mathfrak{T}_2 \mathfrak{T}_1. \tag{14}$$

Consequently, we see that the product of two commutative involutions $\mathfrak{T}_1$ and $\mathfrak{T}_2$ of the first kind is an involution either of the first kind or of the second kind according as $\mathfrak{T}_1 \mathfrak{T}_2 = \mathfrak{T}_2 \mathfrak{T}_1$ or $\mathfrak{T}_1 \mathfrak{T}_2 = -\mathfrak{T}_2 \mathfrak{T}_1$. In particular, for n even, $n = 2p$, say, (11) may be regarded as a product of two involutions of the first kind:
$$W = \begin{pmatrix} 0 & I^{(p)} \\ -I^{(p)} & 0 \end{pmatrix} Z \begin{pmatrix} 0 & -I^{(p)} \\ I & 0 \end{pmatrix}$$

and
$$W = -\begin{pmatrix} 0 & I^{(p)} \\ -I & 0 \end{pmatrix} Z^{-1} \begin{pmatrix} 0 & I^{(p)} \\ I & 0 \end{pmatrix}.$$

For n odd, such a decomposition does not exist.

Further, any involution of the first kind is a product of involutions of the first kind of the signature $(1, n-1)$, which are called *fundamental involutions*.

4. Equivalence of anti-involutions

A little attention should be paid to the equivalence of anti-symplectic transformations. Let $(W_1, W_2) = Q(\bar{Z}_1, \bar{Z}_2)\mathfrak{T}$. From $(U_1, U_2) = Q_1(Z_1, Z_2)\mathfrak{P}$, $(V_1, V_2) = Q_2(W_1, W_2)\mathfrak{P}$, we have
$$(V_1, V_2) = Q_2(W_1, W_2)\mathfrak{P} = Q_2 Q(\bar{Z}_1, \bar{Z}_2)\mathfrak{T}\mathfrak{P} = Q_2 Q \bar{Q}_1^{-1}(U_1, U_2)\mathfrak{P}^{*-1}\mathfrak{T}\mathfrak{P}.$$

Thus we have the following definition.

Definition Two anti-symplectic transformations with matrices $\mathfrak{T}_1$ and $\mathfrak{T}_2$ are said to be equivalent, if there exists a symplectic matrix $\mathfrak{P}$ such that
$$\mathfrak{P}^{*-1}\mathfrak{T}_1\mathfrak{P} = \mathfrak{T}_2. \tag{15}$$

Notice that if (U_1, U_2) and (Z_1, Z_2) (and (V_1, V_2) and (W_1, W_2)) are related antisymplectically, then we have $\mathfrak{P}^{*-1}\mathfrak{T}_1^*\mathfrak{P} = \mathfrak{T}_2$ instead of (15).

5. Relation between anti-involutions and hypercircles and the normal form of the anti-involution

Let $\mathfrak{H}$ be an Hermitian matrix. The points (Z_1, Z_2) of the space for which

$$(Z_1, Z_2)\mathfrak{H}(\bar{Z}_1, \bar{Z}_2)' \tag{16}$$

is positive definite form a hypercircle. $\mathfrak{H}$ is called the matrix of the hypercircle. The skew matrix

$$\mathfrak{H}'\mathfrak{F}\mathfrak{H} \tag{17}$$

is called the discriminantal matrix of $\mathfrak{H}$.

For an anti-involution of the first kind $\mathfrak{T}$, we have a hypercircle with the matrix

$$i\mathfrak{T}\mathfrak{F}, \tag{18}$$

in fact, since $\mathfrak{T}$ is symplectic, $i\mathfrak{T}\mathfrak{F} = i\mathfrak{F}\mathfrak{T}'^{-1} = i\mathfrak{F}\mathfrak{T}^{*\prime} = (i\mathfrak{T}\mathfrak{F})^{*\prime}$, and for an anti-involution of the second kind $\mathfrak{T}$, we have a hypercircle

$$\mathfrak{T}\mathfrak{F}. \tag{19}$$

Their discriminantal matrices are, respectively,

$$(i\mathfrak{T}\mathfrak{F})'\mathfrak{F}(i\mathfrak{T}\mathfrak{F}) = \mathfrak{F}^3 = -\mathfrak{F}$$

and

$$(\mathfrak{T}\mathfrak{F})'\mathfrak{F}(\mathfrak{T}\mathfrak{F}) = -\mathfrak{F}^3 = \mathfrak{F}.$$

Theorem 6 *To each hypercircle with discriminantal matrix $-\mathfrak{F}$ or $\mathfrak{F}$, there corresponds an anti-involution of the first and the second kind respectively.*

Proof (1) Let $\mathfrak{H}$ be an Hermitian matrix satisfying $\mathfrak{H}'\mathfrak{F}\mathfrak{H} = -\mathfrak{F}$. Let

$$\mathfrak{T} = i\mathfrak{H}\mathfrak{F}.$$

Then

$$\mathfrak{T}'\mathfrak{F}\mathfrak{T} = \mathfrak{F}\mathfrak{H}'\mathfrak{F}\mathfrak{H}\mathfrak{F} = -\mathfrak{F}^3 = \mathfrak{F}$$

and

$$\mathfrak{T}^*\mathfrak{T} = -i\mathfrak{H}^*\mathfrak{F}i\mathfrak{H}\mathfrak{F} = -\mathfrak{F}^2 = \mathfrak{J}.$$

Hence $\mathfrak{T}$ is a symplectic matrix and we have an anti-involution of the first kind.

(2) If $\mathfrak{H}'\mathfrak{F}\mathfrak{H} = \mathfrak{F}$, set $\mathfrak{T} = -\mathfrak{H}\mathfrak{F}$. We obtain consequently

$$\mathfrak{T}'\mathfrak{F}\mathfrak{T} = -\mathfrak{F}\mathfrak{H}'\mathfrak{F}\mathfrak{H}\mathfrak{F} = -\mathfrak{F}^3 = \mathfrak{F},$$
$$\mathfrak{T}^*\mathfrak{T} = \mathfrak{H}\mathfrak{F}\mathfrak{H}\mathfrak{F} = -\mathfrak{J},$$

that is, $\mathfrak{T}$ defines an anti-involution of the second kind.

Further, from $\mathfrak{P}^*\mathfrak{T}\mathfrak{P}^{-1} = \mathfrak{T}_1$, we have

$$\mathfrak{P}^*\mathfrak{T}\mathfrak{F}\mathfrak{P}' = \mathfrak{P}^*\mathfrak{T}\mathfrak{P}^{-1}\mathfrak{F} = \mathfrak{T}_1\mathfrak{F}.$$

Therefore, in order to classify anti-involutions, we have to classify their corresponding hypercircles.

Since

$$\mathfrak{H}'\mathfrak{F}\mathfrak{H} - \lambda\mathfrak{F} = \pm\mathfrak{F} - \lambda\mathfrak{F}$$

has characteristic roots either all $+1$ of all -1 and has simple elementary divisors, we have[①] a symplectic matrix $\mathfrak{P}$ such that

$$\mathfrak{P}^*\mathfrak{H}\mathfrak{P}' = \begin{pmatrix} H_1 & 0 \\ 0 & H_2 \end{pmatrix},$$

where H_1 is a diagonal matrix with p terms 1 and q terms -1.

Since the discriminantal matrix is $\pm\mathfrak{F}$, then we have

$$H_1'H_2 = -I, \quad H_2'H_1 = -I$$

for anti-involutions of the first kind, and

$$H_1'H_2 = I, \quad H_2'H_1 = I$$

for anti-involutions of the second kind.

Thus, for anti-involutions of the first kind, the hypercircle is symplectically conjunctive to[②]

$$\begin{pmatrix} I & 0 \\ 0 & -I \end{pmatrix} \tag{20}$$

[①] See the author's paper *On the theory of automorphic functions of a matrix variable*, II, Amer. J. Math., 1944, **66**: 531-563.

[②] Two Hermitian matrices H and K are said to be conjunctive, if there exists a matrix A such that $\bar{A}HA' = K$.

and for the second kind, it is symplectically conjunctive to

$$\begin{pmatrix} H_1 & 0 \\ 0 & H_1 \end{pmatrix}. \tag{21}$$

Therefore, we have the following theorem.

Theorem 7 *An anti-involution of the first kind is equivalent to*

$$W = \bar{Z}^{-1} \tag{22}$$

and an anti-involution of the second kind is equivalent to

$$W = -H\bar{Z}^{-1}H, \tag{23}$$

where H denotes a diagonal matrix with p terms 1 and q terms -1, $p \leqslant q$.

We may also prove that they are all non-equivalent, since their hypercircles are non-equivalent.

An anti-involution, equivalent to (23), is called an *anti-involution of signature* (p,q).

Remark (22) is equivalent to

$$W = -\bar{Z}. \tag{24}$$

In the case $p = q$, (23) is equivalent to

$$W = \begin{pmatrix} 0 & I^{(p)} \\ -I & 0 \end{pmatrix} \bar{Z} \begin{pmatrix} 0 & -I \\ I & 0 \end{pmatrix}. \tag{25}$$

6. Decomposition of involutions into anti-involutions

Now we are going to express involution as product of anti-involutions.

Theorem 8 *Every involution is a product of two commutative anti-involutions of the first kind.*

Before proving Theorem 8, we give the following rules concerning the multiplication of anti-symplectic transformations:

(1) The product of two anti-symplectic transformations with matrices $\mathfrak{T}_1$ and $\mathfrak{T}_2$ is a symplectic transformation with matrix $\mathfrak{T}_1^* \mathfrak{T}_2$;

(2) They are commutative if and only if $\mathfrak{T}_1^* \mathfrak{T}_2 = \pm \mathfrak{T}_2^* \mathfrak{T}_1$.

Proof of Theorem 8 (1) Let

$$\mathfrak{T}_1 = \begin{pmatrix} 0 & iI \\ iI & 0 \end{pmatrix}, \quad \mathfrak{T}_2 = \begin{pmatrix} 0 & H_i \\ H_i & 0 \end{pmatrix}.$$

Then
$$\mathfrak{T}_2^*\mathfrak{T}_1 = \begin{pmatrix} H & 0 \\ 0 & H \end{pmatrix} = \mathfrak{T}_1^*\mathfrak{T}_2,$$

which corresponds to the involution $W = HZH$ of the first kind. Further $\mathfrak{T}_1^*\mathfrak{T}_1 = \mathfrak{T}_2^*\mathfrak{T}_2 = \mathfrak{J}$, hence $\mathfrak{T}_1$ and $\mathfrak{T}_2$ represent two anti-involutions of the first kind.

(2) Let
$$\mathfrak{T}_1 = \begin{pmatrix} iI & 0 \\ 0 & -iI \end{pmatrix}, \quad \mathfrak{T}_2 = \begin{pmatrix} 0 & Ii \\ Ii & 0 \end{pmatrix}.$$

Then $\mathfrak{T}_1^*\mathfrak{T}_1 = \mathfrak{T}_2^*\mathfrak{T}_2 = \mathfrak{J}$. Further

$$\mathfrak{T}_2^*\mathfrak{T}_1 = \begin{pmatrix} 0 & -I \\ I & 0 \end{pmatrix} = -\mathfrak{T}_1^*\mathfrak{T}_2,$$

which corresponds to the involution $Z_1 = -Z^{-1}$ of the second kind.

Theorem 9 *Every anti-involution of the second kind is a product of three mutually commutative anti-involutions of the first kind.*

Proof The anti-involution of the second kind

$$Z_1 = -H\bar{Z}^{-1}H$$

is a product of $Z_1 = -Z_2^{-1}$ and $Z_2 = H\bar{Z}H$, and the former one is a product of two anti-involutions of the first kind:

$$Z_1 = -\bar{Z}_3, \quad Z_3 = \bar{Z}_2^{-1}.$$

The three anti-involutions so obtained are evidently mutually commutative.

Remark Theorems 8 and 9 (and later Theorem 10) suggest that anti-involutions of the first kind can be used as generators in the group of symplectic and anti-symplectic transformations. Does the anti-involution of the second kind play the same role? The answer seems to be negative. In fact, an involution of the second kind cannot be decomposed into a product of two anti-involutions of the second kind. Let $\mathfrak{T}$ be an involution of the second kind. Suppose the contrary, that is, suppose that we have two anti-involutions $\mathfrak{T}_1$ and $\mathfrak{T}_2$ of the second kind such that $\mathfrak{T} = \mathfrak{T}_1^*\mathfrak{T}_2$. Then, we have

$$-\mathfrak{J} = \mathfrak{T}^2 = (\mathfrak{T}_1^*\mathfrak{T}_2)(\mathfrak{T}_1^*\mathfrak{T}_2),$$

that is,

$$\mathfrak{T}_2\mathfrak{T}_1^* = -\mathfrak{T}_1\mathfrak{T}_2^*. \tag{26}$$

Now, we are going to show that, in particular, for $\mathfrak{T}_1 = \mathfrak{F}$ we have no $\mathfrak{T}_2$ satisfying this condition (notice that no generality is lost). From

$$\mathfrak{T}_2 \mathfrak{F} = -\mathfrak{F} \mathfrak{T}_2^*$$

and (by Theorem 3) $\mathfrak{T}_2 \mathfrak{F} = (\mathfrak{T}_2 \mathfrak{F})^{*\prime} = -\mathfrak{F} \mathfrak{T}_2^{*\prime}$, we deduce that $\mathfrak{T}_2$ is symmetric. Further, the equation

$$\mathfrak{T}_2 \mathfrak{T}_2^{*\prime} = \mathfrak{T}_2 \mathfrak{T}_2^* = -\mathfrak{J}$$

is impossible, since the matrix on the left is a positive definite Hermitian matrix, while the matrix on the right is negative definite.

7. Decomposition of symplectic transformations into involutions

First of all we give the normal form of a symplectic matrix $\mathfrak{T}$. We have a symplectic matrix $\mathfrak{P}$ such that $\mathfrak{P}^{-1} \mathfrak{T} \mathfrak{P}$ is a direct sum of symplectic transformations of the forms

$$W^{(\sigma)} = J_{\pm 1}^{(\sigma)} Z^{(\sigma)} J_{\pm 1}^{(\sigma)\prime} + S^{(\sigma)} \tag{27}$$

and

$$W^{(\tau)} = J_\alpha^{(\tau)} Z^{(\tau)} J_\alpha^{(\tau)}, \quad \alpha \neq \pm 1, \tau \text{ odd}, \tag{28}$$

where

$$J_\alpha^{(\tau)} = \begin{pmatrix} \alpha & 0 & \cdots & 0 \\ 1 & \alpha & \cdots & 0 \\ \vdots & \vdots & & \vdots \\ 0 & 0 & \cdots & \alpha \end{pmatrix}, \quad S = \begin{pmatrix} 1 & 0 & \cdots & 0 \\ 0 & 0 & \cdots & 0 \\ \vdots & \vdots & & \vdots \\ 0 & 0 & \cdots & 0 \end{pmatrix}.$$

By a direct sum of two symplectic transformations with matrices

$$\begin{pmatrix} A^{(\sigma)} & B_1 \\ C_1 & D_1 \end{pmatrix}, \quad \begin{pmatrix} A_2^{(\tau)} & B_2 \\ C_2 & D_2 \end{pmatrix},$$

we mean a symplectic transformation with the $2(\sigma + \tau)$-rowed matrix

$$\begin{pmatrix} \begin{pmatrix} A_1 & 0 \\ 0 & A_2 \end{pmatrix} & \begin{pmatrix} B_1 & 0 \\ 0 & B_2 \end{pmatrix} \\ \begin{pmatrix} C_1 & 0 \\ 0 & C_2 \end{pmatrix} & \begin{pmatrix} D_1 & 0 \\ 0 & D_2 \end{pmatrix} \end{pmatrix}.$$

This result, according to valuable information from Professor H. Weyl, is due to Williamson; however, it was also proved by the author independently. Since Williamson's paper is not available in China, the author is obliged to give the preceding result without a necessary quotation.

Theorem 10 *Every symplectic transformation is a product of two involutions of the second kind; consequently, every symplectic transformation is a product of four anti-involutions of the first kind.*

Proof (1) Let

$$J = \begin{pmatrix} 0 & 0 & \cdots & 0 & 1 \\ 0 & 0 & \cdots & 1 & 0 \\ \vdots & \vdots & & \vdots & \vdots \\ 1 & 0 & \cdots & 0 & 0 \end{pmatrix}, \quad \mathfrak{P} = \begin{pmatrix} 0 & J \\ -J & 0 \end{pmatrix}.$$

Evidently, $\mathfrak{P}$ is an involution of the second kind. Let

$$\mathfrak{T} = \begin{pmatrix} J_\alpha & 0 \\ 0 & J_\alpha'^{-1} \end{pmatrix}.$$

Then

$$(\mathfrak{T}\mathfrak{P})^2 = \begin{pmatrix} 0 & J_\alpha J \\ -J_\alpha'^{-1} J & 0 \end{pmatrix}^2 = -\mathfrak{J},$$

since $J_\alpha J = JJ_\alpha'$. Thus $\mathfrak{T}$ is a product of two involutions of the second kind.

(2) Next we consider the case

$$\mathfrak{T} = \begin{pmatrix} J_1 & SJ_1'^{-1} \\ 0 & J_1'^{-1} \end{pmatrix}. \tag{29}$$

Let

$$M^{(n)} = M = (m_{ij}), \quad m_{ij} = \begin{cases} (-1)^{i-1} \begin{pmatrix} i-1 \\ j-1 \end{pmatrix}, & \text{for } i \geqslant j, \\ 0, & \text{for } i < j \end{cases} \tag{30}$$

and $M^2 = (t_{ij})$. Then, by (30),

$$t_{ij} = \sum_{h=1}^{m} m_{ih} m_{hj} = (-1)^{i-1} \sum_{j \leqslant k \leqslant i} (-1)^{k-1} \begin{pmatrix} i-1 \\ k-1 \end{pmatrix} \begin{pmatrix} k-1 \\ j-1 \end{pmatrix} = \begin{cases} 0, & \text{for } i \neq j, \\ 1, & \text{for } i = j. \end{cases}$$

Therefore, we have

$$M^2 = I. \tag{31}$$

Consequently, the matrix

$$\mathfrak{P} = \begin{pmatrix} iM & 0 \\ 0 & -iM' \end{pmatrix}$$

denotes an involution of the second kind. We may verify directly that

$$J_1 M^{(n)} = \begin{pmatrix} 1 & 0 \\ 0 & -M^{(n-1)} \end{pmatrix}.$$

Thus

$$\mathfrak{T}\mathfrak{P} = i \begin{pmatrix} J_1 M & -S J_1'^{-1} M' \\ 0 & -J_1'^{-1} M' \end{pmatrix}$$

is also an involution of the second kind, since $S = J_1 M S M' J_1'$.

(3) Similarly we treat the case

$$\begin{pmatrix} J_{-1} & S J_{-1}'^{-1} \\ 0 & J_{-1}'^{-1} \end{pmatrix}.$$

On account of the result quoted at the beginning of the section, we have the theorem.

8. Geometries induced by anti-involutions

The results of Sections 8, 9, 10, 11 and 12 can be used to introduce several types of geometry; the detailed study will be given elsewhere.

We take an anti-involution $\mathfrak{J}$ of the first kind as an absolute. The group G formed by all symplectic transformations $\mathfrak{T}$ commutative with $\mathfrak{J}$ is called the group of motion. Correspondingly, we have a hypercircle with the matrix $i\mathfrak{J}\mathfrak{F}$. The group of transformations $\mathfrak{T}$ with $\mathfrak{T}\mathfrak{J} = \mathfrak{J}\mathfrak{T}^*$ form a subgroup G_1 whose index in G is equal to 2.

In fact, from

$$\mathfrak{T}_1 \mathfrak{J} = \mathfrak{J} \mathfrak{T}_1^*$$

and

$$\mathfrak{T}_2 \mathfrak{J} = \mathfrak{J} \mathfrak{T}_2^*$$

we deduce $\mathfrak{T}_1 \mathfrak{T}_2 \mathfrak{J} = \mathfrak{T}_1 \mathfrak{J} \mathfrak{T}_2^* = \mathfrak{J} \mathfrak{T}_1 \mathfrak{T}_2^*$.

As the transformation

$$(W_1, W_2) = Q(Z_1, Z_2) \mathfrak{T} \tag{32}$$

of G_1 carries

$$(Z_1, Z_2) i \mathfrak{J} \mathfrak{F} (Z_1, Z_2)^{*\prime} \tag{33}$$

into

$$Q(W_1, W_2) i \mathfrak{T} \mathfrak{J} \mathfrak{F} \mathfrak{T}^{*\prime} (W_1, W_2)^{*\prime} \bar{Q}' = Q(W_1, W_2) i \mathfrak{J} \mathfrak{F} (W_1, W_2)^{*\prime} \bar{Q}'$$

for $\mathfrak{T}$ belonging to G_1, the rank and signature of the Hermitian matrix (33) classify the points of the space into transitive sets. This statement will be proved in the remark of the next section.

Now we take the set of points for which (33) is positive definite. This set is called a *hyperbolic space*. The corresponding geometry is called a *hyperbolic geometry of symmetric matrices*. By Theorem 7, we find that, apart from equivalence, there is one and only one type of hyperbolic geometry. The group G_1 is called the group of motion of the space.

Let P_1 and P_2 be two points of the hyperbolic space. Every motion evidently carries the cross-ratio matrix

$$\{P_1, P_2\} = (P_1, \mathfrak{J}(P_1), P_2, \mathfrak{J}(P_2))$$
$$= (P_1 - P_2)(P_1 - \mathfrak{J}(P_2))^{-1}((\mathfrak{J}(P_1) - P_2)(\mathfrak{J}(P_1) - \mathfrak{J}(P_2))^{-1})^{-1} \quad (34)$$

into a similar matrix. (34) is called the *distance-matrix* between two points. We have the following theorems.

Theorem 11 *The hyperbolic space is transitive.*

Theorem 12 *The distance matrix has simple elementary divisors and positive characteristic roots. Two point-pairs are equivalent if and only if their distance matrices have the same characteristic roots (multiplicities are counted).*

Theorem 13 *The space is symmetric.*

By a symmetric space, we mean that to each point of the space there exists an involution with the point as its isolated fixed point and that every transformation of the space having the point as fixed point is commutative with the involution.

The proof of these three theorems will be given in the next section.

9. Visualization of the hyperbolic geometry

For the sake of concreteness, we take (cf. (24))

$$\mathfrak{J} = i \begin{pmatrix} I & 0 \\ 0 & -I \end{pmatrix}.$$

The hypercircle is now formed by those points $Z = X + iY$ where X and Y are real and Y is positive definite. The group G_1 is formed by the symplectic transformations

$$W = (AZ + B)(CZ + D)^{-1}, \quad (35)$$

where

$$\mathfrak{T} = \begin{pmatrix} A & B \\ C & D \end{pmatrix}$$

is real. This is Siegel's generalization of the Poincaré half-plane.

Sometimes we take, by Theorem 7,

$$\mathfrak{J} = i \begin{pmatrix} 0 & I \\ I & 0 \end{pmatrix}.$$

The space is formed by those points for which $I - Z\bar{Z}$ is positive definite. The group of motions consists of the transformations

$$W = (AZ + B)(\bar{B}Z + \bar{A})^{-1}. \tag{36}$$

This is a generalization of the unit circle in the complex plane.

Theorem 11 is evident, since any point $P = Q + iR$ with positive definite R may be carried into iR by the transformation $W = Z - Q$ of (35) and since there exists a real matrix C such that $CRC' = I$, then $W = C^{-1}ZC'^{-1}$ carries iR into iI.

To prove Theorem 12, we use the second representation. We may take one of the points to be zero. Then $\{Z, 0\} = (Z, \bar{Z}^{-1}, 0, \infty) = Z\bar{Z}$, which is positive definite. We have a unitary matrix U such that

$$UZ\bar{Z}U' = [\lambda_1^2, \cdots, \lambda_n^2], \qquad \lambda_r \geqslant 0$$

and

$$UZU' = [\lambda_1, \cdots, \lambda_n].$$

This gives Theorem 12.

The space is transitive. In order to prove Theorem 13, we have to show that there exists one and only one involution which has 0 as its isolated fixed point and that the subgroup leaving 0 invariant is commutative with it.

The transformation leaving 0 invariant carries also ∞ into itself. Then it takes the form

$$W = AZA'. \tag{37}$$

(37) denotes an involution if and only if $A^2 = \pm I$. The equation $Z = AZA'$ has no other solution but zero if and only if the characteristic roots of the second power matrix of A are different from 1. Further the roots of A are ± 1 and $\pm i$; the only possibility is that $A = \pm iI$ by Theorem 44.4 of Mac-Duffee[1]. Now we have the unique involution

---

[1] *Theory of Matrices*, 1933.

$$W = -Z. \tag{38}$$

It is evidently commutative with all transformations (37).

Remark In order to prove the promised statement of the preceding section, it is sufficient to prove that the points

$$X + Yi, \quad X, \ Y \text{ real},$$

are classified into transitive sets according to the ranks and signatures of Y. This is evident, since $Z_1 = Z - X$ carries the point into Yi and $Z_1 = AZA'$ carries Yi into a diagonal matrix of the form $i[1, \cdots, 1, -1, \cdots, -1, 0, \cdots, 0]$.

10. Elliptic geometries

Now we take $\mathfrak{J}$ to be an anti-involution of the second kind as an absolute. More definitely, we take

$$\mathfrak{J} = \begin{pmatrix} 0 & -H \\ H & 0 \end{pmatrix},$$

where

$$H = [1, \cdots, 1, -1, \cdots, 1],$$

where there are p 1's and q -1's, $q \geqslant q$. The hypercircle has the matrix

$$\begin{pmatrix} H & 0 \\ 0 & H \end{pmatrix}.$$

The space is formed by the points for which

$$Z_1 H \bar{Z}'_1 + Z_2 H \bar{Z}'_2 = (Z_1, Z_2) \mathfrak{J} \mathfrak{F} (Z_1, Z_2)^{*\prime} \tag{39}$$

is of the same signature as H. This geometry will be called the *elliptic geometry of signature* (p, q). The group of motions is defined by the symplectic transformations

$$(Z_1, Z_2) = Q(W_1, W_2) \mathfrak{T}, \tag{40}$$

with

$$\mathfrak{T}\mathfrak{J} = \mathfrak{J}\mathfrak{T}^*, \tag{41}$$

since (40) carries (39) into

$$Q(W_1, W_2) \mathfrak{T} \mathfrak{J} \mathfrak{T}^{*\prime} (W_1, W_2)' \bar{Q}' = Q(W_1, W_2) \mathfrak{J} \mathfrak{F} (W_1, W_2)^{*\prime} \bar{Q}'.$$

In the case $p = q$, the transformations (40) with $\mathfrak{T}\mathfrak{J} = -\mathfrak{J}\mathfrak{T}^*$ instead of (41) also leave the space invariant. Then the group of motions is now constituted by both kinds of transformations.

From (41), we have

$$\mathfrak{T} = \begin{pmatrix} A & B \\ -H\bar{B}H & H\bar{A}H \end{pmatrix}.$$

Theorem 14 *The space is transitive.*

Proof For any point (A, B) of the space making $\bar{A}HA' + \bar{B}HB'$ of signature (p, q) we can find a matrix P such that

$$\bar{P}(\bar{A}HA' + \bar{B}HB')P' = H.$$

Then

$$\mathfrak{P} = \begin{pmatrix} PA & PB \\ -H\bar{P}\bar{B}H & H\bar{P}\bar{A}H \end{pmatrix} \tag{42}$$

is symplectic, in fact

$$PA(PB)' = PAB'P' = (PB)(PA)',$$
$$(-H\bar{P}\bar{B}H)(H\bar{P}\bar{A}H)' = (H\bar{P}\bar{A}H)(-H\bar{P}\bar{B}H)'$$

and

$$PA(H\bar{P}\bar{A}H)' - PB(H\bar{P}\bar{B}H)' = P(AH\bar{A}' + BH\bar{B}')\bar{P}'H = I.$$

The transformation $(Z_1, Z_2) = Q(W_1, W_2)\mathfrak{P}$ is a motion of the space, since

$$\mathfrak{P}^* \begin{pmatrix} H & 0 \\ 0 & H \end{pmatrix} \mathfrak{P}' = \begin{pmatrix} H & 0 \\ 0 & H \end{pmatrix}$$

and it carries $(I, 0)$ into the point (A, B).

The group of stability at the origin, that is, the subgroup of transformations leaving 0 invariant, consists of the transformations of the form

$$W = AZA',$$

where A is a conjunctive automorphism of H, that is, $\bar{A}HA' = H$. We may prove also the following theorem.

Theorem 15 *The elliptic space of signature (p, q) is symmetric.*

As in §9, we may introduce the distance matrix between two points P_1 and P_2 by $\{P_1, P_2\} = (P_1, \mathfrak{J}(P_1), P_2, \mathfrak{J}(P_2))$. Putting $P_1 = Z$ and $P_2 = 0$, we have

$$\{Z, 0\} = ZH\bar{Z}H. \tag{43}$$

Here the problem arises to classify symmetric matrices Z under the group of conjunctive automorphisms of an Hermitian matrix. It would require too much space to solve this problem: hence at present we content ourselves with the special case $H = I$, which is the direct generalization of elliptic geometry.

Theorem 16 *The distance matrix has simple elementary divisors and positive characteristic roots in the elliptic space of signature $(n, 0)$. Two point-pairs are equivalent if and only if their distance matrices have the same characteristic roots (multiplicities are counted).*

Notice that in the elliptic space of signature $(n, 0)$ the group of motions is formed by all unitary symplectic matrices.

11. Geometries induced by involutions

Now we take an involution $\mathfrak{J}$ of the second kind as the absolute; without loss of generality we may take

$$W = -Z. \tag{44}$$

From

$$\begin{pmatrix} A & B \\ C & D \end{pmatrix} \begin{pmatrix} iI & 0 \\ 0 & -iI \end{pmatrix} = \begin{pmatrix} iI & 0 \\ 0 & -iI \end{pmatrix} \begin{pmatrix} A & B \\ C & D \end{pmatrix},$$

we deduce that $B = C = 0$. Thus the transformations of the space are given by

$$Z_1 = AZA', \tag{45}$$

which form a group G_1. If the involution $Z_1 = -Z^{-1}$ is adjoined to G_1, a group G is obtained which contains G_1 as a subgroup of index 2. The geometry of the symmetric matrices under the group G_1 is equivalent to the algebraic problem "congruent classification of matrices". Evidently the space is not transitive. The rank of P is the characteristic invariant for a finite point.

For a pair of finite points P_1 and P_2, we have $(P_1, P_2, 0, \infty) = P_1 P_2^{-1}$. Thus the elementary divisors of $P_1 P_2^{-1}$ characterize the equivalence of a pair of finite points.

Now we take an involution $\mathfrak{J}$ of the first kind with signature (p, q) as an absolute, for example

$$Z_1 = HZH. \tag{46}$$

Now the matrix $\mathfrak{T}$ of the transformations satisfies

$$\begin{pmatrix} A & B \\ C & D \end{pmatrix} \begin{pmatrix} H & 0 \\ 0 & H \end{pmatrix} = \begin{pmatrix} H & 0 \\ 0 & H \end{pmatrix} \begin{pmatrix} A & B \\ C & D \end{pmatrix}.$$

Consequently we have

$$A = \begin{pmatrix} a_1^{(p)} & 0 \\ 0 & a_2^{(q)} \end{pmatrix}, \quad B = \begin{pmatrix} b_1^{(p)} & 0 \\ 0 & b_2^{(q)} \end{pmatrix}, \quad C = \begin{pmatrix} c_1^{(p)} & 0 \\ 0 & c_2^{(q)} \end{pmatrix}, \quad D = \begin{pmatrix} d_1^{(p)} & 0 \\ 0 & d_2^{(q)} \end{pmatrix}.$$

Then $\mathfrak{T}$ may be considered as a direct product of two symplectic matrices

$$\begin{pmatrix} a_1 & b_1 \\ c_1 & d_1 \end{pmatrix}, \quad \begin{pmatrix} a_2 & b_2 \\ c_2 & d_2 \end{pmatrix}.$$

The group so formed is denoted by G_1.

The equation

$$\begin{pmatrix} A & B \\ C & D \end{pmatrix} \begin{pmatrix} H & 0 \\ 0 & H \end{pmatrix} = - \begin{pmatrix} H & 0 \\ 0 & H \end{pmatrix} \begin{pmatrix} A & B \\ C & D \end{pmatrix}$$

cannot hold, except when $p = q$, $n = 2p$. Then we have a group G which is generated by G_1 and the additional transformation

$$W = \begin{pmatrix} 0 & I^{(p)} \\ I & 0 \end{pmatrix} Z \begin{pmatrix} 0 & I \\ I & 0 \end{pmatrix}.$$

For simplicity's sake, we take $p = n - 1$, $q = 1$, that is, the absolute is a fundamental involution.

Theorem 17 *The arithmetic distance of P and $\mathfrak{J}(P)$ characterizes the equivalence of points. More definitely, every point is either equivalent to 0 or equivalent to*

$$\begin{pmatrix} 0 & 0 & \cdots & 0 & 1 \\ 0 & 0 & \cdots & 0 & 0 \\ \vdots & \vdots & & \vdots & \vdots \\ 0 & 0 & \cdots & 0 & 0 \\ 1 & 0 & \cdots & 0 & 0 \end{pmatrix}.$$

Proof Let

$$P = \begin{pmatrix} P_1 & V' \\ V & P \end{pmatrix}, \quad P_1 = P_1^{(n-1)}.$$

Then

$$W = Z - \begin{pmatrix} P_1 & 0 \\ 0 & P \end{pmatrix}$$

carries P into
$$\begin{pmatrix} 0 & v' \\ v & 0 \end{pmatrix}. \tag{47}$$

If $v = 0$, there is nothing to be proved. Otherwise, we have a matrix M such that $vM = (1, 0, \cdots, 0)$, $M = M^{(n-1)}$. Then the transformation
$$W = \begin{pmatrix} M & 0 \\ 0 & 1 \end{pmatrix} Z \begin{pmatrix} M & 0 \\ 0 & 1 \end{pmatrix}$$

carries (47) into the required form.

The non-equivalence of the two points given in Theorem 17 is evident.

It is easy to extend the theorem for arbitrary p and q.

Remarks (1) The geometries so obtained are generalizations of the complex analogy of the Möbius geometry of circles. There is a particularly great variety of geometries over the real field.

(2) We may take several commutative involutions as absolutes; for example, we take $W = -Z$, $W = Z^{-1}$ as absolutes, then the transformations take the form $W = \Gamma Z \Gamma'$, where Γ is an orthogonal matrix. Thus "the orthogonal classification of symmetric matrices" may be considered as the geometry of symmetric matrices with two absolutes.

(3) To each involution $\mathfrak{J}$ of the second kind, we have a symplectic symmetric matrix $\mathfrak{J}\mathfrak{F}$, and conversely. Thus, the geometry of symplectic symmetric matrices may be considered as a geometry of involutions of the second kind. In particular, for $n = 1$, it gives the ordinary treatment of involutions in the complex plane. It may also be extended to the study of anti-involutions, and so on. But the author will not go into the detailed discussion of this problem.

12. Laguerre's geometry of matrices

The transformations
$$W = AZA' + S \tag{48}$$

form a subgroup, which leaves ∞ invariant. The geometry under this group is called *Laguerre's geometry of symmetric matrices, or affine geometry of symmetric matrices.*

Theorem 18 *Finite points are transitive under the group. Two point-pairs are equivalent if and only if their arithmetic distances are the same.*

We may obtain invariants from the "projective" geometry of symmetric matrices by selecting a point to be infinity. For the discussion of the equivalence of the triples of points, we introduce a "simple ratio-matrix", that is,

$$(X_1, X_2, X_3, \infty) = (X_1 - X_2)(X_3 - X_2)^{-1}.$$

Further, the transformations (48) with unitary A form a subgroup. The geometry under this group may be called the Euclidean geometry of symmetric matrices. The transformations (48) with A of determinant I form also a group. The corresponding geometry may be called the special Laguerre geometry.

13. Fixed points of an involution of the second kind

In §13, §14 we determine the fixed points of involutions.

Theorem 19 *An involution of the second kind has two isolated fixed points; they form a nonspecial pair, that is, a pair of points with arithmetic distance n. Conversely, a nonspecial pair of points determines uniquely an involution of the second kind having the points as its isolated fixed points.*

Proof (1) It is sufficient to prove that $\pm iI$ are two isolated solutions of

$$Z^2 = -I. \tag{49}$$

If Z_0 is a non-scalar solution of (49), then $\Gamma Z_0 \Gamma'$ is also a solution for all orthogonal Γ. Thus Z_0 is not an isolated fixed point of $W = -Z^{-1}$.

If Z_0 is scalar, then $Z_0 = \pm iI$. From $(iI + \Delta)^2 = -I$ we have $\Delta(\Delta + 2iI) = 0$. For sufficiently small Δ, $\Delta + 2iI$ is nonsingular, consequently $\Delta = 0$. This establishes that $\pm iI$ are the two isolated fixed points.

(2) Conversely, let 0 and ∞ be two fixed points. Then the involution takes the form

$$W = AZA', \quad A^2 = -I.$$

Using this in connection with (37), we have $W = -Z$.

Theorem 20 *Let P_1 and P_2 be two isolated fixed points of an involution of the second kind; all its fixed points X are given by*

$$r(P_1, X) + r(X, P_2) = n \tag{50}$$

and conversely [①].

Proof (1) All the solutions of (49) are given by

$$Z = \Gamma \begin{pmatrix} -I^{(p)}i & 0 \\ 0 & I^{(q)}i \end{pmatrix} \Gamma',$$

---

① $r(A, B)$ denotes the rank of $A - B$; it is called the arithmetic distance between A and B.

where Γ is orthogonal, since the elementary divisors of $Z - \lambda I$ are all simple. Further

$$r(Z, Ii) = r\left(\begin{pmatrix} -I^{(p)}i & 0 \\ 0 & I^{(q)}i \end{pmatrix}, Ii\right) = p$$

and $r(-Ii, Z) = q$.

(2) Now we let

$$r(Z, iI) = p \quad \text{and} \quad r(Z, -iI) = q.$$

There exists a matrix Q such that

$$QZQ' = \sum{}' J_k^{(C_k)}, \quad QIQ' = \sum{}' J^{(C_k)},$$

where

$$J = \begin{pmatrix} 0 & 0 & \cdots & 0 & 1 \\ 0 & 0 & \cdots & 1 & 0 \\ \vdots & \vdots & & \vdots & \vdots \\ 1 & 0 & \cdots & 0 & 0 \end{pmatrix}, \quad J_k = \begin{pmatrix} 0 & \cdots & 0 & \lambda_k \\ 0 & \cdots & \lambda_k & 1 \\ \vdots & & \vdots & \vdots \\ s\lambda_k & \cdots & 0 & 0 \end{pmatrix}.$$

Direct verification shows that $C_k = 1$ and that $d(Z - \lambda I) = 0$ has i as root of multiplicity p and has $-i$ as root of multiplicity q. Thus we have an orthogonal Q such that

$$Z = Q \begin{pmatrix} I^{(p)}i & 0 \\ 0 & -I^{(q)}i \end{pmatrix} Q'.$$

Theorem 21 Let P_1 and P_2 be two isolated fixed points of an involution of the second kind. Let X be a point satisfying $r(P_1, X) = r(P_2, X) = n$, and let X_0 be the image of X under the involution. Then P_1, P_2, X, X_0 form a harmonic range.

Proof We may take $P_1 = 0, P_2 = \infty$. Then $X_0 = -X$. We have consequently $(P_1, P_2, X, X_0) = (0, \infty, X, -X) = -I$.

14. Fixed elements of an involution of the first kind

In order to determine the fixed elements of an involution of the first kind, we have to introduce the concept of decomposable subspace.

Definition A manifold in the space is said to form a *decomposable subspace* of the type (p, q), if we have a symplectic transformation carrying the manifold to a manifold formed by the points of the form

$$\begin{pmatrix} Z_1^{(p)} & 0 \\ 0 & Z_2^{(q)} \end{pmatrix}, \quad p + q = n, \tag{51}$$

where Z_1 and Z_2 are called *components*.

Theorem 22 Let $p+q=n$. *The fixed points of an involution of signature* (p,q) *form a decomposable subspace of type* (p,q).

Proof We write
$$H = \begin{pmatrix} I^{(p)} & 0 \\ 0 & -I^{(q)} \end{pmatrix}$$
and
$$Z = \begin{pmatrix} Z_{11}^{(p)} & Z_{12} \\ Z_{12}' & Z_{22}^{(q)} \end{pmatrix}.$$
Then
$$HZH - Z = \begin{pmatrix} 0 & -2Z_{12} \\ -2Z_{12}' & 0 \end{pmatrix} = 0,$$
which implies $Z_{12}=0$. Thus we have the theorem.

Theorems 22 and 2 give us a general definition of decomposable subspace:

Let $\mathfrak{K}$ be a skew symmetric symplectic matrix. The points (Z_1, Z_2) satisfying
$$(Z_1, Z_2)\mathfrak{K}(Z_1, Z_2)' = 0 \tag{52}$$
define a decomposable space. It is of the type (p,q), if $d(\mathfrak{K} - \lambda\mathfrak{F}) = 0$ has 1 and -1 as roots of multiplicity $2p$ and $2q$ respectively.

The justification is almost evident, since on multiplying $(W_1, W_2) = Q(Z_1, Z_2)\mathfrak{T}$ on the right by $\mathfrak{F}(Z_1, Z_2)'$, we obtain
$$(W_1, W_2)\mathfrak{F}(Z_1, Z_2)' = Q(Z_1, Z_2)\mathfrak{K}(Z_1, Z_2)'$$
(notice that (Z_1, Z_2) and (W_1, W_2) denote the same point, if and only if $(Z_1, Z_2)\mathfrak{F}(W_1, W_2)' = 0$).

Further we may prove that the involution is uniquely determined by the decomposable subspace.

In the same way we may define the manifold of the fixed points of an involution of the second kind.

15. Fixed points of an anti-involution

Now we are going to find the fixed elements of anti-involutions.

Theorem 23 *The fixed points of an anti-involution of the first kind form a connected piece of dimension* $n(n+1)/2$ *(real parameters)*.

In fact, the fixed points of
$$W = \bar{Z} \tag{53}$$
are the real Z. The theorem is now evident.

Theorem 24 *An anti-involution with signature $(p,q)(p \neq q)$ has no fixed point and that with signature (p,p) has fixed points depending on $n(n+1)/2$ real parameters.*

Proof We have (23). Its fixed points are given by $ZH\bar{Z} = -H$. This is impossible except for $p = q$.

In the case $p = q$, we take the normal form
$$W = \begin{pmatrix} 0 & I^{(p)} \\ I & 0 \end{pmatrix} \bar{Z} \begin{pmatrix} 0 & I \\ I & 0 \end{pmatrix}. \tag{54}$$

Set
$$Z = \begin{pmatrix} Z_1 & Z_2 \\ Z_2' & Z_3 \end{pmatrix}.$$

The fixed points of (54) are given by
$$Z_1 = \bar{Z}_3, \quad Z_2 = \bar{Z}_2'.$$

The number of parameters of Z_1 and Z_2 are $p(p+1)$, p^2 respectively. Thus the total number of parameters is equal to
$$p(p+1) + p^2 = p(2p+1) = n(n+1)/2.$$

16. Number of parameters of involutions

Now we are going to determine the dimensions of the manifolds formed by involutions and anti-involutions.

Theorem 25 *The number of parameters (complex) of involutions of the second kind is equal to $n(n+1)$ and that of involutions of the first kind of signature (p,q) is equal to $4pq$.*

Notice that $4pq \leqslant n^2 < n(n+1)$.

Proof (1) By Theorem 19, we have the first part of the theorem.

(2) Now we consider involutions of the signature (p,q). It is known that the group of $2n$-rowed symplectic matrices depends on $n(2n+1)$ parameters. We are going to find the number of parameters of the subgroup leaving the decomposable subspace
$$\begin{pmatrix} X_1^{(p)} & 0 \\ 0 & X_2^{(q)} \end{pmatrix}$$

invariant. By the result of §11 and Theorem 22, the number of parameters of the subgroup is equal to $p(2p+1) + q(2q+1)$. Therefore the number of parameters of involutions of the signature (p, q) is equal to

$$n(2n+1) - p(2p+1) - q(2q+1) = 2(n^2 - p^2 - q^2) = 4pq,$$

since $n = p + q$.

Now we give also the number of parameters of anti-involutions.

Theorem 26 *Anti-involutions of the first or the second kind each form an $n(2n+1)$ parametric family (real parameters).*

Proof We consider the anti-involution $\mathfrak{T}$ of the first kind. Put

$$\mathfrak{T} = (\mathfrak{J} - \mathfrak{S}\mathfrak{F})(\mathfrak{J} + \mathfrak{S}\mathfrak{F})^{-1}.$$

From $\mathfrak{T}^*\mathfrak{T} = \mathfrak{J}$ and $\mathfrak{T}\mathfrak{F}\mathfrak{T}' = \mathfrak{F}$ we deduce easily that

$$\mathfrak{S} + \mathfrak{S}^* = 0, \quad \mathfrak{S} = \mathfrak{S}'.$$

That is, $\mathfrak{S}$ is pure imaginary and symmetric. The theorem follows.

In the case of anti-involutions of the second kind, we have correspondingly

$$\mathfrak{S}\mathfrak{F}\mathfrak{S}^* = \mathfrak{F}, \quad \mathfrak{S} = \mathfrak{S}'.$$

Putting $\mathfrak{S} = (\mathfrak{J} - \mathfrak{P})(\mathfrak{J} + \mathfrak{P})^{-1}$ again, we have $\mathfrak{P}\mathfrak{F} + \mathfrak{F}\mathfrak{P}^* = 0, \mathfrak{P} = \mathfrak{P}'$, that is,

$$\mathfrak{P} = \begin{pmatrix} S & H \\ H' & -S \end{pmatrix},$$

where S is symmetric and H is Hermitian. The number of parameters is $n(n+1)+n^2 = n(2n+1)$.

17. Dieder manifolds

In this section we study the dimension of dieder manifold.

Definition Let A and B form a nonspecial pair. The locus of the points X satisfying

$$r(A, X) + r(X, B) = n \tag{55}$$

is called a *dieder manifold* of the point-pair (A, B).

As a consequence of Theorem 20, the fixed points of an involution of the second kind form a dieder manifold of its two isolated fixed points. It is called the *dieder manifold of the involution*.

Theorem 27 Let $\mathfrak{S}$ be an involution commutative with $\mathfrak{T}$, an involution of the second kind. Then $\mathfrak{S}$ carries the dieder manifold of $\mathfrak{T}$ into itself.

Proof $\mathfrak{S}$ carries the isolated fixed points (A, B) of $\mathfrak{T}$ to (C, D). Then (C, D) are isolated fixed points of $\mathfrak{S}\mathfrak{T}\mathfrak{S}^{-1}$. Since $\mathfrak{S}\mathfrak{T}\mathfrak{S}^{-1} = \pm\mathfrak{T}$, $\mathfrak{S}$ either keeps A and B fixed or permutes A and B. In both cases the dieder manifold is invariant.

Definition The part of the dieder manifold satisfying

$$r(A, X) = p \tag{56}$$

is called the *component of index p* with respect to A.

Theorem 28 *The component of index p is of dimension $2pq$.*

Proof It is now more convenient to use the homogeneous coordinate-system. We may assume that A and B are $(0, I)$ and $(I, 0)$ respectively. The component of index p is constituted by the points

$$(Z_1, Z_2), \tag{57}$$

where Z_1 is of rank p and Z_2 of rank q. Evidently

$$Z_1 = \begin{pmatrix} I^{(p)} & a \\ 0 & 0 \end{pmatrix}, \quad Z_2 = \begin{pmatrix} 0 & 0 \\ b & I^{(q)} \end{pmatrix} \tag{58}$$

are points on the component. The number of parameters is $2pq$. Thus we have to establish the following two facts:

(1) We have no Q, different from the identity, such that

$$Q\left(\begin{pmatrix} I^{(p)} & a \\ 0 & 0 \end{pmatrix}, \begin{pmatrix} 0 & 0 \\ b & I^{(q)} \end{pmatrix}\right) = \left(\begin{pmatrix} I^{(p)} & a_1 \\ 0 & 0 \end{pmatrix}, \begin{pmatrix} 0 & 0 \\ b_1 & I^{(q)} \end{pmatrix}\right); \tag{59}$$

(2) In general, each point on the component is of the form (58).

To prove (1), we let

$$Q = \begin{pmatrix} q_{11} & q_{12} \\ q_{21} & q_{22} \end{pmatrix}.$$

Comparing the elements on both sides of (59), we have $q_{11} = I^{(p)}$, $q_{12} = 0$, $q_{21} = 0$, $q_{22} = I^{(q)}$. Thus (1) is true.

(2) The transformation leaving $(0, I)$ and $(I, 0)$ invariant is of the form

$$W_1 = QZ_1A, \quad W_2 = QZ_2D, \quad AD' = I.$$

Thus the points on the components are given by

$$W_1 = Q \begin{pmatrix} I^{(p)} & 0 \\ 0 & 0 \end{pmatrix} \quad A = Q \begin{pmatrix} a_{11}^{(p)} & a_{12} \\ 0 & 0 \end{pmatrix},$$

$$W_2 = Q \begin{pmatrix} 0 & 0 \\ 0 & I^{(q)} \end{pmatrix} \quad A'^{-1} = Q \begin{pmatrix} 0 & 0 \\ \alpha_{21} & \alpha_{22} \end{pmatrix},$$

where

$$A = \begin{pmatrix} a_{11} & a_{12} \\ a_{21} & a_{22} \end{pmatrix}, \quad A'^{-1} = \begin{pmatrix} \alpha_{11} & \alpha_{12} \\ \alpha_{21} & \alpha_{22} \end{pmatrix}.$$

We take the general case $d(a_{11}) \neq 0$, $d(d_{22}) \neq 0$. Thus

$$\begin{pmatrix} a_{11}^{-1} & 0 \\ 0 & a_{22}^{-1} \end{pmatrix} \left(\begin{pmatrix} a_{11} & a_{12} \\ 0 & 0 \end{pmatrix}, \begin{pmatrix} 0 & 0 \\ \alpha_{21} & \alpha_{22} \end{pmatrix} \right)$$

gives the required form. The conditions of inequality are irrelevant in counting the parameters.

Consequently, two components of indices r and s of a dieder manifold are not equivalent topologically, if $r \neq s$ and $r + s \neq n$. As a consequence of Theorem 27, we have the following theorem.

Theorem 29 *A dieder manifold is of dimension $2[n/2](n - [n/2])$ where $[x]$ denotes the integral part of x.*

18. Commutative involutions of the second kind

Without loss of generality the problem to be solved can be stated as follows:

Given an involution of the second kind, say (12), find all involutions of the second kind commutative with it.

Let $\mathfrak{T}$ be the matrix of the required involution, then $\mathfrak{T}^2 = -\mathfrak{T}$ and

$$\mathfrak{T} \begin{pmatrix} iI & 0 \\ 0 & -iI \end{pmatrix} = \pm \begin{pmatrix} iI & 0 \\ 0 & -iI \end{pmatrix} \mathfrak{T}. \tag{60}$$

(1) Taking the upper sign, we have

$$\mathfrak{T} = \begin{pmatrix} A & B \\ C & D \end{pmatrix}, \quad B = C = 0.$$

That is, the involution takes the form

$$W = AZA', \quad A^2 = -I. \tag{61}$$

Now we are going to find its isolated fixed points. There exists a nonsingular matrix Γ such that $\Gamma A \Gamma^{-1} = iH$, where $H = [1, \cdots, 1, -1, \cdots, -1]$, where there are p 1's and q -1's. Let
$$W_0 = \Gamma W \Gamma', \quad Z_0 = \Gamma Z \Gamma',$$
we have
$$W_0 = -H Z_0 H \tag{62}$$
and notice that (12) takes the form $W_0 = -Z_0$. We drop the subscript 0.

Using homogeneous coordinates, we have
$$(Z_1, Z_2) = Q(W_1, W_2) \begin{pmatrix} iH & 0 \\ 0 & -iH \end{pmatrix}. \tag{63}$$

The transformation
$$\mathfrak{T} = \begin{pmatrix} I^p & iI_q \\ iI_q & I^p \end{pmatrix}, \quad I^p = \begin{pmatrix} I^{(p)} & 0 \\ 0 & 0 \end{pmatrix}, \quad I_q = \begin{pmatrix} 0 & 0 \\ 0 & I^{(q)} \end{pmatrix} \tag{64}$$
carries (63) into
$$\mathfrak{T} \begin{pmatrix} iH & 0 \\ 0 & -iH \end{pmatrix} \mathfrak{T}^{-1} = \begin{pmatrix} I^p & iI_q \\ iI_q & I^p \end{pmatrix} \begin{pmatrix} iH & 0 \\ 0 & -iH \end{pmatrix} \begin{pmatrix} I^p & -iI_q \\ -iI_q & I^p \end{pmatrix}$$
$$= \begin{pmatrix} iI & 0 \\ 0 & -iI \end{pmatrix}.$$

The last matrix represents (12), and it has 0 and ∞ as its isolated fixed points. Thus the isolated fixed points of (63) are $(I^p, iI_q), (iI_q, I^p)$ which lie on the dieder manifold of (12).

(2) Taking the lower sign of (60), we have $A = D = 0$. Then
$$\mathfrak{T} = \begin{pmatrix} 0 & B \\ -B'^{-1} & 0 \end{pmatrix}.$$
Since
$$\begin{pmatrix} 0 & B \\ -B'^{-1} & 0 \end{pmatrix}^2 = \begin{pmatrix} -BB'^{-1} & 0 \\ 0 & * \end{pmatrix} = -\mathfrak{J},$$
B is symmetric ($=S$, say). Then we have
$$W = SZ^{-1}S. \tag{65}$$

It has $\pm S$ as its two isolated fixed points. Since $0, \infty, S, -S$ form a harmonic range, we have the following theorem.

Theorem 30 *If an involution of the second kind is commutative with a fixed involution of the same kind, its isolated fixed points either lie on the dieder manifold of the fixed involution or separate the isolated fixed points of the fixed involution harmonically.*

The proof of Theorem 29 suggests also the following theorem.

Theorem 31 *The involution of the second kind commutative with a fixed involution of the second kind depends on $n(n+1)/2$ parameters (complex).*

Proof The case (2) gives us $n(n+1)/2$ parameters, as S is symmetric. The case (1) gives $2[n/2](n-[n/2])$ ($< n(n+1)/2$) parameters as shown by the following:

Lemma *The solution of the matrix equation*

$$A^2 = I \tag{66}$$

depends on $2[n/2](n-[n/2])$ parameters.

In fact, let A have 1 as characteristic root of multiplicity p, and -1 of multiplicity q. The most general solution is given by

$$\Gamma^{-1}[1,\cdots,1,-1,\cdots,-1]\Gamma,$$

where Γ is nonsingular.

Further, if $\Gamma^{-1}[1,\cdots,1,-1,\cdots,-1]\Gamma = [1,\cdots,1,-1,\cdots,-1]$, then

$$\Gamma = \begin{pmatrix} \gamma_1^{(p)} & 0 \\ 0 & \gamma_2^{(q)} \end{pmatrix},$$

which depends on $p^2 + q^2$ variables. Thus for a fixed P, A depends on

$$n^2 - p^2 - q^2 = 2pq$$

parameters. This expression has its maximum for $p = [n/2]$.

Since

$$2[n/2](n-[n/2]) \leqslant 2(n/2)^2 < n(n+1)/2,$$

we have Theorem 30.

19. Number of parameters of involutions commutative with a given involution

Now we push Theorem 31 a little further.

Theorem 32   *The involutions commutative with a fixed involution of the second kind depend on $n(n+1)/2$ parameters.*

Proof   We need only to consider the involutions of the first kind commutative $W = -Z$. Then, we have either

$$Z_1 = AZA', \quad A^2 = I, \tag{67}$$

or, for even n,

$$Z_1 = KZ^{-1}K, \tag{68}$$

where K is skew symmetric. (67) depends on $[n/2](n - [n/2])$ parameters by the lemma of the last section. (68) depends on $n(n-1)/2$ parameters. Both do not exceed $n(n+1)/2$.

Theorem 33   *The involutions commutative with a fixed involution of the first kind of signature (p, q) depend on $p(p+1) + q(q+1)$ parameters.*

Proof   We take $Z_1 = HZH, H = [1, \cdots, 1, -1, \cdots, -1]$ as the fixed involution. Let

$$\mathfrak{T} = \begin{pmatrix} A & B \\ C & D \end{pmatrix}$$

be an involution commutative with it, then we have

$$\begin{pmatrix} A & B \\ C & D \end{pmatrix} \begin{pmatrix} H & 0 \\ 0 & H \end{pmatrix} = \pm \begin{pmatrix} H & 0 \\ 0 & H \end{pmatrix} \begin{pmatrix} A & B \\ C & D \end{pmatrix}. \tag{69}$$

(1) Taking the upper sign, we have

$$A = \begin{pmatrix} a_1^{(p)} & 0 \\ 0 & a_2 \end{pmatrix}, \quad B = \begin{pmatrix} b_1^{(p)} & 0 \\ 0 & b_2 \end{pmatrix}, \quad C = \begin{pmatrix} c_1^{(p)} & 0 \\ 0 & c_2 \end{pmatrix}, \quad D = \begin{pmatrix} d_1^{(p)} & 0 \\ 0 & d_2 \end{pmatrix},$$

where

$$\begin{pmatrix} a_1 & b_1 \\ c_1 & d_1 \end{pmatrix}, \quad \begin{pmatrix} a_2 & b_2 \\ c_2 & d_2 \end{pmatrix}$$

are involutions of the same kind. By Theorem 25, the number of parameters is

$$p(p+1) + q(q+1).$$

(2) We take the lower sign in (69). This is possible only when $p = q$. Then

$$A = \begin{pmatrix} 0^{(p)} & a_1 \\ a_2 & 0 \end{pmatrix}, \quad B = \begin{pmatrix} 0 & b_1 \\ b_2 & 0 \end{pmatrix}, \quad C = \begin{pmatrix} 0 & c_1 \\ c_2 & 0 \end{pmatrix}, \quad D = \begin{pmatrix} 0 & d_1 \\ d_2 & 0 \end{pmatrix}.$$

Let

$$T_1 = \begin{pmatrix} a_1 & b_1 \\ c_1 & d_1 \end{pmatrix}, \quad T_2 = \begin{pmatrix} a_2 & b_2 \\ c_2 & d_2 \end{pmatrix}.$$

We have $T_1 T_2 = I^{(p)}$ and T_1 is symplectic. Thus the number of parameters of T_1 is equal to $p(2p+1)$ and T_2 is determined uniquely. Since

$$p(2p+1) < 2p(p+1)$$

we have the theorem.

Notice that

$$p(p+1) + q(q+1) = p^2 + q^2 + p + q$$
$$\geqslant (p+q)^2/2 + p + q > n(n+1)/2. \tag{70}$$

20. Further study of commutative involutions

First of all, let us classify commutative involutions.

Definition Two commutative involutions of the second kind, $\mathfrak{T}_1$ and $\mathfrak{T}_2$, which have isolated fixed points forming a harmonic range, are said to be commutative nondegenerately. Otherwise, we say they are commutative degenerately. For the degenerate case, the arithmetic distance between an isolated fixed point of $\mathfrak{T}_1$ and the other of $\mathfrak{T}_2$ is either p or q. We assume that $0 < p \leqslant n/2$. Then p is called the *arithmetic distance* of $\mathfrak{T}_1$ and $\mathfrak{T}_2$.

Analytically, $\mathfrak{T}_1$ and $\mathfrak{T}_2$ are commutative nondegenerately if and only if

$$\mathfrak{T}_1 \mathfrak{T}_2 = -\mathfrak{T}_2 \mathfrak{T}_1 \tag{71}$$

(see (2) of §18). In the present section we are going to establish the following theorem.

Theorem 34 Let $n = 2^\sigma \tau$, where τ is odd. There are $\sigma + 3$ involutions of the second kind any two of which are commutative nondegenerately, and $\sigma + 3$ is the maximal number.

In other words, *there are $\sigma + 3$ nonspecial pairs of points such that any two pairs form a harmonic range.*

We prove the present theorem together with the following:

Theorem 35 Let $n = 2^\sigma \tau$, where τ is odd. Let $P(n), \Sigma(n)$ and $K(n)$ denote the greatest integer s for which there exist s n-rowed symplectic, symmetric and skew-symmetric matrices $T_1, \cdots, T_s$ such that

$$T_i^2 = -I, \quad T_i T_j = -T_j T_i, \quad i \neq j, \tag{72}$$

respectively. Then

$$P(n) = \sigma + 2 \ (\sigma \geqslant 1), \tag{73}$$

$$\Sigma(n) = \sigma + 1 \ (\sigma \geqslant 0) \tag{74}$$

and

$$K(n) = \sigma \ (\sigma \geqslant 1). \tag{75}$$

Notice that (73) implies Theorem 35 and that $P(n)$ and $K(n)$ are defined only for even n.

We establish (73), (74) and (75) as consequences of the equalities

$$P(2n) = 2 + \Sigma(n), \tag{76}$$

$$\Sigma(2n) = 2 + K(n) \tag{77}$$

and

$$K(2n) = 1 + K(n) \tag{78}$$

and the three initial equalities

$$K(2\tau) = 1, \tag{79}$$

$$\Sigma(\tau) = 1 \tag{80}$$

and

$$\Sigma(2\tau) = 2 \tag{81}$$

for odd τ.

We are going to prove (76) first. We start with the involution (12). By (2) of §18, the involutions commutative with (12) are of the form (65), where S is a symmetric matrix. There is a matrix Γ such that $\Gamma'\Gamma = S^{-1}$. We put $\Gamma'W\Gamma$ and $\Gamma'Z\Gamma$ instead of W and Z, then (12) takes its original form and (65) becomes

$$W = Z^{-1}. \tag{82}$$

Hence any pair of nondegenerately commutative involutions may be carried simultaneously to (12) and (82).

Now we consider an involution of the second kind commutative with both (12) and (82). Since it commutes with (12), it is of the form (65); since it commutes with (82), we have $S^2 = -I$. Let

$$W = S_i Z^{-1} S_i, \quad 1 \leqslant i \leqslant P(2n) - 2,$$

be involutions commutative nondegenerately. Then we have $S_i^2 = -I$, $S_i S_j^{-1} = -S_j S_i^{-1}$, that is,

$$S_i S_j = -S_j S_i, \quad S_i^2 = -I. \tag{83}$$

Therefore, we have

$$P(2n) - 2 \leqslant \Sigma(n).$$

On the other hand, if we have $\Sigma(n)$ symmetric matrices satisfying (83), we may find $2 + \Sigma(n)$ symplectic matrices with the required property (in the following we shall not repeat this statement at similar occasions). Therefore we have (76).

From (83), it follows that S_1 has simple elementary divisors and $\pm i$ are its characteristic roots. We have an orthogonal matrix Γ such that

$$\Gamma S_1 \Gamma' = \begin{pmatrix} I^{(a)} i & 0 \\ 0 & -I^{(b)} i \end{pmatrix}.$$

We may put

$$S_1 = \begin{pmatrix} I^{(a)} i & 0 \\ 0 & -I^{(b)} i \end{pmatrix}. \tag{84}$$

If there exists an S_2 such that

$$S_1 S_2 = -S_2 S_1, \quad S_1 = -S_2 S_1 S_2^{-1},$$

then S_1 and $-S_1$ have the same characteristic equation. This is only possible for $a = b$. Therefore, we have (80). Now we take $n = 2a$.

From $S_1 S_l + S_l S_1 = 0, S_l^2 = -I$ we deduce that

$$S_l = \begin{pmatrix} 0 & i A_l^{(a)} \\ i A_l^{-1} & 0 \end{pmatrix}, \quad l \geqslant 2, \tag{85}$$

where

$$A_l A_l' = I. \tag{86}$$

Putting

$$\begin{pmatrix} A_2' & 0 \\ 0 & I^{(a)} \end{pmatrix} S_k \begin{pmatrix} A_2 & 0 \\ 0 & I^{(a)} \end{pmatrix}$$

instead of S_k, we find that (84) remains unaltered and S_2 takes the form

$$S_2 = \begin{pmatrix} 0 & iI \\ iI & 0 \end{pmatrix}. \tag{87}$$

If (85) is commutative with (87) nondegenerately, for $l \geqslant 3$, then the $A_l(l \geqslant 3)$ are skew symmetric and satisfy $A_l^2 = -I^{(a)}$, $A_l A_k = -A_k A_l$. In case a is odd, there is no such A, we have therefore (81). Otherwise, we have (77).

Now we start again with $(2n)$-rowed skew symmetric matrices

$$\mathfrak{K}_1, \cdots, \mathfrak{K}_s$$

satisfying

$$\mathfrak{K}_i \mathfrak{K}_j = -\mathfrak{K}_j \mathfrak{K}_i, \quad \mathfrak{K}_i^2 = -\mathfrak{J}. \tag{88}$$

We may take

$$\mathfrak{K}_1 = \begin{pmatrix} 0 & I \\ -I & 0 \end{pmatrix} = \mathfrak{F}.$$

From $\mathfrak{K}_i \mathfrak{K}_j = -\mathfrak{K}_j \mathfrak{K}_i$, we find

$$\mathfrak{K}_k = \begin{pmatrix} K_{1k} & K_{2k} \\ K_{2k} & -K_{1k} \end{pmatrix},$$

where K_{1k} and K_{2k} are skew symmetric matrices.

We introduce the transformation

$$\mathfrak{K}_k^\dagger = \frac{i}{2} \begin{pmatrix} I & iI \\ I & -iI \end{pmatrix} \mathfrak{K}_k \begin{pmatrix} I & I \\ iI & -iI \end{pmatrix}. \tag{89}$$

Then $\mathfrak{K}_1^\dagger = \mathfrak{F}$ and

$$\mathfrak{K}_k^\dagger = \begin{pmatrix} L_k & 0 \\ 0 & M_k \end{pmatrix},$$

where

$$L_k = -K_{2k} + iK_{1k}, \tag{90}$$

$$M_k = K_{2k} + iK_{1k}. \tag{91}$$

$\mathfrak{K}_k^\dagger (k \geqslant 2)$ can exist only when n is even. Thus we have (79). From (88), we deduce that

$$L_i M_i = I \tag{92}$$

and
$$L_i M_j + L_j M_i = 0, \quad i \geqslant 2, \; j \geqslant 2. \tag{93}$$
Consequently, we have $L_2, \cdots, L_s$ such that
$$L_i L_j^{-1} + L_j L_i^{-1} = 0. \tag{94}$$
We have Γ such that $\Gamma L_2 \Gamma' = F$, where
$$F = \begin{pmatrix} 0 & I^{(a)} \\ -I & 0 \end{pmatrix}, \quad a = 2n.$$
We may take $L_2 = F$. Let
$$N_i = F^{-1} L_i, \quad 3 \leqslant i \leqslant s.$$
Then, from (94),
$$N_i^2 = L_2^{-1} L_i L_2^{-1} L_i = -I, \tag{95}$$
$$F N_i + N_i F = L_i + F^{-1} L_i F = 0 \tag{96}$$
and
$$N_i N_j + N_j N_i = L_2^{-1} L_j L_2^{-1} L_i + L_2^{-1} L_i L_2^{-1} L_j = -(L_j^{-1} L_i + L_i^{-1} L_j) = 0. \tag{97}$$

Further
$$N_i' = L_i' F'^{-1} = L_i F^{-1} = -F^{-1} L_i = -N_i.$$
Thus we have $s-1$ n-rowed skew symmetric matrices $F, N_3, \cdots, N_s$ satisfying (95), (96) and (97). Thus, we have (78).

Remark If we restrict the T's to be orthogonal in the theorem, the maximum s is equal to $K(n) = \sigma$. In fact, from
$$T^2 = -I, \quad TT' = I,$$
we deduce immediately $T = -T'$.

21. Parameters of involutions commutative with a pair of involutions

Now we establish the characteristic distinction of all sorts of commutative involutions of the second kind.

Theorem 36 *Given two nondegenerately commutative involutions of the second kind, the involutions of the second kind commutative with both depend on $[n/2]$ $(n - [n/2])$ parameters (complex).*

Proof (1) Now we take, without loss of generality, the two fixed involutions as (12) and (11).

The involutions, nondegenerately commutative with both given involutions, are given by
$$W = SZ^{-1}S, \quad S^2 = -I. \tag{98}$$
We have an orthogonal matrix Γ such that $\Gamma S\Gamma' = Hi$, $H = [1, \cdots, 1, -1, \cdots, -1]$. The totality of orthogonal matrices Γ depends on $n(n-1)/2$ parameters. Further from $\Gamma H \Gamma' = H$, we deduce
$$\Gamma = \begin{pmatrix} t^{(p)} & 0 \\ 0 & t^{(q)} \end{pmatrix},$$
which depends on $p(p-1)/2 + q(q-1)/2$ parameters. Thus S depends on
$$(n(n-1) - p(p-1) - q(q-1))/2 = pq \leqslant [n/2](n - [n/2])$$
parameters.

(2) Consider the involutions commutative degenerately with one of the fixed involutions, say $W = -Z$. Then they are of the form (61). If they commute with $W = -Z^{-1}$, either degenerately or not, then
$$W = AZ^{-1}A' = (AZA')^{-1} = A'^{-1}ZA^{-1}$$
and we deduce $A'A = \pm I$. Then
$$A' = -A'A^2 = \mp A,$$
that is, A is either symmetric or skew-symmetric. If it is symmetric, the number of symmetric involutions A has been counted in (1).

Now we assume that A is skew-symmetric and orthogonal. Consequently n is even, equal to $2p$, say. Since $A^2 = -I$, A has $\pm i$ as its characteristic roots and its elementary divisors are simple. Since
$$d(A - \lambda I) = d(A - \lambda I)' = d(-A - \lambda I),$$
the multiplicities of i and $-i$ are equal.

The pair of matrices
$$\begin{pmatrix} 0 & I^{(p)} \\ -I^{(p)} & 0 \end{pmatrix}, \quad \begin{pmatrix} I^{(p)} & 0 \\ 0 & I^{(p)} \end{pmatrix}$$
has the prescribed elementary divisors. Thus we have an orthogonal matrix Γ such that
$$\Gamma A \Gamma' = \begin{pmatrix} 0 & I \\ -I & 0 \end{pmatrix}.$$

The totality of orthogonal matrices depends on $n(n-1)/2$ parameters and symplectic orthogonal matrices of order $2p$ depend on p^2 parameters. Thus the number of parameters is

$$n(n-1)/2 - n^2/4 < ((n-1)/2)(n-((n-1)/2)) \leqslant [n/2](n-[n/2])$$

and the theorem follows.

Theorem 37 *The involutions of the second kind commutative with two commutative involutions of the second kind of arithmetic distance p depend on $p(p+1)/2 + q(q+1)/2$ parameters.*

Proof We may write both fixed involutions as (12) and

$$W = -HZH, \tag{99}$$

where $H = [1, \cdots, 1, -1, \cdots, -1]$.

(1) The involutions commutative degenerately with (12) are given by (61). If they commute with (99), we have $AH = \pm HA$. Consequently, we have

$$A = \begin{pmatrix} a_1^{(p)} & 0 \\ 0 & a_2^{(q)} \end{pmatrix}, \quad \text{or} \quad A = \begin{pmatrix} 0 & b_1^{(p)} \\ b_2^{(p)} & 0 \end{pmatrix},$$

the second case can appear only when $p = n/2$.

① The number of parameters of

$$\begin{pmatrix} a_1 & 0 \\ 0 & a_2 \end{pmatrix}, \quad a_1^2 = -I^{(p)}, \quad a_2^2 = -I^{(q)}$$

is equal to $2[p/2](p - [p/2]) + 2[q/2](q - [q/2])$, by the lemma of §18.

② From $A^2 = -1$, we deduce

$$\begin{pmatrix} 0 & b_1 \\ b_2 & 0 \end{pmatrix}, \quad b_1 b_2 = -I^{(p)}.$$

Thus the number of parameters is equal to p^2, which is less than $p(p+1)$.

(2) The involutions commuting with (12) nondegenerately are of the form

$$W = SZ^{-1}S.$$

If they commute with (99), we have $HS = \pm SH$. Consequently, we obtain

$$S = \begin{pmatrix} s_1 & 0 \\ 0 & s_2 \end{pmatrix}, \quad \text{or} \quad S = \begin{pmatrix} 0 & t \\ t' & 0 \end{pmatrix}.$$

The number of parameters in the first case is

$$p(p+1)/2 + q(q+1)/2,$$

and in the second case, $n = 2p$, is $p^2 < p(p+1)$.

Finally, we have the theorem, since

$$2[p/2](p - [p/2]) + 2[q/2](q - [q/2]) < p(p+1)/2 + q(q+1)/2.$$

22. Automorphisms of the group of symplectic transformations

Finally, we are going to prove the following theorem.

Theorem 38 *A topological automorphism of the group of symplectic transformations is either an inner automorphism or an anti-symplectic transformation.*

More explicitly: Let $\mathfrak{G}$ be the group of symplectic transformations. It may be visualized as the totality of symplectic matrices $\mathfrak{T}$, but we have to identify $\mathfrak{T}$ and $-\mathfrak{T}$. $\mathcal{A}$ is called an automorphism over $\mathfrak{G}$, written as

$$\mathcal{A}(\mathfrak{T}) = \mathfrak{T}^\dagger, \tag{100}$$

if it builds up a one-to-one and bi-continuous relation between $\mathfrak{T}$ and $\mathfrak{T}^\dagger$, and if

$$\mathcal{A}(\mathfrak{T}_1 \mathfrak{T}_2) = \mathfrak{T}_1^\dagger \mathfrak{T}_2^\dagger. \tag{101}$$

The conclusion of the theorem is that we have either a symplectic matrix $\mathfrak{P}$ such that

$$\mathcal{A}(\mathfrak{T}) = \mathfrak{P}\mathfrak{T}\mathfrak{P}^{-1} \tag{102}$$

or a symplectic matrix $\mathfrak{P}$ such that

$$\mathcal{A}(\mathfrak{T}) = (\mathfrak{P}^\dagger \mathfrak{T})^\dagger \mathfrak{P}^{-1} = \mathfrak{P}\mathfrak{T}^\dagger \mathfrak{P}^{-1}, \tag{103}$$

for all $\mathfrak{T}$.

Proof Evidently the automorphism $\mathcal{A}$ carries an involution into an involution and a pair of commutative involutions into a pair of commutative involutions. The involutions commutative with an involution of the first kind of signature (p,q) form a manifold of dimension $p(p+1) + q(q+1)$, by Theorem 33, and those commutative with an involution of the second kind form a manifold of dimension $n(n+1)/2$ by Theorem 32. Since the numbers

$$n(n+1)/2, \quad p(p+1) + q(q+1), \quad 1 \leqslant q \leqslant n/2$$

are never equal, and since a topological transformation leaves the dimension invariant, $\mathcal{A}$ leaves the kind and signature of involutions invariant.

Further, the manifold formed by the involutions of the second kind permuting with a pair of commutative involutions of arithmetic distance p is of dimension

$$p(p+1)/2 + q(q+1)/2$$

by Theorem 37. They are all different for $0 < p \leqslant n/2$. Moreover the manifold formed by the involutions of the second kind permuting with a pair of nondegenerate commutative involutions is of dimension $[n/2](n - [n/2])$, by Theorem 36. Therefore the "degeneracy" of a pair of commutative involutions is also invariant under $\mathcal{A}$.

Each involution of the second kind is determined uniquely by its two isolated fixed points. Thus, we use (A, B) to denote an involution of the second kind possessing A and B as its two isolated fixed points.

We now prove that two involutions with a common isolated fixed point, and the other two fixed points forming a nonspecial pair, are carried by $\mathcal{A}$ into involutions with the same property. In fact, let

$$(A, B), \quad (A, C) \tag{104}$$

be the two involutions under consideration, and suppose that $\mathcal{A}$ carries them into

$$(P, Q), \quad (R, S). \tag{105}$$

The transformations commutative degenerately with both involutions (104) form a group isomorphic to the orthogonal group. In fact we may take $A = 0, B = \infty$ and $C = I$; the conclusion follows immediately.

Suppose that at least one of the arithmetic distances $r(P, R), r(P, S), r(Q, R)$ and $r(Q, S)$ is different from 0 and n. Then the subgroup formed by the transformations commutative degenerately with (P, Q) and (R, S) is reducible. In fact, we may let, in homogeneous coordinates,

$$P = (I, 0), \quad Q = (0, I), \quad R = \left(\begin{pmatrix} I & 0 & 0 \\ 0 & I & 0 \\ 0 & 0 & 0 \end{pmatrix}, \begin{pmatrix} 0 & 0 & 0 \\ 0 & I & 0 \\ 0 & 0 & I \end{pmatrix} \right).$$

Direct verification asserts that the group is formed by the elements of the form:

$$\begin{pmatrix} A & 0 \\ 0 & A'^{-1} \end{pmatrix}, \quad A = \begin{pmatrix} A_{11} & 0 & 0 \\ A_{21} & A_{22} & 0 \\ A_{31} & A_{32} & A_{33} \end{pmatrix}, \quad A_{22} A'_{22} = I.$$

Since the orthogonal group cannot be isomorphic to a reducible group, we have the conclusion that either our assertion is true or that

$$r(P,R) = r(P,S) = r(Q,R) = r(Q,S) = n.$$

The latter case cannot happen. In fact, there is no pair of matrices (X,Y) to separate both (A,B) and (A,C) harmonically, and this may be justified easily by putting $A = 0, B = \infty$ and $C = I$. On the other hand, there exists a nonspecial pair (X,Y) to separate (P,Q) and (R,S) both harmonically. In fact, we may take $P = 0, Q = \infty, R = I$; then the pair $(S_1, -S_1)$ separates both pairs harmonically when $S_1^2 = S$. Thus we have established our assertion.

Let $\Sigma(A)$ be the set of involutions having a common fixed point A. Then, by $\mathcal{A}$, it is mapped into another set with the same property. In fact, if $(A,B), (A,C), (r(B,C) = n)$ are mapped into $(A^\dagger, B^\dagger), (A^\dagger, C^\dagger)$; a third pair (A,D) must be mapped into either $(A^\dagger, D^\dagger)$ or $(B^\dagger, C^\dagger)$. The latter case cannot happen, since $\mathcal{A}$ cannot map a fourth one. We may remove the assumption $r(B,C) = n$ by means of a consideration of continuity.

Suppose $\Sigma(A)$ is carried into $\Sigma(A^\dagger)$ by $\mathcal{A}$. Therefore $\mathcal{A}$ induces a mapping of the space of symmetric matrices: $(A \to A^\dagger)$.

Let $r(A,B) = p$. The points C satisfying

$$r(A,C) = n, \quad r(B,C) = q$$

form a manifold of dimension $nq - q(q+1)/2$. Let D be the image of B under the involution (A,C). Then $(A,C), (B,D)$ are two commutative involutions with the arithmetic distance q. Suppose $\mathcal{A}$ carries $\Sigma(A)$ and $\Sigma(B)$ into $\Sigma(A^\dagger)$ and $\Sigma(B^\dagger)$. Then we find that $r(A^\dagger, B^\dagger) = p$ or q. Since $np - p(p+1)/2 \neq nq - q(q+1)/2$ for $p \neq q$, we have $r(A^\dagger, B^\dagger) = p$.

Therefore the automorphism $\mathcal{A}$ induces a mapping on the space of symmetric matrices possessing the following properties:

(1) It is topological;
(2) It keeps harmonic separation invariant;
(3) It keeps arithmetic distance invariant.

The theorem follows from the generalization of von Staudt's theorem for symmetric matrices.

Geometries of matrices. III. Fundamental theorems in the geometries of symmetric matrices[*]

1. Introduction

The author gave, in the paper I_1[①], a discussion of the redundancy of the conditions which appeared in the generalization of von Staudt's theorem for symmetric matrices of order 2. The author then could also give a proof for matrices of all even orders but was unable to prove the result in its full generality, and further the proof depended essentially on some results about involutions. The author has now found a simpler proof which holds for any order greater than 2 and which is independent of the results about involutions, Nevertheless, the proof is carried out by means of mathematical induction upon the result of I_1, which depends essentially on the theory of involutions.

Several theorems for geometries keeping an involution as an absolute have also been obtained.

The second part of the paper is concerned with analytic mappings. The projective space of symmetric matrices may also be considered as the extended space of several complex variables as defined by Osgood[②]. So far as the author is aware, the completeness of the group of automorphic mappings of an extended space has been established only for two special cases, namely the space of function theory and the complex projective space. If we assume that the group is topological, we may deduce the result from a theorem of E. Catan[③] for semi-simple groups. In this paper, the problem is solved without any restriction besides analyticity.

As an application of the previous result in combination with the continuity theo-

---

[*] Presented to the Society, April 27, 1946; received by the editors May 30, 1945. Reprinted from *Transactions of the American Mathematical Society*, 1947, **61**: 229-255.

[①] Papers I, I_1, and II under the same title were published in *Trans. Amer. Math. Soc.*, 1945, **57**: 441–481, 482–490; 1947, **61**: 193–228. They will be referred to in the present paper simply by their numbers.

[②] *Lehrbuch der Funktionentheorie*, 1929, **2**.

[③] *La théorie des groupes finis et continus el l'analysis situs*. Mémorial des sciences mathématiques, Paris, 1930: 51.

rem due to Levi[①] and with a result due to the author[②], we solve also the corresponding problem for the group of automorphs of the elliptic space. It should be remarked that the corresponding problem for the hyperbolic space was solved by C. L. Siegel[③] in a recent important paper.

2. Arithmetic distance

Let $\mathfrak{S}$ and $\mathfrak{T}$ be two sets of points in the "projective" space of symmetric matrices. The arithmetic distance between $\mathfrak{S}$ and $\mathfrak{T}$ is defined to be the least upper bound of the arithmetic distance between any two points S and T, where S and T belong to $\mathfrak{S}$ and $\mathfrak{T}$ respectively. It is denoted by $r(\mathfrak{S}, \mathfrak{T})$.

In particular for $\mathfrak{S} = \mathfrak{T}$, the distance $r(\mathfrak{S}, \mathfrak{S})$ is called the arithmetic diameter of the set $\mathfrak{S}$.

Given a positive integer ρ, a set $\mathfrak{S}$ is called a *maximal set of rank ρ*, if $\mathfrak{S}$ is of arithmetic diameter ρ, and if any set properly containing $\mathfrak{S}$ is of arithmetic diameter greater than ρ.

Theorem 1 *A normal subspace of rank ρ is a maximal set of rank ρ.*

Proof Since arithmetic distance is invariant, we may take the normal subspace to consist of the points
$$\begin{pmatrix} X^{(\rho)} & 0 \\ 0 & 0 \end{pmatrix}.$$

For any vector v and any number a, not both zero, we have $X^{(\rho)}$ such that
$$\begin{pmatrix} X^{(\rho)} & v' \\ v & a \end{pmatrix}$$

is a nonsingular $(\rho+1)$-rowed matrix. The theorem is now evident.

Is the converse of Theorem 1 true? It is true for $\rho = 1$, but in general we have the following "gegenbeispiel":

The set of all matrices of rank 1 forms a maximal set $\mathfrak{S}$ of rank 2, but does not form a normal subspace. In fact, the equation
$$d\left(\begin{pmatrix} x & y & z \\ y & u & v \\ z & v & w \end{pmatrix} - \begin{pmatrix} a^2 & ab & ac \\ ab & b^2 & bc \\ ac & bc & c^2 \end{pmatrix} \right) = 0$$

[①] E. E. Levi. *Annales de Mathématiques Pures et Appliquées*, 1910, **17**(3).
[②] L. K. Hua. *Trans. Amer. Math. Soc.*, 1946, **59**: 508–523.
[③] C. L. Siegel. *Amer. J. Math.*, 1943, **65**: 1–86.

for all a, b, c implies that
$$\begin{pmatrix} x & y & z \\ y & u & v \\ z & v & w \end{pmatrix}$$
is of rank 1. This fact shows that $\mathfrak{S}$ is a maximal set. On the other hand, we have $r(\mathfrak{S}, 0) = 1$, which cannot hold in the normal subspace.

3. Arithmetic property of the normal subspace

Now we extend the concept of dieder manifold a little further:

Definition Let P_1 and P_2 be two points. The points X satisfying
$$r(P_1, X) + r(X, P_2) = r(P_1, P_2)$$
are said to form a *dieder manifold spanned by* P_1 *and* P_2.

Theorem 2 *A maximal set of rank ρ is a normal space if and only if it contains all dieder manifolds spanned by any two points of the set.*

Proof (1) Let
$$P = 0, \quad Q = [1, \cdots, 1, 0, \cdots, 0]$$
(where the 1's are ρ in number); we shall prove that, if
$$r(P, X) + r(X, Q) = r(P, Q),$$
then X takes the form
$$\begin{pmatrix} X_1^{(\rho)} & 0 \\ 0 & 0 \end{pmatrix}.$$
We consider the pair of matrices Q and X. There exists a nonsingular matrix Γ[①] such that
$$\Gamma^{-1} X \Gamma'^{-1} = [1, \cdots, 1, 0, \cdots, 0], \quad 0 \leqslant p \leqslant \rho$$
(where the 1's are p in number), and
$$\Gamma^{-1} Q \Gamma'^{-1} = Q.$$
From the second equation, we find that
$$\Gamma = \begin{pmatrix} A & B \\ 0 & C \end{pmatrix}, \quad AA' = I.$$

[①] See Turnbull and Aitken. *Theory of canonical matrices*, 1931: 135.

Then
$$X = \Gamma \begin{pmatrix} I^{(p)} & 0 \\ 0 & 0 \end{pmatrix} \Gamma'.$$
Consequently, we have the assertion.

Therefore a normal subspace contains the dieder manifold spanned by any pair of points of the subspace.

(2) Without loss of generality, we may assume that the maximal set contains two points
$$P = 0, \quad Q = [1, \cdots, 1, 0, \cdots, 0]$$
(where the 1's are ρ in number). So it contains
$$P_i = [1, \cdots, 1, 0, 1, \cdots, 1, 0, \cdots, 0], \quad 1 \leqslant i \leqslant \rho$$
(where the first zero in the brackets appears in the ith place, the second in the $(\rho - 1)$th), and
$$P_{ij} = P_i P_j, \quad P_{ijk} = P_i P_j P_k,$$
and so on.

Let
$$X = (a_{rs})_{1 \leqslant r, s \leqslant n}$$
be a point of the set. Then
$$r(X, P_i) \leqslant \rho, \quad r(X, P_{ij}) \leqslant \rho, \quad r(X, P_{ijk}) \leqslant \rho,$$
and so on. Consequently, the matrix
$$\begin{pmatrix} a_{11} - \varepsilon_1 & a_{12} & \cdots & a_{1\rho} & a_{1\,\rho+1} & \cdots & a_{1n} \\ a_{12} & a_{22} - \varepsilon_2 & \cdots & a_{2\rho} & a_{2\,\rho+1} & \cdots & a_{2n} \\ \vdots & \vdots & & \vdots & \vdots & & \vdots \\ a_{1\rho} & a_{2\rho} & \cdots & a_{\rho\rho} - \varepsilon_\rho & a_{\rho\,\rho+1} & \cdots & a_{\rho n} \\ a_{1\,\rho+1} & a_{2\,\rho+1} & \cdots & a_{\rho\,\rho+1} & a_{\rho+1\,\rho+1} & \cdots & a_{\rho+1\,n} \\ \vdots & \vdots & & \vdots & \vdots & & \vdots \\ a_{1n} & a_{2n} & \cdots & a_{\rho n} & a_{\rho+1\,n} & \cdots & a_{nn} \end{pmatrix}$$
is always of rank ρ for $\varepsilon_1, \cdots, \varepsilon_\rho$ arbitrarily taken from 0 and 1. In particular,
$$\begin{vmatrix} a_{11} - \varepsilon_1 & a_{12} & \cdots & a_{1\rho} & a_{1i} \\ a_{12} & a_{22} - \varepsilon_2 & \cdots & a_{2\rho} & a_{2i} \\ \vdots & \vdots & & \vdots & \vdots \\ a_{1\rho} & a_{2\rho} & \cdots & a_{\rho\rho} - \varepsilon_\rho & a_{\rho i} \\ a_{1i} & a_{2i} & \cdots & a_{\rho i} & a_{ii} \end{vmatrix} = 0$$

for $\rho < i \leqslant n$ and any choice of ε. Putting $\varepsilon_1 = 0$ and $\varepsilon_1 = 1$ and subtracting the results, we have

$$\begin{vmatrix} a_{22} - \varepsilon_2 & \cdots & a_{2\rho} & a_{2i} \\ \vdots & & \vdots & \vdots \\ a_{2\rho} & \cdots & a_{\rho\rho} - \varepsilon_\rho & a_{\rho i} \\ a_{2i} & \cdots & a_{\rho i} & a_{ii} \end{vmatrix} = 0.$$

Repeating the same process, we have finally

$$\begin{vmatrix} a_{\rho\rho} - \varepsilon_\rho & a_{\rho i} \\ a_{\rho i} & a_{ii} \end{vmatrix} = 0$$

for $\varepsilon_\rho = 0$ and 1. Consequently $a_{ii} = 0$ and $a_{\rho i} = 0$. Varying ρ, we find that $a_{ji} = 0$ for all $1 \leqslant j \leqslant \rho$. Further varying i from $\rho + 1$ to n, we have the assertion.

Therefore the elements of the set are of the form

$$\begin{pmatrix} X^{(\rho)} & 0 \\ 0 & 0 \end{pmatrix}.$$

By Theorem 1 and (1) we have the second part of our theorem.

Since pairs of points with arithmetic distance ρ $(1 \leqslant \rho \leqslant n)$ form a transitive set, the proof of the previous theorem establishes also the following:

Theorem 3 *Given two points with arithmetic distance ρ, there is one and only one normal subspace of rank ρ which contains both.*

4. Proof of the fundamental theorem

Theorem 4 *A continuous mapping carrying symmetric matrices into symmetric matrices and leaving arithmetic distance invariant is either a symplectic transformation or an anti-symplectic transformation.*

Proof (1) The theorem was proved for matrices of order 2 in I_1. Now we shall establish the general theorem by induction on the order of matrices. We thus suppose that the order of matrices is not less than 3.

(2) Let

$$\Gamma(Z) = Z_1 \tag{1}$$

be a mapping satisfying our conditions. The points of the form

$$\begin{pmatrix} *^{(n-1)} & 0 \\ 0 & 0 \end{pmatrix}$$

form a normal subspace of rank $n-1$. Since the arithmetic distance is invariant, the set of points

$$\Gamma \begin{pmatrix} *^{(n-1)} & 0 \\ 0 & 0 \end{pmatrix}$$

forms also a normal subspace of rank $n-1$. Since the totality of all normal subspaces of rank $n-1$ form a transitive set under the group of symplectic transformations, we may therefore assume that Γ satisfies

$$\Gamma \begin{pmatrix} W^{(n-1)} & 0 \\ 0 & 0 \end{pmatrix} = \begin{pmatrix} W_1^{(n-1)} & 0 \\ 0 & 0 \end{pmatrix}. \tag{2}$$

It induces a continuous mapping on the $(n-1)$-rowed matrices W and it keeps arithmetic distance invariant. Therefore by the hypothesis of induction we find either a symplectic mapping

$$W = (\alpha W_1 + \beta)(\gamma W_1 + \delta)^{-1}, \quad \alpha = \alpha^{(n-1)}, \tag{3}$$

and so on, or an anti-symplectic mapping

$$\overline{W} = (\alpha W_1 + \beta)(\gamma W_1 + \delta)^{-1}. \tag{4}$$

The symplectic transformation $Z = (AZ_1 + B)(CZ_1 + D)^{-1}$ with

$$A = \begin{pmatrix} \alpha & 0 \\ 0 & 1 \end{pmatrix}, \quad B = \begin{pmatrix} \beta & 0 \\ 0 & 0 \end{pmatrix}, \quad C = \begin{pmatrix} \gamma & 0 \\ 0 & 0 \end{pmatrix}, \quad D = \begin{pmatrix} \delta & 0 \\ 0 & 1 \end{pmatrix}$$

carries the transformation Γ inducing (3) into a new one satisfying

$$\Gamma \begin{pmatrix} W^{(n-1)} & 0 \\ 0 & 0 \end{pmatrix} = \begin{pmatrix} W^{(n-1)} & 0 \\ 0 & 0 \end{pmatrix}. \tag{5}$$

Applying the same method to those Γ inducing (4), we conclude, in both cases, that we may assume, without loss of generality, that Γ satisfies (5).

(3) Since

$$\Gamma \begin{pmatrix} 0^{(n-1)} & 0 \\ 0 & 1 \end{pmatrix}$$

is a matrix of rank 1, we may let

$$\Gamma \begin{pmatrix} 0^{(n-1)} & 0 \\ 0 & 1 \end{pmatrix} = (a_1, \cdots, a_n)'(a_1, \cdots, a_n).$$

Since the arithmetic distance between

$$\begin{pmatrix} I^{(n-1)} & 0 \\ 0 & 0 \end{pmatrix}, \quad \begin{pmatrix} 0^{(n-1)} & 0 \\ 0 & 1 \end{pmatrix}$$

is equal to n, we have $a_n \neq 0$. The transformation $Z_1 = A'ZA$ with

$$A = \begin{pmatrix} I^{(n-1)} & 0 \\ -va_n^{-1} & a_n^{-1} \end{pmatrix}, \quad v = (a_1, \cdots, a_{n-1})$$

carries (5) into itself and $(a_1, \cdots, a_n)'(a_1, \cdots, a_n)$ into

$$\begin{pmatrix} 0^{(n-1)} & 0 \\ 0 & 1 \end{pmatrix}.$$

Therefore, we may assume further that

$$\Gamma \begin{pmatrix} 0^{(n-1)} & 0 \\ 0 & 1 \end{pmatrix} = \begin{pmatrix} 0^{(n-1)} & 0 \\ 0 & 1 \end{pmatrix}. \tag{6}$$

(4) The transformation Γ leaves

$$\begin{pmatrix} 0^{(1)} & 0 & 0 \\ 0 & I^{(n-2)} & 0 \\ 0 & 0 & 0 \end{pmatrix}, \quad \begin{pmatrix} 0 & 0 & 0 \\ 0 & 0^{(n-2)} & 0 \\ 0 & 0 & 1 \end{pmatrix}$$

invariant, therefore, by Theorem 2, we find

$$\Gamma \begin{pmatrix} 0 & 0 \\ 0 & W^{(n-1)} \end{pmatrix} = \begin{pmatrix} 0 & 0 \\ 0 & W_1^{(n-1)} \end{pmatrix}.$$

By the supposition of induction, we have either

$$W^{(n-1)} = (\alpha W_1 + \beta)(\gamma W_1 + \delta)^{-1} \tag{7}$$

or

$$\overline{W}^{(n-1)} = (\alpha W_1 + \beta)(\gamma W_1 + \delta)^{-1}, \tag{8}$$

where

$$\begin{pmatrix} \alpha & \beta \\ \gamma & \delta \end{pmatrix}$$

is a $2(n-1)$-rowed symplectic matrix.

According to (5), the transformation (7) and (8) leaves every point of the form

$$W = \begin{pmatrix} X^{(n-2)} & 0 \\ 0 & 0 \end{pmatrix} \qquad (9)$$

invariant; (8) cannot hold.

Since (7) leaves every point of the form (9) invariant, we have

$$\alpha = \begin{pmatrix} \pm I^{(n-2)} & a'_2 \\ 0 & a_4 \end{pmatrix}, \quad \beta = 0,$$

$$\gamma = \begin{pmatrix} 0 & c'_2 \\ c_3 & c_4 \end{pmatrix}, \quad \delta = \begin{pmatrix} \pm I^{(n-2)} & 0 \\ \mp a_2 a_4^{-1} & a_4^{-1} \end{pmatrix}.$$

Further, (7) carries

$$\begin{pmatrix} 0 & 0 \\ 0 & 1^{(1)} \end{pmatrix}$$

into itself; we have $a_2 = 0$.

Let

$$A = \begin{pmatrix} \pm 1 & 0 \\ 0 & \alpha^{(n-1)} \end{pmatrix}, \quad B = \begin{pmatrix} 0 & 0 \\ 0 & \beta^{(n-1)} \end{pmatrix},$$

$$C = \begin{pmatrix} 0 & 0 \\ 0 & \gamma^{(n-1)} \end{pmatrix}, \quad D = \begin{pmatrix} \pm 1 & 0 \\ 0 & \delta^{(n-1)} \end{pmatrix}.$$

The transformation

$$Z = (AZ_1 + B)(CZ_1 + D)^{-1}$$

carries Γ into a new one satisfying (5) and

$$\Gamma \begin{pmatrix} 0 & 0 \\ 0 & W^{(n-1)} \end{pmatrix} = \begin{pmatrix} 0 & 0 \\ 0 & W^{(n-1)} \end{pmatrix}, \qquad (10)$$

since the transformation leaves every point

$$\begin{pmatrix} X_1^{(n-1)} & 0 \\ 0 & 0 \end{pmatrix}$$

invariant.

(5) Since

$$\Gamma \begin{pmatrix} 1 & 0 \\ 0 & 0^{(n-1)} \end{pmatrix} = \begin{pmatrix} 1 & 0 \\ 0 & 0^{(n-1)} \end{pmatrix}, \quad \Gamma \begin{pmatrix} 0^{(n-1)} & 0 \\ 0 & 1 \end{pmatrix} = \begin{pmatrix} 0^{(n-1)} & 0 \\ 0 & 1 \end{pmatrix},$$

we have

$$\Gamma \begin{pmatrix} x_{11} & 0 & \cdots & 0 & x_{1n} \\ 0 & 0 & \cdots & 0 & 0 \\ \vdots & \vdots & & \vdots & \vdots \\ 0 & 0 & \cdots & 0 & 0 \\ x_{1n} & 0 & \cdots & 0 & x_{nn} \end{pmatrix} = \begin{pmatrix} x'_{11} & 0 & \cdots & 0 & x'_{1n} \\ 0 & 0 & \cdots & 0 & 0 \\ \vdots & \vdots & & \vdots & \vdots \\ 0 & 0 & \cdots & 0 & 0 \\ x'_{1n} & 0 & \cdots & 0 & x'_{nn} \end{pmatrix}$$

and by hypothesis of induction, we have

$$\begin{pmatrix} x_{11} & x_{1n} \\ x_{1n} & x_{nn} \end{pmatrix} = \left(\alpha \begin{pmatrix} x'_{11} & x'_{1n} \\ x'_{1n} & x'_{nn} \end{pmatrix} + \beta \right) \left(\gamma \begin{pmatrix} x'_{11} & x'_{1n} \\ x'_{1n} & x'_{nn} \end{pmatrix} + \delta \right)^{-1},$$

which is a symplectic transformation (notice that we omit the "anti-symplectic" case by the same reason given in (3)).

Since it leaves every point

$$\begin{pmatrix} x & 0 \\ 0 & 0 \end{pmatrix}, \quad \begin{pmatrix} 0 & 0 \\ 0 & y \end{pmatrix}$$

invariant, we have either

$$\alpha = I^{(2)}, \quad \beta = 0, \quad \gamma = \begin{pmatrix} 0 & c \\ c & 0 \end{pmatrix}, \quad \delta = I^{(2)}$$

or

$$\alpha = \begin{pmatrix} 1 & 0 \\ 0 & -1 \end{pmatrix}, \quad \beta = 0, \quad \gamma = \begin{pmatrix} 0 & c \\ -c & 0 \end{pmatrix}, \quad \delta = \begin{pmatrix} 1 & 0 \\ 0 & -1 \end{pmatrix}.$$

For the first case the transformation with

$$A = I, \quad B = 0, \quad C = \begin{pmatrix} 0 & 0 & \cdots & 0 & c \\ 0 & 0 & \cdots & 0 & 0 \\ \vdots & \vdots & & \vdots & \vdots \\ 0 & 0 & \cdots & 0 & 0 \\ c & 0 & \cdots & 0 & 0 \end{pmatrix}, \quad D = I$$

and for the second case the transformation with

$$A = [1, 1, \cdots, 1, -1], \quad B = 0, \quad C = \begin{pmatrix} 0 & 0 & \cdots & 0 & c \\ 0 & 0 & \cdots & 0 & 0 \\ \vdots & \vdots & & \vdots & \vdots \\ 0 & 0 & \cdots & 0 & 0 \\ -c & 0 & \cdots & 0 & 0 \end{pmatrix}, \quad D = A,$$

carry Γ into a new one which satisfies (5), (10) and

$$\Gamma \begin{pmatrix} x_{11} & 0 & \cdots & 0 & x_{1n} \\ 0 & 0 & \cdots & 0 & 0 \\ \vdots & \vdots & & \vdots & \vdots \\ 0 & 0 & \cdots & 0 & 0 \\ x_{1n} & 0 & \cdots & 0 & x_{nn} \end{pmatrix} = \begin{pmatrix} x_{11} & 0 & \cdots & 0 & x_{1n} \\ 0 & 0 & \cdots & 0 & 0 \\ \vdots & \vdots & & \vdots & \vdots \\ 0 & 0 & \cdots & 0 & 0 \\ x_{1n} & 0 & \cdots & 0 & x_{nn} \end{pmatrix}. \quad (11)$$

(6) In particular, we conclude that we may assume that

$$\Gamma([\lambda_1, \cdots, \lambda_{n-1}, 0]) = [\lambda_1, \cdots, \lambda_{n-1}, 0],$$
$$\Gamma([0, \lambda_2, \cdots, \lambda_n]) = [0, \lambda_2, \cdots, \lambda_n]$$

and

$$\Gamma([\lambda_1, 0, \cdots, 0, \lambda_n]) = [\lambda_1, 0, \cdots, 0, \lambda_n].$$

Now we are going to prove that

$$\Gamma([\lambda_1, \cdots, \lambda_n]) = [\lambda_1, \cdots, \lambda_n]. \quad (12)$$

Since

$$r([\lambda_1, \cdots, \lambda_n], [\lambda_1, \cdots, \lambda_{n-1}, 0]) = 1,$$

we have

$$\Gamma([\lambda_1, \cdots, \lambda_n]) = [\lambda_1, \cdots, \lambda_{n-1}, 0] + (a_1, \cdots, a_n)'(a_1, \cdots, a_n).$$

Since

$$r([\lambda_1, \cdots, \lambda_n], [0, \lambda_2, \cdots, \lambda_n]) = 1,$$

we have $a_2 = a_3 = \cdots = a_{n-1} = 0$. Further, since

$$r([\lambda_1, \lambda_2, \cdots, \lambda_n], [\lambda_1, 0, \cdots, 0, \lambda_n])$$
$$= r([\lambda_1, \cdots, \lambda_{n-1}, 0] + (a_1, 0, \cdots, 0, a_n)'(a_1, 0, \cdots, 0, a_n), [\lambda_1, 0, \cdots, 0, \lambda_n]),$$

we obtain that the matrix

$$\begin{pmatrix} a_1^2 & a_1 a_n \\ a_1 a_n & a_n^2 - \lambda_n \end{pmatrix}$$

is of rank zero, that is $a_1 = 0$ and $a_n^2 = \lambda_n$. Thus we have the assertion.

(7) Let

$$\Gamma(Z) = Z_1, \quad Z = (z_{ij}), \quad Z_1 = (z'_{ij}).$$

Since (12) holds, the equation

$$d(Z - [\lambda_1, \cdots, \lambda_n]) = 0$$

(consider Z as fixed and vary $\lambda_1, \cdots, \lambda_n$) implies

$$d(Z_1 - [\lambda_1, \cdots, \lambda_n]) = 0$$

and vice versa. This implies

$$d(Z - [\lambda_1, \cdots, \lambda_n]) = d(Z_1 - [\lambda_1, \cdots, \lambda_n])$$

identically in $\lambda_1, \cdots, \lambda_n$. Consequently, we have

$$z_{ii} = z'_{ii}, \quad z_{ii}z_{jj} - z_{ij}^2 = z'_{ii}z'_{jj} - z'^2_{ij}$$

and

$$z_{ij} = \pm z'_{ij}.$$

The transformation

$$[1, -1, 1, \cdots, 1]Z[1, -1, 1, \cdots, 1]$$

carries z_{ii} into itself and z_{12} into $-z_{12}$. Thus we may assume that

$$z_{1i} = z'_{1i}.$$

By the same consideration given in the proof of Theorem 8 of I, we have

$$\Gamma(Z) = Z.$$

The theorem therefore follows.

5. Affine geometry of symmetric matrices

Theorem 5 *A continuous mapping carrying finite points into finite points, infinite points into infinite points, and keeping arithmetic distance invariant is either an affine mapping*

$$W = AZA' + S \tag{13}$$

or an anti-affine mapping

$$\overline{W} = AZA' + S. \tag{13'}$$

This is an immediate consequence of Theorem 4.

6. Möbius geometry of symmetric matrices

As in II, we let $\mathfrak{J}$ be a fundamental involution as an absolute, for example:
$$W = HZH, \tag{14}$$
where $H = [1, 1, \cdots, 1, -1]$. The arithmetic distance between P and $\mathfrak{J}(P)$ is either 0 or 2. In case $P = \mathfrak{J}(P)$ we say that P is a *point-matrix*.

Theorem 6 *A continuous mapping carrying point-matrices into point-matrices, non-point-matrices into non-point-matrices and keeping arithmetic distance invariant is a Möbius mapping or an anti-Möbius mapping.*

Proof We take $\mathfrak{J}$ in the form of (14), which carries matrices of the form
$$\begin{pmatrix} Z_1^{(n-1)} & 0 \\ 0 & z \end{pmatrix} \tag{15}$$
into themselves. Thus the mappings under consideration carry the set defined by (15) onto itself.

Without loss of generality we may assume that the mappings keep 0 and ∞ invariant; by Theorem 4, it takes either the form
$$W = AZA'$$
or
$$\overline{W} = AZA'.$$

For the first case, we put
$$A = \begin{pmatrix} A_1 & \alpha_1' \\ \alpha_2 & a \end{pmatrix}$$
and we have
$$\begin{pmatrix} W_1 & 0 \\ 0 & w \end{pmatrix} = \begin{pmatrix} A_1 & \alpha_1' \\ \alpha_2 & a \end{pmatrix} \begin{pmatrix} Z_1 & 0 \\ 0 & z \end{pmatrix} \begin{pmatrix} A_1' & \alpha_2' \\ \alpha_1 & a \end{pmatrix}.$$
Consequently, we have
$$A_1 Z_1 \alpha_2' + \alpha_1' z a = 0$$
for all Z_1 and z. It follows that

(i) $\qquad\qquad\qquad\alpha_1 = 0$ or $a = 0$

and

(ii) $\qquad\qquad\qquad\alpha_2 = 0$ or $A_1 = 0$.

The case with $\alpha_1 = 0, \alpha_2 = 0$ is what we are looking for. The other cases cannot happen by the nonsingularity of A, except when $n = 2$, $a = A_1 = 0$. For this case we have also a Möbius mapping.

A similar method may be used for the second case.

7. Manifold at infinity

Let M be an idempotent symmetric matrix, that is, $M^2 = M$. Then $I - M$ is also an idempotent symmetric matrix, since $(I - M)^2 = I - 2M + M^2 = I - M$. Further M and $I - M$ annihilate each other, namely

$$M(I - M) = (I - M)M = 0.$$

From an idempotent symmetric matrix M, we can construct a symplectic transformation

$$W = (MZ + (I - M))(-(I - M)Z + M)^{-1}.$$

Now we consider those transformations with diagonal M. They are 2^n in number, including the identity. The $2^n - 1$ non-identity transformations are called the *fundamental semi-involutions* of the space of symmetric matrices. They are of fundamental importance in the study of the extended space.

For the sake of later use, we give the explicit expression of the semi-involution with

$$M = \begin{pmatrix} I^{(r)} & 0 \\ 0 & 0 \end{pmatrix}.$$

Let

$$Z = \begin{pmatrix} Z_{11}^{(r)} & Z_{12} \\ Z_{12}' & Z_{22} \end{pmatrix},$$

we have

$$W = \begin{pmatrix} Z_{11} - Z_{12} Z_{22}^{-1} Z_{12}' & -Z_{12} Z_{22}^{-1} \\ -Z_{22}^{-1} Z_{12}' & -Z_{22}^{-1} \end{pmatrix}. \tag{16}$$

Theorem 7 *Every point at infinity can be carried into a finite point by one of the $2^n - 1$ semi-involutions.*

Proof Suppose that

$$(Z_1, Z_2)$$

is the homogeneous coordinate of a point at infinity, with Z_2 of rank r. There exist two permutation matrices[①] P and Q such that

$$PZ_2Q = \begin{pmatrix} Z_{11}^{(r)} & * \\ * & * \end{pmatrix},$$

---

① Corresponding to a permutation

$$\begin{pmatrix} 1 & 2 & \cdots & n \\ i_1 & i_2 & \cdots & i_n \end{pmatrix},$$

we have a permutation matrix which is the matrix of the linear transformation $x_p' = x_{i_p}$ ($1 \leqslant p \leqslant n$). Evidently, we have $P' = P^{-1}$.

where $Z_{11}^{(r)}$ is nonsingular. We have a nonsingular R such that

$$RPZ_2Q = \begin{pmatrix} Z_{11} & Z_{12} \\ 0 & 0 \end{pmatrix}.$$

Let

$$RP(Z_1, Z_2) \begin{pmatrix} Q'^{-1} & 0 \\ 0 & Q \end{pmatrix} = (W_1, W_2).$$

Then

$$W_2 = \begin{pmatrix} Z_{11} & Z_{12} \\ 0 & 0 \end{pmatrix}.$$

Let

$$W_1 = \begin{pmatrix} W_{11} & W_{12} \\ W_{21} & W_{22} \end{pmatrix}$$

and

$$T = \begin{pmatrix} I & -Z_{11}^{-1} Z_{12} \\ 0 & I \end{pmatrix}.$$

We have immediately

$$W_2 T = \begin{pmatrix} Z_{11} & 0 \\ 0 & 0 \end{pmatrix}$$

and

$$W_1 T'^{-1} = \begin{pmatrix} * & * \\ 0 & W_{22} \end{pmatrix}.$$

The zero at the left lower corner is obtained because (W_1, W_2) is a symmetric pair. Then W_{22} is nonsingular, since (W_1, W_2) is a nonsingular pair.

Let

$$M = \begin{pmatrix} I^{(r)} & 0 \\ 0 & 0 \end{pmatrix}.$$

Then

$$(W_1, W_2) \begin{pmatrix} M & I - M \\ -(I - M) & M \end{pmatrix} = (P_1, P_2),$$

where

$$P_1 = W_1(I - M) + W_2 M = \begin{pmatrix} 0 & W_{12} \\ 0 & W_{22} \end{pmatrix} + \begin{pmatrix} Z_{11} & Z_{12} \\ 0 & 0 \end{pmatrix},$$

which is evidently nonsingular.

Now we consider

$$\begin{pmatrix} Q & 0 \\ 0 & Q \end{pmatrix} \begin{pmatrix} M & I-M \\ -(I-M) & M \end{pmatrix} \begin{pmatrix} Q' & 0 \\ 0 & Q' \end{pmatrix} = \begin{pmatrix} M_1 & I-M_1 \\ -(I-M_1) & M_1 \end{pmatrix},$$

$$M_1 = QMQ^{-1},$$

which is a semi-involution, since $Q' = Q^{-1}$. The theorem is therefore established.

Since

$$d(-(I-M)Z + M) \tag{17}$$

is a principal minor of Z, we have the following theorem.

Theorem 8 *The manifold at infinity is carried by the $2^n - 1$ semi-involutions into manifolds defined by equating the principal minors of Z to zero.*

8. Group of analytic automorphs of the space

Now we come to the second part of the paper. The projective space of symmetric matrices may be regarded as an extended space of several complex variables as defined by Osgood. The aim of the present section is to establish the following theorem.

Theorem 9 *An analytic mapping carrying the extended space onto itself is a symplectic mapping.*

Proof (1) Let

$$W = \Gamma(Z) \tag{18}$$

be an analytic mapping carrying the projective space of symmetric matrices onto itself. By a theorem due to Osgood[1], the mapping is birational; consequently, (18) may be written as

$$w_{ij} = p_{ij}(Z)/q(Z), \tag{19}$$

where $p_{ij}(Z)$ and $q(Z)$ are $2^{-1}n(n+1) + 1$ polynomials without common divisor.

(2) There is a point S at which

$$p_{ij}(S) \neq 0, \quad q(S) \neq 0.$$

The transformation $Z = Z_1 + S$ carries (18) into a new one with

$$p_{ij}(0) \neq 0, \quad q(0) \neq 0. \tag{20}$$

The transformation

$$W = \Gamma(-Z_1^{-1}) \tag{21}$$

---

[1] Ibid. p.295.

maps also the space onto itself, and

$$w_{ij} = p_{ij}(-Z_1^{-1})(d(Z_1))^\lambda / q(-Z_1^{-1})(d(Z_1))^\lambda,$$

where λ is the least integer making all numerators and denominators integral. By (20), $p_{ij}(-Z_1^{-1})(d(Z_1))^\lambda$ and $q(-Z_1^{-1})(d(Z_1))^\lambda$ are of degree $n\lambda$. Consider the Jacobian of (21). Let Δ and Δ_1 be the inverses of the Jacobians of (18) and (21) respectively. Notice that Δ and Δ_1 are polynomials, for otherwise there would exist some point for which the Jacobians vanish. Then we have

$$\Delta_1(Z_1) = \Delta(-Z_1^{-1})(d(Z_1))^{n+1},$$

since the Jacobian of $Z = -Z_1^{-1}$ is $(d(Z_1))^{-(n+1)}$. Since $q(0) \neq 0$, we have $\Delta(0) \neq 0$. Consequently $\Delta_1(Z_1)$ is a polynomial of degree $n(n+1)$.

Without loss of generality, we may now assume that p_{ij} and q are polynomials of degree $n\lambda$ and that the inverse of the Jacobian of (18) is a polynomial of degree $n(n+1)$ and its term of highest degree is equal to a constant multiple of $(d(Z))^{n+1}$. The highest terms of p_{ij} and q are also constant multiples of $(d(Z))^\lambda$.

(3) We decompose q into irreducible factors

$$q = q_1^{\lambda_1} \cdots q_l^{\lambda_l}. \tag{22}$$

Now we are going to establish that $(q_1 \cdots q_l)^{n+1}$ divide Δ. By some easy transformation we may suppose that q_1 does not divide p_{nn}. We consider the product of the transformation (16) with $r = n - 1$ and (18). The inverse of the Jacobian of the new mapping is equal to

$$(p_{nn}/q)^{n+1}\Delta,$$

which is a polynomial. Thus q_1^{n+1} divides Δ. Therefore we have the assertion.

(4) Since $d(Z)$ is irreducible and

$$\Delta(Z) = c(d(Z))^{n+1} + \cdots,$$

we have immediately that $l = 1$. Therefore

$$q(Z) = (ad(Z) + \cdots)^\lambda.$$

Further (16) with $r = n - 1$ carries p_{nn} into the denominator, then we have also

$$p_{nn}(Z) = (a_{nn}d(Z) + \cdots)^\lambda.$$

The mapping cannot be one-to-one except when $\lambda = 1$.

Now we may assume that

$$q(Z) = ad(Z) + \cdots, \tag{23}$$

$$p_{ij}(Z) = a_{ij}d(Z) + \cdots \tag{24}$$

and

$$\Delta(Z) = c(q(Z))^{n+1}. \tag{25}$$

(5) Now we shall prove that $p_{ii}p_{jj} - p_{ij}^2$ is divisible by q. In fact, by (16) with $r = n-2$, we have

$$W = \begin{pmatrix} * & & * & \\ * & \dfrac{q}{p_{n-1\,n-1}p_{nn} - p_{n\,n-1}^2} & \begin{pmatrix} p_{nn} & -p_{n\,n-1} \\ -p_{n\,n-1} & p_{n-1\,n-1} \end{pmatrix} \end{pmatrix}.$$

If q does not divide $p_{n-1\,n-1}p_{nn} - p_{n\,n-1}^2$, the manifold $q = 0$ is mapped into a manifold of dimension not greater than $n(n+1) - 3$, which is impossible. Similarly, every three-rowed principal minor of (p_{ij}) is divisible by q.

(6) Since

$$\Delta(-Z^{-1})(d(Z))^{n+1} = \Delta_1(Z),$$

we have

$$q(-Z^{-1})(d(Z)) = q_1(Z),$$

where $q_1(Z)$ is a polynomial. By means of (16) with $r = n-1$, we may find that

$$p_{ii}(-Z^{-1})d(Z)$$

is also a polynomial. Further, by means of (16) with $r = n-2$, we find that

$$\frac{p_{ii}(-Z^{-1})p_{jj}(-Z^{-1}) - p_{ij}^2(-Z^{-1})}{q(-Z^{-1})}d(z)$$

is also a polynomial. Thus

$$p_{ij}^2(-Z^{-1})(d(Z))^2$$

is a polynomial, and so is $p_{ij}(-Z^{-1})d(Z)$. Therefore we have to find the polynomial $p(Z)$ of degree n such that

$$p(-Z^{-1})d(Z)$$

is a polynomial.

(7) The answer to the question raised in (6) is that if we put

$$p(Z) = \sum_{k=0}^{n} p^{(k)}(Z),$$

where $p^{(k)}$ is a homogeneous polynomial of degree k, then $p^{(k)}(Z)$ is a linear combination of the k-rowed minors of Z. This will be proved in the next section, owing to its independent interest.

(8) Now we have, instead of (23) and (24), the following expressions

$$q(Z) = ad(Z) + \sum_{k=0}^{n-1} q^{(k)}, \quad a \neq 0,$$

$$p_{ij}(Z) = a_{ij}d(Z) + \sum_{k=0}^{n-1} p_{ij}^{(k)},$$

where $q^{(k)}$ and $p_{ij}^{(k)}$ are linear combinations of the k-rowed minors of Z. We may let $a = 1$. Let z_{ij}^* be the cofactor of z_{ij}, then

$$q^{(n-1)} = \sum_{1 \leq i \leq j \leq n} c_{ij} z_{ij}^*.$$

There exists S such that

$$q(Z - S)$$

contains no term of order $n - 1$. Thus, we may assume that $q^{(n-1)} = 0$.

Further the transformation

$$X = W - (a_{ij})$$

carries (18) into a new one with $a_{ij} = 0$. Up to the present, (18) takes the form

$$q(Z) = d(Z) + \sum_{k=0}^{n-2} q^{(k)} \tag{26}$$

and

$$p_{ij}(Z) = \sum_{k=0}^{n-1} p_{ij}^{(k)}, \tag{27}$$

where $q^{(k)}$ and $p_{ij}^{(k)}$ are linear combinations of the k-rowed minors of Z. Δ is given by (25).

Since the singular matrices form a manifold of dimension $n(n+1) - 2$, there is a nonsingular matrix Z carried into a nonsingular matrix W. Without loss of generality we may assume that

$$I = \Gamma(I). \tag{28}$$

(9) We write

$$p_{ij}^{(n-1)} = \sum_{1 \leq s \leq t \leq n} a_{ij,st} z_{st}^*, \quad 1 \leq i; j \leq n. \tag{29}$$

We shall prove that (29) forms a system of independent equations. In fact, the degree of the Jacobian of (18) with (26) and (27) is, by direct verification, not greater than

$$(-2)\frac{n(n+1)}{2} - 1 = -n(n+1) - 1.$$

the last "-1" appears, as the highest terms are dependent. Then Δ is of degree not less than $n(n+1)+1$, which is impossible. We have therefore the assertion.

(10) Now we consider the matrix

$$(p_{ij}^{(n-1)}) = \sum_{1 \leqslant s,t \leqslant n} A_{st} z_{st}^*, \quad A_{st}' = A_{st}. \tag{30}$$

By (5), we have the consequence that $d(Z)$ divides

$$\begin{vmatrix} p_{ii}^{(n-1)} & p_{ij}^{(n-1)} \\ p_{ij}^{(n-1)} & p_{jj}^{(n-1)} \end{vmatrix} \quad \text{and} \quad \begin{vmatrix} p_{ii}^{(n-1)} & p_{ij}^{(n-1)} & p_{ik}^{(n-1)} \\ p_{ij}^{(n-1)} & p_{jj}^{(n-1)} & p_{jk}^{(n-1)} \\ p_{ik}^{(n-1)} & p_{jk}^{(n-1)} & p_{kk}^{(n-1)} \end{vmatrix}.$$

Therefore if $d(Z) = 0$, we find that (30) is of rank not greater than 1. In particular A_{ii} are of rank 1, as that A_{ii} cannot be of rank 0 has been shown in (9).

We write

$$A_{ii} = (a_{i1}, \cdots, a_{in})'(a_{i1}, \cdots, a_{in}).$$

By (28), we have

$$\sum_{i=1}^{n} A_{ii} = I.$$

Let

$$A = (a_{ij}),$$

then

$$A'A = I.$$

Thus $A\Gamma(Z)A'$ carries (30) into a new one with

$$A_{ii} = [0, \cdots, 0, 1, 0, \cdots, 0]$$

(where there are $i-1$ zeros preceding the one).

We write

$$p_{11}^{(n-1)} = z_{11}^* + \sum_{1 \leqslant i < j \leqslant n} a_{ij} z_{ij}^*, \quad p_{22}^{(n-1)} = z_{22}^* + \sum_{1 \leqslant i < j \leqslant n} c_{ij} z_{ij}^*.$$

Since $d(Z)$ divides
$$p_{11}^{(n-1)}p_{22}^{(n-1)} - (p_{12}^{(n-1)})^2,$$
we write
$$p_{11}^{(n-1)}p_{22}^{(n-1)} - (p_{12}^{(n-1)})^2 = d(Z)g(Z).$$
By the lemma given in the next section, $p_{11}^{(n-1)}p_{22}^{(n-1)} - (p_{12}^{(n-1)})^2$ is a linear combination of the two-rowed minors of
$$(z_{ij}^*).$$
We write
$$p_{12}^{(n-1)} = \sum_{1 \leqslant i < j \leqslant n} b_{ij} z_{ij}^*.$$
If p_{12} contains z_{ij}^* $(i,j \neq 1,2)$ then $p_{11}p_{22}$ must contain $x_{ii}^* x_{jj}^*$, which is impossible. Therefore
$$p_{12}^{(n-1)} = \pm z_{12}^*.$$
Consequently,
$$p_{11}^{(n-1)}p_{22}^{(n-1)} - (p_{12}^{(n-1)})^2 = z_{11}^* \sum c_{ij} z_{ij}^* + z_{ij}^* \sum a_{ij} z_{ij}^* + \left(\sum a_{ij} z_{ij}^*\right)\left(\sum c_{ij} z_{ij}^*\right)$$
is a multiple of $d(Z)$. This is possible only when $c = a = 0$. Thus we have
$$p_{ii}^{(n-1)} = z_{ii}^*, \quad p_{ij}^{(n-1)} = \pm z_{ij}^*.$$
The transformation
$$[1,-1,1,\cdots,1]Z^*[1,-1,\cdots,1]$$
carries z_{12}^* into $-z_{12}^*$, therefore we may assume that
$$p_{1i}^{(n-1)} = z_{1i}^*.$$
Since
$$\begin{vmatrix} z_{11}^* & z_{12}^* & z_{13}^* \\ z_{12}^* & z_{22}^* & -z_{23}^* \\ z_{13}^* & -z_{23}^* & z_{33}^* \end{vmatrix}$$
is not divisible by $d(Z)$, we have
$$p_{ij}^{(n-1)} = z_{ij}^*.$$

Up to the present, we arrive at the conclusion that we may let

$$q(Z) = d(Z) + \sum_{k=0}^{n-2} q^{(k)} \tag{31}$$

and

$$p_{ij}(z) = z_{ij}^* + \sum_{k=0}^{n-2} p_{ij}^{(k)}. \tag{32}$$

(11) We write

$$p_{11}p_{22} - p_{12}^2 \equiv 0 \,(\mathrm{mod}\, q)$$

in a more precise form

$$p_{11}p_{22} - p_{12}^2 = q\psi, \tag{33}$$

where

$$\psi = \psi^{(n-2)} + \psi^{(n-3)} + \cdots$$

and $\psi^{(k)}$ is a homogeneous polynomial of degree k.

Comparing the terms of degree $2n-3$ in (33), we have

$$z_{11}^* p_{22}^{(n-2)} + z_{22}^* p_{11}^{(n-2)} - 2 z_{12}^* p_{12}^{(n-2)} = d(Z)\psi^{(n-3)}.$$

By the result which will be proved in (12) we have

$$p_{11}^{(n-2)} = p_{12}^{(n-2)} = p_{22}^{(n-2)} = \psi^{(n-3)} = 0.$$

Now we are going to prove that $p_{11}^{(\sigma)}, p_{12}^{(\sigma)}, p_{22}^{(\sigma)}$ and $\psi^{(\sigma-1)}$ are all zero. Suppose that the assertion is true for $\sigma > \rho$. Then

$$z_{11}^* p_{22}^{(\rho)} + z_{22}^* p_{11}^{(\rho)} - 2 z_{12}^* p_{12}^{(\rho)} = d(Z)\psi^{(\rho-1)}.$$

By (12) we have $p_{11}^{(\rho)} = p_{12}^{(\rho)} = p_{22}^{(\rho)} = \psi^{(\sigma-1)} = 0$. Thus we have finally

$$q(Z) = d(Z), \quad p_{ij}(Z) = z_{ij}^*.$$

The theorem is proved completely, except the verification of the following assertion.

(12) For $n - 2 \geqslant \sigma > 0$, from

$$z_{11}^* p_{22}^{(\sigma)} + z_{22}^* p_{11}^{(\sigma)} - 2 z_{12}^* p_{12}^{(\sigma)} = d(Z)\psi^{(\sigma-1)},$$

we can deduce that $p_{11}^{(\sigma)} = p_{12}^{(\sigma)} = p_{22}^{(\sigma)} = \psi^{(\sigma-1)} = 0$.

To prove this assertion, we make a transformation $Z = W^{-1}$. Then
$$z_{ij}^* = w_{ij}/d(W), \quad p_{ij}^{(\sigma)}(Z) = r_{ij}^{(n-\sigma)}(W)/d(W),$$
$$\psi^{(\sigma-1)}(Z) = \phi^{(n-\sigma+1)}(W)/d(W),$$
we have
$$w_{11}r_{22}^{(n-\sigma)} + w_{22}r_{11}^{(n-\sigma)} - 2w_{12}r_{12}^{(n-\sigma)} = \phi^{(\tau)}(W),$$
where $\phi^{(\tau)}$ is a sum of $\tau = (n - \sigma + 1)$-rowed minors. Let ϕ_1 be a τ-rowed minor contained in ϕ. Putting all elements of W other than those contained in ϕ equal to zero, we find that ϕ_1, a determinant of order τ, may be expressed as
$$w_{12}r_{22} + w_{22}r_{11} - 2w_{12}r_{12},$$
where r_{11}, r_{12}, r_{22} are sums of $(\tau - 1)$-rowed minors of ϕ_1. This is impossible for $n > 3$ as shown by the fact that the determinant vanishes identically for $w_{11} = w_{12} = w_{22} = 0$. Thus we suppose that $\tau = 3$. Since w_{22} contains no term with factors w_{11} and w_{13}, we have
$$r_{22} = c_1(w_{22}w_{33} - w_{23}^2)$$
and similarly
$$r_{11} = c_2(w_{11}w_{33} - w_{13}^2).$$
The equality
$$c_1(w_{22}w_{33} - w_{23}^2)w_{11} + c_2(w_{11}w_{33} - w_{13}^2)w_{22} - 2w_{12}r_{12}$$
$$= w_{11}w_{22}w_{33} - w_{23}^2 w_{11} - w_{13}^2 w_{22} + \cdots$$
implies
$$c_1(w_{22}w_{33} - w_{23}^2)w_{11} + c_2(w_{11}w_{33} - w_{13})^2 w_{23}$$
$$\equiv w_{11}w_{22}w_{33} - w_{23}^2 w_{11} - w_{13}^2 w_{22} \pmod{w_{12}},$$
which is evidently impossible.

9. A result concerning adjugate

Now we are going to verify the assertion stated in (7) of the previous section. Let
$$X = (x_{ij})$$
denote an n-rowed symmetric matrix and let
$$Y = (y_{ij})$$

be its in verse. Now we consider the following problem: find the polynomial $f(Y)$ in y_{ij} satisfying

$$f(Y)d(X) = g(X), \tag{34}$$

where $g(X)$ is a polynomial in x_{ij}. Since $d(X)$ is a homogeneous polynomial, we need only consider the homogeneous part of $f(Y)$ as well as $g(X)$. Thus our problem is reduced to finding homogeneous $f(Y)$ satisfying (34). Let l be the degree of $f(Y)$ and

$$f(Y) = \sum_{p_{11}+\cdots+p_{nn}=l} p(p_{11},\cdots,p_{nn})y_{11}^{p_{11}}\cdots y_{nn}^{p_{nn}}. \tag{35}$$

Then $g(X)$ is of degree $n-l$ and can be written as

$$g(X) = \sum_{q_{11}+\cdots+q_{nn}=n-l} q(q_{11},\cdots,q_{nn})x_{11}^{q_{11}}\cdots x_{nn}^{q_{nn}}. \tag{36}$$

Let

$$X^* = (x_{ij}^*) = d(X)Y.$$

Then (34) may be written as

$$\sum_{p_{11}+\cdots+p_{nn}=k} p(p_{11},\cdots,p_{nn})x_{11}^{*p_{11}}\cdots x_{nn}^{*p_{nn}}$$
$$=(d(X))^{l-1}\sum_{q_{11}+\cdots+q_{nn}=n-l} q(q_{11},\cdots,q_{nn})x_{11}^{q_{11}}\cdots x_{nn}^{q_{nn}}. \tag{37}$$

Notice that the relation (37) is a reciprocal one, in fact, from (37), we deduce also

$$\sum_{q_{11}+\cdots+q_{nn}=n-l} q(q_{11},\cdots,q_{nn})x_{11}^{*q_{11}}\cdots x_{nn}^{*q_{nn}}$$
$$=(d(X))^{n-l-1}\sum_{p_{11}+\cdots+p_{nn}=n} p(p_{11},\cdots,p_{nn})x_{11}^{p_{11}}\cdots x_{nn}^{p_{nn}}. \tag{38}$$

Does (37) have a solution? It is known[1] that a minor of order l satisfies our requirement. The purpose of the present section is to establish the converse, namely:

Theorem 10 (37) *holds if and only if* (36) *is a linear combination of the* $(n-l)$-*rowed minors. Consequently,* (34) *holds if and only if* $g(X)$ *is a linear combination of the minors of* X.

Proof The theorem is evidently true for $l=0,1,n$ and $n-1$. Consequently the theorem is true for $n=1,2$, and 3.

---

[1] Wedderburn. Lectures on matrices. *Amer. Math. Soc. Colloquium Publications*, vol. 17, p.67, formula (20).

Let $n \geq 4$ and $n/2 \leq l \leq n-2$. We write
$$g(X) = g_1(X) + g_0(X),$$
where $g_1(X)$ vanishes for $x_{11} = x_{12} = \cdots = x_{1n} = 0$. Putting
$$X = \begin{pmatrix} 1 & 0 \\ 0 & X_1 \end{pmatrix},$$
we find, by comparing the homogeneous parts, that $g_0(X) = g_0(X_1)$ satisfies (37) with $l-1, n-1$ instead of l, n. By hypothesis of induction, $g_0(X)$ is a linear combination of $(n-l)$-rowed minors of X_1.

Let
$$g_1(X) = g_2(X) + g_{1,0}(X),$$
where $g_2(X)$ vanishes for
$$x_{11} = x_{12} = \cdots = x_{1n} = 0$$
and for
$$x_{12} = x_{22} = \cdots = x_{2n} = 0.$$
Then $g_{10}(X)$ is also a linear combination of the $(n-l)$-rowed minors of X.

Proceeding successively, we have
$$g(X) = g_0(X) + g_{10}(X) + \cdots + g_{n0}(X) + \psi(X),$$
where $\psi(X)$ vanishes for
$$x_{11} = x_{12} = \cdots = x_{1n} = 0,$$
for
$$x_{12} = x_{22} = \cdots = x_{2n} = 0,$$
$$\cdots\cdots\cdots$$
and for
$$x_{1n} = x_{2n} = \cdots = x_{nn} = 0.$$
Since the degree of $\psi(X)$ is $n - l \leq n/2$, it is only possible for $l = n/2$, and $\psi(X)$ contains only terms of the form
$$c x_{12} x_{34} \cdots x_{n-1\,n}.$$
The term $(x_{11}\cdots x_{nn})^{l-1}(x_{12}x_{34}\cdots x_{n-1\,n})$ cannot appear on the left side of (37). Therefore we have the theorem.

10. Elliptic geometry

Now we let
$$\xi = \begin{pmatrix} H & 0 \\ 0 & H \end{pmatrix},$$
where $H = [1, \cdots, 1, -1, \cdots, -1]$ with p positive 1's and q negative 1's, $p + q = n$.

The points (W_1, W_2) making
$$\overline{(W_1, W_2)} \xi (W_1, W_2)' \tag{39}$$
of the same signature as H form the elliptic space with signature (p, q). The symplectic transformation
$$(Z_1, Z_2) = Q(W_1, W_2)F, \quad \overline{F} \xi F' = \xi$$
is called a motion of the space.

Theorem 11 *Any nonsingular symmetric pair of matrices (W_1, W_2) making (39) nonsingular belongs to the elliptic space of signature (p, q).*

Proof We have a nonsingular matrix Γ such that
$$\overline{\Gamma(W_1, W_2)} \xi (W_1, W_2)' \Gamma' = H_0, \tag{40}$$
where H_0 is a diagonal matrix with p' positive 1's and q' negative 1's.

Let
$$(W_1^*, W_2^*) = \Gamma(W_1, W_2).$$

Construct the symplectic matrix
$$F = \begin{pmatrix} W_1^* & W_2^* \\ -H_0 \overline{W}_2^* H & H_0 \overline{W}_1^* H \end{pmatrix}.$$

We may verify directly that
$$\overline{F} \begin{pmatrix} H & 0 \\ 0 & H \end{pmatrix} F' = \begin{pmatrix} H_0 & 0 \\ 0 & H_0 \end{pmatrix}.$$

Owing to the invariance of signature, we have $H = H_0$. The theorem follows.

From Theorem 11, we have that the elliptic space of signature (p, q) is formed by all nonsingular symmetric pairs of matrices (W_1, W_2) except those lying on
$$d(\overline{(W_1, W_2)} \xi (W_1, W_2)') = 0.$$

From Theorem 11, we may easily find that

$$d(\overline{(W_1,W_2)}\xi(W_1,W_2)'H) \geq 0 \tag{41}$$

for any nonsingular pair of matrices (W_1, W_2) and that the equality holds on a manifold of dimension not greater than $n(n+1) - 2$. Now we shall go further to establish the following theorem.

Theorem 12 *If*

$$d(\overline{(W_1,W_2)}\xi(W_1,W_2)') = 0,$$

then (W_1, W_2) lies on a manifold of dimension not greater than $n(n+1) - 3$.

Proof By the consideration of semi-involutions, we have only to consider the nonhomogeneous expression, that is, we are going to prove that the symmetric matrices Z satisfying

$$d(H + ZH\overline{Z}) = 0 \tag{42}$$

form a manifold of dimension not greater than $n(n+1) - 3$.

The equation (42) implies that $HZ^{-1}H\overline{Z}^{-1}$ has at least a negative root, if Z is nonsingular. By a theorem due to the author[①] those Z form a manifold of dimension not greater than $n(n+1) - 3$.

The equation $d(Z) = 0$ does not imply $d(H + ZH\overline{Z}) = 0$ identically. The manifold defined by

$$d(Z) = d(H + ZH\overline{Z}) = 0$$

is therefore of dimension not greater than $n(n+1) - 3$.

Theorem 13 (Fundamental theorem of the elliptic space) *An analytic automorph of the elliptic space of signature (p, q) is a motion of the space.*

Proof Let

$$W = \Gamma(Z), \quad \Gamma = (f_{ij})$$

by an analytic mapping of the elliptic space onto itself. Then $f_{ij}(Z)$ is analytic in the whole extended space of symmetric matrices except possibly on a manifold of dimension not greater than $n(n+1) - 3$. By the continuity theorem[②] of functions of several complex variables, $f_{ij}(Z)$ is analytic in the whole extended space. Its inverse mapping is also analytic everywhere. The theorem follows from Theorem 12.

[①] Ibid. Theorem 6.

[②] Levi, ibid., or Satz 17, Folgerung 1 of Behnke and Thullen. *Theorie der Funktionen mehrerer komplexer Veränderlichen.* Julius Springer, 1934.

11. Hyperbolic space

The hyperbolic space is formed by the symmetric matrices Z making $I - Z\overline{Z}$ positive definite. A symplectic mapping carrying the space onto itself is called a motion of the space.

Theorem 14 (Siegel) *An analytic mapping carrying the hyperbolic space onto itself is a motion of the space.*

For completeness, we give here a proof which is different from that due to Siegel.

Proof (1) The space is transitive, hence we need only to consider the transformations keeping 0 invariant. By a theorem for circular regions proved by H. Cartan[①], the mapping is a linear one. Let

$$W = \sum_{1 \leqslant r \leqslant s \leqslant h} A_{rs} z_{rs}, \quad A_{rs} = A'_{rs} \tag{43}$$

be the mapping. Since $I - Z\overline{Z} = 0$ is the intersection of all the algebraic surfaces bounding the space, the mapping (43), therefore, carries unitary symmetric matrices into unitary symmetric matrices.

Since any unitary symmetric matrix can be expressed as UU' where U is unitary, we may assume that (43) carries I into itself. Consequently, we have

$$\sum_{r=1}^{n} A_{rr} = I. \tag{44}$$

Putting $Z = [e^{i\theta_1}, \cdots, e^{i\theta_n}]$, we have

$$I = W\overline{W} = \left(\sum_{r=1}^{n} A_{rr} e^{i\theta_r} \right) \left(\sum_{r=1}^{n} \overline{A}_{rr} e^{-i\theta_r} \right)$$

for any real θ. Consequently we obtain

$$A_{rr} \overline{A}_{ss} = 0 \quad \text{for } r \neq s. \tag{45}$$

(2) From (44) and (45), we have

$$A_{rr} = A_{rr} \left(\sum_{s=1}^{n} \overline{A}_{ss} \right) = A_{rr} \overline{A}_{rr} = \left(\sum_{s=1}^{n} A_{ss} \right) \overline{A}_{rr} = \overline{A}_{rr}.$$

Then A_{rr} are real and $A_{rr} A_{ss} = 0, A_{rr}^2 = A_{rr}$. Therefore, we have a real orthogonal T such that

$$TA_{rr}T' = [0, \cdots, 0, 1, 0, \cdots, 0]$$

---

① *Journal de Mathématique*, 1931, **10**(9), Theorem 6.

(where the 1 in the brackets is in the rth place). Without loss of generality we assume that
$$A_{rr} = [0, \cdots, 0, 1, 0, \cdots, 0] \qquad (46)$$
(where the 1 in brackets is in the rth place).

(3) Now we put
$$Z = \begin{pmatrix} 0 & e^{i\theta} \\ e^{i\theta} & 0 \end{pmatrix} + [1, \cdots, 1].$$

From (44) and $I = W\overline{W}$, we have
$$A_{12}\overline{A}_{12} + e^{i\theta}A_{12}(I - A_{11} - A_{22}) + (I - A_{11} - A_{22})e^{-i\theta}\overline{A}_{12}$$
$$+ ((I - A_{11} - A_{22})^2 = I,$$
for all θ. Thus we have
$$A_{12} = \alpha^{(2)} \dotplus 0,$$
where $\alpha\bar{\alpha} = I^{(2)}$.

(4) Further we let
$$Z = \begin{pmatrix} e^{i\tau}/2^{1/2} & 1/2^{1/2} \\ 1/2^{1/2} & -e^{-i\tau}/2^{1/2} \end{pmatrix} \dotplus 0^{(n-2)}.$$

From (44) and $I - W\overline{W} = 0$, we deduce immediately that
$$I^{(2)} = 2^{-1}\left(\begin{pmatrix} e^{i\tau} & 0 \\ 0 & -e^{-i\tau} \end{pmatrix} + \alpha\right)\left(\begin{pmatrix} e^{-i\tau} & 0 \\ 0 & -e^{i\tau} \end{pmatrix} + \bar{\alpha}\right)$$
$$= 2^{-1}I^{(2)} + 2^{-1}\left(\alpha\begin{pmatrix} e^{-i\tau} & 0 \\ 0 & -e^{i\tau} \end{pmatrix} + \begin{pmatrix} e^{i\tau} & 0 \\ 0 & -e^{-i\tau} \end{pmatrix}\bar{\alpha}\right) + 2^{-1}\alpha\bar{\alpha}.$$

Consequently
$$\alpha\begin{pmatrix} e^{-i\tau} & 0 \\ 0 & -e^{i\tau} \end{pmatrix} + \begin{pmatrix} e^{i\tau} & 0 \\ 0 & -e^{-i\tau} \end{pmatrix}\bar{\alpha} = 0,$$
for all τ. Then
$$\alpha = \pm\begin{pmatrix} 0 & 1 \\ 1 & 0 \end{pmatrix}.$$
Similarly A_{rs} is determined completely apart from a sign. We may assume that A_{1s} takes the positive sign. Since
$$\begin{pmatrix} z_{11} & z_{12} & z_{13} \\ z_{12} & z_{22} & z_{23} \\ z_{13} & z_{23} & z_{33} \end{pmatrix}$$

being unitary does not imply that

$$\begin{pmatrix} z_{11} & z_{12} & z_{13} \\ z_{12} & z_{22} & -z_{23} \\ z_{13} & -z_{23} & z_{33} \end{pmatrix}$$

is unitary, the theorem is now established.

Finally, the author gives the following theorem which shows the importance of the notion of the characteristic roots of the distance-matrix.

Theorem 15 *A mapping carrying the hyperbolic space "$I - Z\overline{Z}$ being positive definite" into itself, and keeping the characteristic roots of the distance matrix between two points invariant, is either a hyperbolic motion or a hyperbolic motion combined with a reflexion $Z = \overline{W}$.*

Proof The proof is comparatively simple, and it contains some repetition of our old argument. The author gives only the main procedure of the proof.

(1) The distance matrix $\mathfrak{D}(A,B)$ of two points A and B of the space is defined by

$$\mathfrak{D}(A,B) = (A-B)(A-\overline{B}^{-1})^{-1}((\overline{A}^{-1}-B)(\overline{A}^{-1}-\overline{B}^{-1})^{-1})^{-1}.$$

In particular for $B = 0$, we have

$$\mathfrak{D}(A,0) = A\overline{A}'.$$

(2) Owing to the transitivity, we need only consider the mapping $W = \Gamma(Z)$ keeping 0 invariant. Thus both Hermitian matrices $W\overline{W}$ and $Z\overline{Z}$ are conjunctive. In particular the mapping keeps the rank of Z invariant.

(3) Let

$$\Gamma([1/2, 0, \cdots, 0]) = (a_1, \cdots, a_n)'(a_1, \cdots, a_n).$$

By a slight modification, we may let

$$\Gamma([1/2, 0, \cdots, 0]) = [1/2, 0, \cdots, 0].$$

From the invariance of the characteristic roots of distance-matrix we find

$$\Gamma([\lambda_1, 0, \cdots, 0]) = [\lambda_1, 0, \cdots, 0],$$

for all real λ_1. Without loss of generality, we may modify Γ such that

$$\Gamma([0, \lambda_2, 0, \cdots, 0]) = [0, \lambda_2, 0, \cdots, 0],$$

$$\cdots\cdots\cdots$$

$$\Gamma([0, 0, \cdots, 0, \lambda_n]) = [0, 0, \cdots, 0, \lambda_n]$$

for all real $\lambda_2, \cdots, \lambda_n$.

Next, we may show that for any real diagonal Λ we have
$$\Gamma(\Lambda) = \Lambda$$
and that for all real X, we have
$$\Gamma(X) = X.$$

(4) Since $W = \Gamma(Z)$ and
$$r(X, W) = r(X, Z),$$
we find that
$$|d(X - W)|^2 = |d(X - Z)|^2$$
for all real X. We deduce that
$$W = Z \quad \text{or} \quad \overline{Z}.$$

Some "Anzahl" theorems for groups of prime power orders*

Abstract

We define a group $\mathcal{G}$ of order p^n (p being a prime) to be of rank α if the maximum of the orders of all elements of $\mathcal{G}$ is equal to $p^{n-\alpha}$. First the author introduces a "pseudo base", which is the main weapon of the paper. Then the author proves the "Anzahl" theorem that if $p \geqslant 3, n \geqslant 2\alpha + 1$, then (i) $\mathcal{G}$ contains one and only one subgroup of order p^m ($2\alpha + 1 \leqslant m \leqslant n$) of rank α; (ii) $\mathcal{G}$ contains p^α cyclic subgroups of order p^m ($\alpha < m < n - \alpha + 1$); and (iii) $\mathcal{G}$ contains $p^{m+\alpha}$ elements satisfying $G^{p^m} = 1$. Next the author determines completely the ranks of all major subgroups and the exact number of the major subgroups of each rank. More precisely, the rank of a major subgroup of index p^β is either $\alpha - \beta + 1$ or $\alpha - \beta$, and for $p \geqslant 3$ and $n \geqslant 2\alpha + 1$, the number of those of rank $\alpha - \beta + 1$ and the number of those of rank $\alpha - \beta$ are equal to $\phi_{d-1,\beta-1}$ and $p^\beta \phi_{d-1,\beta}$ respectively, where d is the number of mutually independent generators of the group and for $0 \leqslant \tau \leqslant \nu$,

$$\phi_{\nu,\tau} = \frac{(p^\nu - 1) \cdots (p^{\nu-\tau+1} - 1)}{(p^\tau - 1) \cdots (p - 1)}.$$

Then the author deduces a more precise formula than the enumeration principle due to Hall. This is the second weapon of the paper. By these two weapons, the following Anzahl theorem is obtained: if $p \geqslant 3$ and $n \geqslant 2\alpha+1$, then the number of its subgroups of order p^m ($2\alpha - \beta < m < n - \beta + 1$) and of rank $\alpha - \beta$ with $0 \leqslant \beta \leqslant d$ is congruent to $p^\beta \phi_{d-1,\beta}$ (mod $p^{d-\beta+1}$). For $\beta = d$, $p^\beta \phi_{d-1,\beta}$ is to be replaced by zero. In the later part of the paper, the author selects the following typical one: for $p \geqslant 3$ if $\mathcal{G}$ contains p^β cyclic subgroups of order p^m for a fixed m satisfying $m > \beta + 1$, it is of rank β.

Introduction

For convenience, we shall define a group $\mathcal{G}$ of order p^n (p being a prime) to be of rank δ if the maximum of the orders of all elements G of $\mathcal{G}$ is equal to $p^{n-\delta}$.

---

* Reprinted from *The Science Reports of National Tsing Hua University*, Series A, 1947, 4 (4–6): 313–327.

Throughout this paper, we shall always denote by p an odd prime.

In §1, by introducing a "pseudo-base", we shall prove the following "Anzahl" theorem:

If $\mathcal{G}$ is of order p^n ($p \geqslant 3$, $n \geqslant 2\alpha + 1$) of rank α, then (i) $\mathcal{G}$ contains one and only one subgroup of order p^m ($2\alpha + 1 \leqslant m \leqslant n$) of rank α; (ii) $\mathcal{G}$ contains p^α cyclic subgroups of order p^m ($\alpha < m < n - \alpha + 1$); and (iii) $\mathcal{G}$ contains $p^{m+\alpha}$ elements satisfying
$$G^{p^m} = 1.$$

The second and the third statements of the above theorem yield better results than the well-known theorems due to Miller and Kulakoff respectively.

In §2, the ranks of all the major subgroups and the exact number of the major subgroups of each rank will be determined completely; more precisely, in a group $\mathcal{G}$ of order p^n of rank α, the rank of a major subgroup of index p^β is either $\alpha - \beta + 1$ or $\alpha - \beta$, and for $p \geqslant 3$ and $n \geqslant 2\alpha + 1$, the number of those of rank $\alpha - \beta + 1$ and the number of those of rank $\alpha - \beta$ are equal to $\phi_{d-1,\beta-1}$ and $p^\beta \phi_{d-1,\beta}$ respectively, where d is the number of generators of the group and for $0 \leqslant \tau \leqslant \nu$,
$$\phi_{\nu,\tau} = \frac{(p^\nu - 1) \cdots (p^{\nu - \tau + 1} - 1)}{(p^\tau - 1) \cdots (p - 1)} = \frac{(p^\nu - 1) \cdots (p^\nu - p^{\tau - 1})}{(p^\tau - 1) \cdots (p^\tau - p^{\tau - 1})}.$$

By means of these results we shall establish a more precise formula than the enumeration principle due to Hall.

Some precise results concerning the number of subgroups of order p^m in $\mathcal{G}$ will be given in §3, the chief theorem is the following:

Let $\mathcal{G}$ be a group of order p^n and of rank α with $p \geqslant 3$ and $n \geqslant 2\alpha + 1$. Then the number of its subgroups of order p^m ($2\alpha - \beta < m < n - \beta + 1$) and of rank $\alpha - \beta$ with $0 \leqslant \beta \leqslant d$ is congruent to $p^\beta \phi_{d-1,\beta}$ (mod $p^{d-\beta+1}$). For $\beta = d$, $p^\beta \phi_{d-1,\beta}$ has to be replaced by 0.

Among the theorems of §4, we shall state the following typical one which is a generalization of a result due to Miller and is in a certain sense a converse of a result of §1:

A group of order p^n ($p \geqslant 3$) which contains p^β cyclic subgroups of order p^m for a fixed m satisfying $m > \beta + 1$ is of rank β.

For $\beta = 0$, the theorem is trivial, and for $\beta = 1$, we have the case discussed by Miller.

The author is indebted to Mr. H. F. Tuan for his revision of the manuscripts. Some new results of Kulakoff's type have been obtained by him.

§1. **Theorem 1** *Let $\mathcal{G}$ be a group of order p^n and of rank α with $p \geq 3$ and $n \geq 2\alpha + 1$, then we have the following conclusions:*

(i) *There exist $\alpha + 1$ elements $A, A_1, \cdots, A_\alpha$ of $\mathcal{G}$ such that every element G of $\mathcal{G}$ can be expressed uniquely as*
$$G = A_\alpha^{\lambda_\alpha} \cdots A_1^{\lambda_1} A^\lambda, \quad 1 \leq \lambda_\alpha \leq p, \cdots, 1 \leq \lambda_1 \leq p,\ 1 \leq \lambda \leq p^{n-\alpha},$$
where the order of A_δ does not exceed p^δ and that of A is exactly $p^{n-\alpha}$ (pseudo-base);

(ii) *The equality*
$$(B_1 B_2) p^\alpha = B_1^{p^\alpha} B_2^{p^\alpha}$$
holds for any pair of elements B_1 and B_2 of $\mathcal{G}$;

(iii) *The commutators of $\mathcal{G}$ are of order $\leq p^\alpha$, and the simple 3-fold commutators of $\mathcal{G}$ are of order $\leq p^{\alpha-1}$; and*

(iv) *A^{p^α} belongs to the central of $\mathcal{G}$.*

Proof For $\alpha = 0$ the theorem is trivial. For $\alpha = 1$, the theorem is true[①]. Assume that $\alpha > 1$ and that the theorem is true for $\alpha - 1$. $\mathcal{G}$ has a maximal subgroup $\mathfrak{M}$ of rank $\alpha - 1$. By hypothesis of induction, since $(n-1) \geq 2(\alpha-1)+1$, the elements M of $\mathfrak{M}$ are uniquely expressible as

$$M = A_{\alpha-1}^{\lambda_{\alpha-1}} \cdots A_1^{\lambda_1} A^\lambda, \quad 1 \leq \lambda_{\alpha-1} \leq p, \cdots, 1 \leq \lambda_1 \leq p, \cdots, 1 \leq \lambda \leq p^{n-\alpha}, \quad \text{(i')}$$

where A_δ is of order $\leq p^\delta$ and A is of order $p^{n-\alpha}$, and the following relations

$$(M_1 M_2)^{p^{\alpha-1}} = M_1^{p^{\alpha-1}} M_2^{p^{\alpha-1}}, \tag{ii'}$$

$$(M_1, M_2)^{p^{\alpha-1}} = 1 \quad \text{and} \quad (M_1, M_2, M_3)^{p^{\alpha-2}} = 1 \tag{iii'}$$

hold for arbitrary elements M_1, M_2 and M_3 of $\mathfrak{M}$, and further

$$A^{p^{\alpha-1}} \text{ belongs to the central of } \mathfrak{M}. \tag{iv'}$$

Let B be any element of $\mathcal{G}$ but not in $\mathfrak{M}$. Then $\mathcal{G} = \{\mathfrak{M}, B\}$. Since B^p belongs to $\mathfrak{M}$, by (iv') we have
$$A^{p^{\alpha-1}} B^p = B^p A^{p^{\alpha-1}}.$$

Let
$$B^{-1} A B = AC,$$

---

① See, e.g., Zassenhaus H. Lehrbuch der Gruppentheorie I, 1937: 114–115.

where C belengs to $\mathfrak{M}$, since $\mathfrak{M}$ is normal in $\mathcal{G}$, By (ii'), we have

$$B^{-1}A^{p^{\alpha-1}}B = A^{p^{\alpha-1}}C^{p^{\alpha-1}}.$$

Further, $C^{p^{\alpha-1}}$ belongs to the cyclic subgroup $\{A\}$, for otherwise the group $\{C, A\}$ would be greater than $\mathfrak{M}$. We can let

$$C^{p^{\alpha-1}} = A^{p^{\alpha-1}b},$$

since the order of C cannot exceed that of A. Then

$$B^{-1}A^{p^{\alpha-1}}B = A^{p^{\alpha-1}(1+b)}.$$

Consequently by (iv'),

$$A^{p^{\alpha-1}} = B^{-p}A^{p^{\alpha-1}}B^p = A^{p^{\alpha-1}(1+b)^p}.$$

Therefore

$$(1+b)^p \equiv 1 \pmod{p^{n-2\alpha+1}}. \tag{1}$$

For $n \geqslant 2\alpha + 1$, we have then

$$b \equiv 0 \pmod{p^{n-2\alpha}}. \tag{2}$$

Thus

$$C^{p^{\alpha-1}} = A^{p^{n-\alpha-1}c}.$$

i.e. $(A, B) = C$ is of order $\leqslant p^\alpha$, and it follows that (B, A^p) is of order $\leqslant p^{\alpha-1}$. We then obtain (iv) by $(B, A^{p^\alpha}) = 1$ and (iv').

Further, the orders of the commutators (B, A_i), $i = 1, \cdots, \alpha - 1$, are $\leqslant p^{\alpha-1}$, since let

$$A_i^{-1}B^{-1}A_iB = C_i \quad (C_i \text{ in } \mathfrak{M}),$$

we have

$$1 = B^{-1}A_i^{p^{\alpha-1}}B = (A_iC_i)^{p^{\alpha-1}} = C_i^{p^{\alpha-1}}$$

by (ii') and (i'). Let M be any element of $\mathfrak{M}$, and let $M = M_1A^\lambda$ with $M_1 = A_{\alpha-1}^{\lambda_{\alpha-1}} \cdots A_1^{\lambda_1}$. Then since

$$(B, M) = (B, M_1A^\lambda) = (B, A^\lambda)A^{-\lambda}(B, M_1)A^\lambda,$$

we have

$$(B, M)^{p^\alpha} = (B, A^\lambda)^{p^\alpha}A^{-p^\alpha\lambda}(B, M_1)^{p^\alpha}A^{p^\alpha\lambda} = 1.$$

Further, if both B and B' are in $\mathcal{G}$ but not in $\mathfrak{M}$, then we have an element M of $\mathfrak{M}$ such that $B' = B^\lambda M$. Therefore

$$(B, B') = B^{-1}M^{-1}B^{-\lambda}BB^\lambda M = (B, M)$$

is of order $\leqslant p^\alpha$.

Now let K be any commutator, then K belongs to $\mathfrak{M}$ and is of order $\leqslant p^\alpha$. Thus (K, C) is of order $\leqslant p^{\alpha-1}$ for any C of $\mathfrak{M}$ by (iii'). Finally we can write

$$K = M_1 A^{p^{n-2\alpha}\lambda'} \text{ with } M_1 = A_{\alpha-1}^{\lambda_{\alpha-1}} \cdots A_1^{\lambda_1}.$$

Then, for any B of $\mathcal{G}$ not of $\mathfrak{M}$, we have

$$(K, B) = A^{-p^{n-2\alpha}\lambda'}(M_1, B)A^{p^{n-2\alpha}\lambda'}(A^{p^{n-2\alpha}\lambda'}, B).$$

By (ii') and since $(M, B)^{p^{\alpha-1}} = 1$, we have

$$(K, B)^{p^{\alpha-1}} = 1, \quad \text{for } n \geqslant 2\alpha + 1.$$

Hence the proof of (iii) is complete.

It is easy to see by mathematical induction that

$$(B_1 B_2)^p = B_1^p B_2^p K_1 K_2 \cdots K_t,$$

where among these K's $\dfrac{1}{2}p(p-1)$ are (B_1, B_2) and the others are higher fold commutators. By (ii') we then have for $p \geqslant 3$

$$(B_1 B_2)^{p^\alpha} = B_1^{p^\alpha} B_2^{p^\alpha} K_1^{p^{\alpha-1}} \cdots K_t^{p^{\alpha-1}} = B_1^{p^\alpha} B_2^{p^\alpha} (B_1, B_2)^{\frac{1}{2}p^\alpha(p-1)} = B_1^{p^\alpha} B_2^{p^\alpha}.$$

Therefore (ii) is true.

If B is of order $\leqslant p^\alpha$, then $B = A_\alpha$ satisfies our requirement of (i). If B is of order p^β ($\beta > \alpha$), then

$$B^p = A_{\alpha-1}^{\mu_{\alpha-1}} \cdots A_1^{\mu_1} A^{p^{n-\alpha-\beta+1}\mu} \text{ with } p + \mu.$$

Therefore

$$B^{p^\alpha} = A^{p^{n-\beta}}\mu$$

by (ii') and (i'). Let $A_\alpha = B^{-1} A^{p^{n-\alpha-\beta}}\mu$, then

$$A_\alpha^{p^\alpha} = B^{-p^\alpha} A^{p^{n-\beta}}\mu = 1$$

by (ii). (i) is thus proved, since the uniqueness is a trivial consequence of the relative orders of A_αs'.

Remark For $n = 2\alpha$, (1) implies $b \equiv 0 \pmod{p}$. The theorem is still ture.

Theorem 2 (Generalization of a result due to Kulakoff) *The number of elements of order $\leqslant p^m$ ($\alpha \leqslant m \leqslant n - \alpha$) of a group of order p^n ($p \geqslant 3, n \geqslant 2\alpha + 1$) of rank α is equal to $p^{m+\alpha}$.*

Proof By Theorem 1 (i) and (ii), the elements of order $\leqslant p^m$ are of the form

$$A_\alpha^{\lambda_\alpha} \cdots A_1^{\lambda_1} A^{p^{n-\alpha-m}\mu}, \quad 1 \leqslant \lambda_\alpha \leqslant p, \cdots, 1 \leqslant \lambda_1 \leqslant p, \ 1 \leqslant \mu \leqslant p^m.$$

There are $p^{m+\alpha}$ such elements.

Theorem 3 (Generalization of a result due to Miller) *The number of cyclic subgroups of order p^m ($\alpha < m < n - \alpha + 1$) of a group of order p^n ($p \geqslant 3, n \geqslant 2\alpha + 1$) of rank α is equal to p^α.*

Proof The number of elements of order p^m of the group $\mathcal{G}$ is $p^\alpha \phi(p^m)$. Therefore we have the theorem.

Theorem 4 *The group of order p^n ($p \geqslant 3$, $n \geqslant 2\alpha + 1$) of rank α contains one and only one subgroup of order p^m ($n \geqslant m \geqslant 2\alpha$) of rank α.*

Proof The set of elements

$$M = A_\alpha^{\lambda_\alpha} \cdots A_1^{\lambda_1} A^{p\lambda'}, \quad 0 \leqslant \lambda_\alpha \leqslant p-1, \cdots, 0 \leqslant \lambda_1 \leqslant p-1, \ 0 \leqslant \lambda' \leqslant p^{n-\alpha-1}-1 \quad (3)$$

forms a group, since (M, A^P) is of order $\leqslant p^{\alpha-1}$, i.e $(M, A^p) = M_1 A^{\lambda p}$ with $M_1 = A_{\alpha-1}^{\lambda_{\alpha-1}} \cdots A_1^{\lambda_1}$. Further (3) contains all elements of order $\leqslant p^{n-\alpha-1}$. The theorem is then proved by induction.

Remark For $n = 2\alpha$, Theorem 4 is then trivial.

§2. Let $\mathcal{G}$ be any p-group of order p^n and of rank α and let $\mathfrak{M}_\beta$ denote a typical major subgroup of index p^β in $\mathcal{G}$ with $0 \leqslant \beta \leqslant d = d(\mathcal{G})$, where d is the number of mutually independent generating elements of $\mathcal{G}$. In particular, $\mathcal{G} = \mathfrak{M}_0$, $D = \mathfrak{M}_d =$ the principal subgroup of $\mathcal{G}$.

Theorem 5 *If $\mathcal{G}$ is of rank α, then the rank of $\mathcal{H}$, which is a subgroup of $\mathcal{G}$ of index p^β, is $\geqslant \alpha - \beta$ and $\leqslant \alpha$.*

Proof 1) Evidently

$$H^{p^{n-\alpha}} = 1$$

for all elements H of $\mathcal{H}$. Thus the rank of $\mathcal{H}$ is $\geqslant (n - \beta) - (n - \alpha) = \alpha - \beta$.

2) Evidently G^{p^β} belongs to $\mathscr{H}$ for all elements G of $\mathcal{G}$. If $\mathscr{H}$ is of rank γ, then

$$(G^{p^\beta})^{p^{n-\beta-\gamma}} = G^{p^{n-\gamma}} = 1.$$

Therefore $n - \alpha \leqslant n - \gamma$ and $\gamma \leqslant \alpha$.

Theorem 6 *If $\mathcal{G}$ is of rank α, then the rank of any $\mathfrak{M}_\beta$ is $\leqslant \alpha - \beta + 1$.*

Proof Suppose $\mathfrak{M}_\beta$ is of rank γ. Then

$$(G^p)^{p^{n-\beta-\gamma}} = 1$$

for all elements G of $\mathcal{G}$, since G^p belongs to $\mathfrak{M}_\beta$. Therefore $n - \beta - \gamma + 1 \geqslant n - \alpha$ and $\gamma \leqslant \alpha - \beta + 1$.

According to Theorems 5 and 6, we can classify the major subgroups $\mathfrak{M}_\beta$ into two classes.

I. Those $\mathfrak{M}_\beta$ of rank $\alpha - \beta + 1$; the typical one will be denoted by $\mathfrak{M}_{\beta_1}$;

II. Those $\mathfrak{M}_\beta$ of rank $\alpha - \beta$; the typical one will be denoted by $\mathfrak{M}_{\beta_2}$.

Theorem 4 asserts that there is one and only one $\mathfrak{M}_{1,1}$ provided $p \geqslant 3$, $n \geqslant 2\alpha + 1$.

Theorem 7 *Let $p \geqslant 3$ and $n \geqslant 2\alpha + 1$. The subgroups of $\mathcal{G}$ which lie between $\mathfrak{M}_{1,1}$ and $\mathscr{D}$ are those of I and conversely.*

Proof If $\mathfrak{M}_\beta$ lies in $\mathfrak{M}_{1,1}$, which is of rank α, the rank of $\mathfrak{M}_\beta$ is $\geqslant \alpha - (\beta - 1) = \alpha - \beta + 1$, by Theorem 5. By Theorem 6, the rank of $\mathfrak{M}_\beta$ is equal to $\alpha - \beta + 1$. Hence $\mathfrak{M}_\beta$ lies always in I.

Next we shall prove that every $\mathfrak{M}_{\beta_1}$ lies between $\mathfrak{M}_{11}$ and $\mathscr{D}$. This is trivial for $\beta = 1$. Hence we suppose $\beta \geqslant 2$. There exists an $\mathfrak{M}_{\beta-1}$ which contains $\mathfrak{M}_{\beta,1}$. If $\mathfrak{M}_{\beta-1}$ belongs to I, i.e. $\mathfrak{M}_{\beta-1} = \mathfrak{M}_{\beta-1,1}$ the theorem follows by induction. Hence suppose that $\mathfrak{M}_{\beta-1}$ belongs to II and we write $\mathfrak{M}_{\beta-1} = \mathfrak{M}_{\beta-1,2}$. The intersection of $\mathfrak{M}_{1,1}$ and $\mathfrak{M}_{\beta-1,2}$ is of order $p^{n-\beta}$ and of rank $\alpha - \beta + 1$, since $\mathfrak{M}_{\beta-1,2}$ is not a subgroup of $\mathfrak{M}_{1,1}$ by the direct part of the present theorem and

$$\frac{(\text{order of } \mathfrak{M}_{1,1})(\text{order of } \mathfrak{M}_{\beta-1,2})}{\text{order of } \mathfrak{M}_{1,1} \wedge \mathfrak{M}_{\beta-1,2}} = \text{order of } \mathcal{G}$$

or

$$\frac{p^{n-1} \cdot p^{n-\beta+1}}{\text{order of } \mathfrak{M}_{1,1} \wedge \mathfrak{M}_{\beta-1,2}} = p^n.$$

Now $\mathfrak{M}_{\beta-1,2}$ of order $p^{n-\beta+1}$ and rank $\alpha - (\beta - 1)$, has one and only one subgroup of rank $\alpha - \beta + 1$ and of order $p^{n-\beta}$ by Theorem 4 for $n - \beta + 1 \geqslant 2\alpha - \beta + 2 \geqslant 2(\alpha - \beta + 1) + \beta - 1 \geqslant 2(\alpha - \beta + 1) + 1$, the intersection should be $\mathfrak{M}_{\beta,1}$. Thus

$$\mathfrak{M}_{1,1} \wedge \mathfrak{M}_{\beta-1,2} = \mathfrak{M}_{\beta,1}.$$

Consequently $\mathfrak{M}_{\beta,1}$ is a subgroup of $\mathfrak{M}_{1,1}$. The theorem is thus completely proved.

Corollary 7.1 *If $p \geqslant 3$ and $n \geqslant 2\alpha + 1$, then the principal subgroup of a group of order p^n and of rank α is of rank $\alpha - d + 1$.*

Corollary 7.2 *If $p \geqslant 3$ and $n \geqslant 2\alpha + 1$, then a group of order p^n with a cyclic principal subgroup of index p^d is of rank $\alpha = d - 1$. In other words, the number of members of a minimal base is equal to its rank plus one.*

Corollary 7.3 *If $p \geqslant 3$ and $n \geqslant 2\alpha + 1$, the number of elements of a minimal base of a group of order p^n and of rank α does not exceed $\alpha + 1$.*

Theorem 8 *Let $p \geqslant 3$ and $n \geqslant 2\alpha + 1$. The number of $\mathfrak{M}_{\beta,1}$ is equal to $\phi_{d-1,\beta-1}$ and that of $\mathfrak{M}_{\beta,2}$ is equal to $p^\beta \phi_{d-1,\beta}$.*

Proof The number of $\mathfrak{M}_{\beta,1}$ is equal to the number of subgroups of order $p^{n-\beta}$ which lie between $\mathfrak{M}_{1,1}$ and $\mathscr{D}$. Now $\dfrac{\mathfrak{M}_{1,1}}{\mathscr{D}}$ is Abelian of type $\underbrace{(1,1,\cdots,1)}_{d-1}$, we have the first statement of the theorem. The second follows immediately since

$$\phi_{d,\beta} - \phi_{d-1,\beta-1} = p^\beta \phi_{d-1,\beta}.$$

Theorem 9 *Let $p \geqslant 3$ and $n > 2\alpha + 1$. Then*

$$d(\mathfrak{M}_{\beta,1}) \geqslant d(\mathcal{G}) - \beta + 1$$

and

$$d(\mathfrak{M}_{\beta,2}) \geqslant d(\mathcal{G}) - \beta.$$

Proof It is evident that $\mathscr{D}(\mathcal{G}) \geqslant \mathscr{D}(\mathfrak{M}_\beta)$, hence we have the inequality

$$d(\mathfrak{M}_\beta) \geqslant d(\mathcal{G}) - \beta.$$

In particular, the last inequality is trivial and is true for arbitrary n. To prove the first inequality, we have merely to show that for $n > 2\alpha + 1$,

$$\mathscr{D}(\mathcal{G}) \neq \mathscr{D}(\mathfrak{M}_{1,1}).$$

For suppose $\mathscr{D}(\mathcal{G}) = \mathscr{D}(\mathfrak{M}_{1,1})$, then Corollary 7.1 shows that $\mathscr{D}(\mathfrak{M}_{1,1})$ would be of rank $\alpha - (d-1) + 1 = \alpha - d + 2$, since $n - 1 \geqslant 2\alpha + 1$ for $n > 2\alpha + 1$. But the rank of $\mathscr{D}(\mathcal{G})$ is $\alpha - d + 1$. Thus $\mathscr{D}(\mathcal{G}) \neq \mathscr{D}(\mathfrak{M}_{1,1})$, and the theorem is proved.

Corollary 9.1 *Let $\mathcal{G}$ be a group of order p^n and of rank α with $p \geqslant 3$ and $n \geqslant 2\alpha + 1$. If $\mathcal{G}$ has a cyclic principal subgroup of index p^d, then each major subgroup*

$\mathfrak{M}_{\gamma,i}$ has also a cyclic principal subgroup, and for $n > 2\alpha + 1$, there is one and only one maximal subgroup which has also a cyclic principal subgroup of index p^d.

This is an immediate consequence of Corollaries 7.2 and 7.3 together with Theorem 9. The converse of the corollary is not true as is shown by the following "Gegenbeispiel":

$$A_1^{p^{n-2}} = 1, \quad A_2^p = 1, \quad (A_1, A_2)^p = 1, \quad ((A_1, A_2), A_1) = 1, \quad ((A_1, A_2), A_2) = 1.$$

Theorem 10 (Modified form of Hall's enumeration principle)   Let $\mathcal{G}$, $\mathfrak{M}_{\beta,1}$ and $\mathfrak{M}_{\beta,2}$ have the same meaning as defined above. Let $\mathfrak{R}$ be any class whose members are elements or subgroups or (still more generally) sets of elements of $\mathcal{G}$, and let each member of $\mathfrak{R}$ belong to at least one maximal subgroup of $\mathcal{G}$. Let $n(\mathfrak{M}_{\beta,i})$ denote the number of members of $\mathfrak{R}$ which belong to $\mathfrak{M}_{\beta,i}$. Then

$$n(\mathfrak{M}_0) - \sum_{(\mathfrak{M}_{1,1})} n(\mathfrak{M}_{1,1}) - \sum_{(\mathfrak{M}_{1,2})} n(\mathfrak{M}_{1,2}) + p \left(\sum_{(\mathfrak{M}_{2,1})} n(\mathfrak{M}_{2,1}) + \sum_{(\mathfrak{M}_{2,2})} n(\mathfrak{M}_{2,2}) \right)$$
$$- p^3 \left(\sum_{(\mathfrak{M}_{3,1})} n(\mathfrak{M}_{3,1}) - \sum_{(\mathfrak{M}_{3,2})} n(\mathfrak{M}_{3,2}) \right) + \cdots + (-1)^d p^{\frac{1}{2}(d-1)} n(\mathfrak{M}_d) = 0,$$

the sum $\sum_{(\mathfrak{M}_{\beta,i})}$ being taken over all the subgroups $\mathfrak{M}_{\beta,i}$ of $\mathcal{G}$.

The derivation of this theorem from that due to Hall is immediate.

§3. Theorem 11   Let $\mathcal{G}$ be a group of order p^n and of rank $\alpha \geqslant 1$ (consequently $d \geqslant 2$), with $p \geqslant 3$ and $n \geqslant 2\alpha + 1$. Then the number of its subgroups of order p^m ($n > m \geqslant 2\alpha$) and of rank $\alpha - 1$ is congruent to

$$p\phi_{d-1,1} = p + p^2 + \cdots + p^{d-1} \pmod{p^d}.$$

Proof   For $m = n - 1$, the subgroups considered are those of $\mathfrak{M}_{1,2}$, which are $p\phi_{d-1,1}$ in number, hence the theorem is true for $m = n-1$. In particular, the theorem is true for $n = 2\alpha + 1$. We shall therefore assume that $n > 2\alpha + 1$ and $n - 1 > m \geqslant 2\alpha$ and apply mathematical induction on n.

Taking $\mathfrak{R}$ to be the set of subgroups of order p^m and of rank $\alpha - 1$, we have, by Theorems 10, 4 and 5

$$n(\mathfrak{M}_0) = n(\mathfrak{M}_{1,1}) + \sum_{(\mathfrak{M}_{1,2})} n(\mathfrak{M}_{1,2}) - p \sum_{(\mathfrak{M}_{2,1})} n(\mathfrak{M}_{2,1}).$$

By Theorems 8 and 9 and hypothesis of induction, we have

$$n(\mathfrak{M}_0) \equiv 1 \cdot p\phi_{d(\mathfrak{M}_{1,1})-1,1} + p\phi_{d-1,1} \cdot 1 - p\phi_{d-1,1} \cdot 1 \pmod{p^{d(\mathfrak{M}_{1,1})}}$$
$$\equiv p\phi_{d-1,1} \pmod{p^d}.$$

Theorem 12 *In a group $\mathcal{G}$ of order p^n and of rank α ($p \geqslant 3$ and $n \geqslant 2\alpha + 1$), the number of subgroups of order $p^{n-\beta}$ ($d \geqslant \beta \geqslant 0$) and of rank $\alpha - \beta$ is congruent to $p^\beta \phi_{d-1,\beta} \pmod{p^{d-\beta+1}}$. Thus the number of nonmajor subgroups of order $p^{n-\beta}$ and of rank $\alpha - \beta$ in $\mathcal{G}$ is divisible by $p^{\alpha-\beta+1}$(for $\beta = d$, $p^\beta \phi_{d-1,\beta}$ is to be replaced by 0).*

Proof There is no subgroup of rank $\alpha - \beta$ of order greater than $p^{n-\beta}$ by Theorem 5.

The theorem is trivial for $\beta = 0$ and it is true for $\beta = 1$ by Theorem 8. We shall therefore assume $\beta > 1$ and apply induction on β. Taking $\mathfrak{R}$ to be the set of all subgroups of order $p^{n-\beta}$ and of rank $\alpha - \beta$, we have, by Theorems 10 and 5 (hence $\mathfrak{M}_{\gamma,1}$ contains no subgroup of rank $(\alpha - \gamma + 1) - (\beta - \gamma + 1)$ of order $> p^{(n-\gamma)-(\beta-\gamma+1)} = p^{n-\beta-1}$),

$$n(\mathfrak{M}_0) = \sum_{(\mathfrak{M}_1)} n(\mathfrak{M}_1) - p \sum_{(\mathfrak{M}_2)} n(\mathfrak{M}_2) + p^3 \sum_{(\mathfrak{M}_3)} n(\mathfrak{M}_3) - \cdots$$
$$- (-1)^{\beta+1} p^{\frac{1}{2}\beta(\beta+1)} \sum_{(\mathfrak{M}_{\beta+1})} n(\mathfrak{M}_{\beta+1}) - \cdots$$
$$= \sum_{(\mathfrak{M}_{1,2})} n(\mathfrak{M}_{1,2}) - p \sum_{(\mathfrak{M}_{2,2})} n(\mathfrak{M}_{2,2}) + p^3 \sum_{(\mathfrak{M}_{3,2})} n(\mathfrak{M}_{3,2}) - \cdots$$
$$- (-1)^\beta p^{\frac{1}{2}\beta(\beta-1)} \sum_{(\mathfrak{M}_{\beta,2})} n(\mathfrak{M}_{\beta,2}).$$

Now $\mathfrak{M}_{\gamma,2}$ is a group of order $p^{n-\gamma}$ and of rank $\alpha - \gamma$, and $n(\mathfrak{M}_{\gamma,2})$ is equal to the number of subgroups of order $p^{(n-\gamma)-(\beta-\gamma)}$ and of rank $(\alpha-\gamma)-(\beta-\gamma)$ with $\beta-\gamma < \beta$, further $n-\gamma \geqslant 2(\alpha-\gamma)+1$ and $0 \leqslant \beta-\gamma \leqslant d-\gamma \leqslant d(\mathfrak{M}_{\gamma,2})$, we have, by hpyothesis of induction and by Theorems 8 and 9

$$- (-1)^\gamma p^{\frac{1}{2}\gamma(\gamma-1)} \sum_{(\mathfrak{M}_{\gamma,2})} n(\mathfrak{M}_{\gamma,2})$$
$$\equiv - (-1)^\gamma p^{\frac{1}{2}\gamma(\gamma-1)} \sum_{(\mathfrak{M}_{\gamma,2})} p^{\beta-\gamma} \phi_{d(\mathfrak{M}_{\gamma,2})-1,\beta-\gamma} \pmod{p^\delta}$$
$$\left(\delta = \min_{(\mathfrak{M}_{\gamma,2})} d(\mathfrak{M}_{\gamma,2}) - \beta + \gamma + 1 + \frac{1}{2}\gamma(\gamma-1)\right)$$
$$\equiv - (-1)^\gamma p^{\frac{1}{2}\gamma(\gamma-1)} \sum_{(\mathfrak{M}_{\gamma,2})} p^{\beta-\gamma} \phi_{d-\gamma-1,\beta-\gamma} \left(\bmod\ p^{d-\beta+1+\frac{1}{2}\gamma(\gamma-1)}\right)$$

$$\equiv -(-1)^\gamma p^{\frac{1}{2}\gamma(\gamma-1)} \sum_{(\mathfrak{M}_{\gamma,2})} p^{\beta-\gamma}\phi_{d-\gamma-1,\beta-\gamma} \pmod{p^{d-\beta+1}}$$
$$\equiv -(-1)^\gamma p^{\frac{1}{2}\gamma(\gamma-1)} p^\gamma \phi_{d-1,\gamma} p^{\beta-\gamma}\phi_{d-\gamma-1,\beta-\gamma} \pmod{p^{d-\beta+1}}$$
$$\equiv -(-1)^\gamma p^{\frac{1}{2}\gamma(\gamma-1)} p^\beta \phi_{d-1,\gamma}\phi_{d-\gamma-1,\beta-\gamma} \pmod{p^{d-\beta+1}}.$$

Consequently
$$n(\mathfrak{M}_0) \equiv p^\beta\left(\phi_{d-1,1}\phi_{d-2,\beta-1} - p\phi_{d-1,2}\phi_{d-3,\beta-2} + \cdots - (-1)^\beta p^{\frac{1}{2}\beta(\beta-1)}\phi_{d-1,\beta}\right)$$
$$\equiv p^\beta \phi_{d-1,\beta} \pmod{p^{d-\beta+1}},$$

and the theorem is thus proved.

Corollary 12.1 *Let $\mathcal{G}$ have a cyclic principal subgroup of index p^α (consequently it is of rank $\alpha = d-1$). Then the number of subgroups of order $p^{n-\beta}$ ($d \geqslant \beta \geqslant 0$) and of rank $\alpha - \beta$ is equal to $p^\beta \phi_{d-1,\beta}$.*

Proof Since for this case, $d(\mathfrak{M}_{\gamma,2}) = d - \gamma$, we have everywhere equalities in place of former congruences.

Theorem 13 *Let $\mathcal{G}$ be a group of order p^n and of rank α with $p \geqslant 3$ and $n \geqslant 2\alpha + 1$. The number of its subgroups of order p^m ($2\alpha - \beta < m < n - \beta + 1$) and of rank $\alpha - \beta$ ($0 \leqslant \beta \leqslant d$) is congruent to $p^\beta \phi_{d-1,\beta}$ (mod $p^{d-\beta+1}$) (for $\beta = d$, $p^\beta \phi_{d-1,\beta}$ has to be replaced by 0).*

Proof Theorem 12 asserts that the theorem is true for $m = n - \beta$. Thus, in particular, the theorem is true for $n = 2\alpha + 1$. By Theorems 4 and 11, the theorem is true for $\beta = 0$ and $\beta = 1$ respectively. Hence suppose $n > 2 + 1$, $m < n - \beta$ and $\beta > 1$, and prove by induction on n and β. Let $\Re$ be the set of all subgroups of order p^m of rank $\alpha - \beta$. By Theorems 8, 9 and 10 and hypothesis of induction, we then have as before

$$n(\mathfrak{M}_0) \equiv n(\mathfrak{M}_{1,1}) + \sum_{(\mathfrak{M}_{1,2})} n(\mathfrak{M}_{1,2}) - p \sum_{(\mathfrak{M}_{2,1})} n(\mathfrak{M}_{2,1}) - p \sum_{(\mathfrak{M}_{2,2})} n(\mathfrak{M}_{2,2}) + \cdots$$
$$\equiv p^\beta(\phi_{d-1,\beta} + \phi_{d-1,1}\phi_{d-1,\beta-1} - \phi_{d-1,1}\phi_{d-1,\beta-1} + \cdots$$
$$- (-1)^\gamma p^{\frac{1}{2}\gamma(\gamma-1)}\phi_{d-\gamma-1,\beta-\gamma}\phi_{d-1,\gamma}$$
$$- (-1)^{\gamma+1} p^{\frac{1}{2}(\gamma-1)\gamma}\phi_{d-\gamma-1,\beta-\gamma}\phi_{d-1,\gamma} + \cdots)$$
$$\equiv p^\beta \phi_{d-1,\beta} \pmod{p^{d-\beta+1}}.$$

hence the theorem follows.

Corollary 13.1 *Let $\mathcal{G}$ have a cyclic principal subgroup of index p^d (consequently it is of rank $\alpha = d-1$). Then the number of its subgroups of order p^m with*

$2\alpha - \beta < m < n - \beta + 1$ and of rank $\alpha - 3$ with $0 \leqslant \beta \leqslant d$ is equal to $p^\beta \phi_{d-1,\beta}$.

Proof Since $d(\mathfrak{M}_{\gamma,1}) = \alpha - \gamma + 2 = d - \gamma + 1$ and $d(\mathfrak{M}_{\gamma,2}) = d - \gamma$, we have everywhere equalities in place of the former congruences.

Corollary 13.2 *In a group of order p^n and of rank α with $p \geqslant 3$ and $n \geqslant 2\alpha + 1$, the number of subgroups of order p^m and of rank $\alpha - \beta$ with $0 \leqslant \beta \leqslant \alpha$ and $2\alpha - \beta < m < n - \beta + 1$ is congruent to p^β (mod $p^{\beta+1}$).*

Proof For $0 \leqslant \beta \leqslant \left[\dfrac{d}{2}\right]$, the corollary is true by Theorem 13. Hence we can assume that $\beta > \left[\dfrac{d}{2}\right]$ and apply induction on m.

For $m = n - \beta$, we have then

$$n(\mathfrak{M}_0) = n(\mathfrak{M}_{1,1}) + \sum n(\mathfrak{M}_{1,2}) - p\sum n(\mathfrak{M}_{2,1}) - p\sum n(\mathfrak{M}_{2,2}) + \cdots$$
$$= 0 + p\phi_{d-1,1}(p^{\beta-1} + \cdots) - 0 - p \cdot p^2 \phi_{d-2,2}(p^{\beta-2} + \cdots) + \cdots$$
$$\equiv p^\beta \pmod{p^{\beta+1}}.$$

Hence the corollary is true for $m = n - \beta$ and is thus true for $n = 2\alpha + 1$. We can then assume $n > 2\alpha + 1$, also $m < n - \beta$, and apply induction on n.

Now as before

$$n(\mathfrak{M}_0) = n(\mathfrak{M}_{1,1}) + \sum n(\mathfrak{M}_{1,2}) - p\sum n(\mathfrak{M}_{2,1}) - p\sum n(\mathfrak{M}_{2,2}) + \cdots$$
$$\equiv p^\beta + p\phi_{d-1,1}p^{\beta-1} - p\phi_{d-1,1}p^{\beta-1} - p \cdot p^2 \phi_{d-1,2}p^{\beta-2} + \cdots \pmod{p^{\beta+1}}$$
$$\equiv p^\beta \pmod{p^{\beta+1}},$$

hence the corollary.

By the above corollary and Theorem 5, we have

Corollary 13.3 *In a group of order p^n and of rank α with $p \geqslant 3$, the number of subgroups of order p^m and of rank $\alpha - \beta$ with $2\alpha - \beta < n$ is always congruent to 0 (mod p^β).*

§4. Theorem 14 *The number of subgroups of order p^m of rank β ($p \geqslant 3$, $m > 2\beta + 1$) in a group of order p^n ($n \geqslant m$) of rank α ($\alpha > \beta$) is congruent to 0 (mod p).*

Proof For $m = n$, the theorem is trivial, hence we assume $m < n$. For $n \leqslant \alpha + \beta + 1$, the theorem is true, since the group would then contain no subgroup of order p^m of rank β ($m > 2\beta + 1$) by Theorem 5. For $\alpha = \beta + 1$, we have $n > m > 2\beta + 1 = 2\alpha - 1$, and therefore the theorem is true by Theorem 11. Hence assume $n > \alpha + \beta + 1$ and $\alpha > \beta + 1$ and apply induction on n and β. By the enumeration principle, we have therefore

$$n(\mathfrak{M}_0) \equiv \sum n(\mathfrak{M}_{1,1}) + \sum n(\mathfrak{M}_{1,2}) \pmod{p};$$

the by hypothesis of induction, $n(\mathfrak{M}_0) \equiv 0 \pmod{p}$.

Theorem 15 *The number of cyclic subgroups of order p^m ($p \geqslant 3, m > \beta$) in a group of order p^n ($n \geqslant m$) of rank $\alpha \geqslant \beta$ is congruent to $0 \pmod{p^\beta}$.*

Proof For $m = n$, the theorem is trival, hence we assume $m < n$. For $n \leqslant \alpha+\beta$, the theorem is true since the group would then contain no cyclic subgroups of such orders by Theorem 5. For $\alpha = \beta$, the theorem is true for $\beta < m \leqslant n - \alpha$ (then $n \geqslant m + \alpha > \alpha + \beta \geqslant 2\alpha$) by Theorem 3 and for $m > n - \alpha$ by Theorem 5. Hence assume $n > \alpha + \beta$ and $\alpha > \beta$ and apply induction on n and β. By the enumeration principle, we have

$$n(\mathfrak{M}_0) \equiv \sum n(\mathfrak{M}_{1,1}) + \sum n(\mathfrak{M}_{1,2}) - p \sum n(\mathfrak{M}_{2,1}) - p \sum n(\mathfrak{M}_{2,2}) + \cdots.$$

By hypothesis of induction, we have then

$$n(\mathfrak{M}_0) \equiv -(-1)^{\alpha-\beta+1} p^{\frac{1}{2}(\alpha-\beta+1)(\alpha-\beta)} \sum n(\mathfrak{M}_{\alpha-\beta+1,2}) - (-1)^{\alpha-\beta+2}$$
$$\times p^{\frac{1}{2}(\alpha-\beta+2)(\alpha-\beta+1)} \sum n(\mathfrak{M}_{\alpha-\beta+2,1}) + \cdots \pmod{p^\beta}.$$

Now for $\gamma \geqslant 2\alpha - \beta + 1$, we have $m >$ (rank of $\mathfrak{M}_{\gamma,i}$); hence by Theorems 3 and 5,

$$n(\mathfrak{M}_{\gamma,1}) \equiv 0 \pmod{p^{\alpha-\gamma+1}}, \quad n(\mathfrak{M}_{\gamma,2}) \equiv 0 \pmod{p^{\alpha-\gamma}}$$

and for $\gamma \geqslant 2\alpha - \beta + 1$, we have also

$$\frac{1}{2}\gamma(\gamma - 1) + \alpha - \gamma = \frac{1}{2}\gamma(\gamma - 3) + \alpha \geqslant \beta,$$

so the theorem is proved.

Theorem 16 *If a group of order p^n ($p \geqslant 3$) contains one and only one subgroup of rank β of order p^m for a fixed m, $n \geqslant m > 2\beta + 1$, then it is of rank β.*

Proof If the rank of the group is less than β, then by Theorem 5, it contains no subgroup of rank β. And if the rank of the group is greater than β, then by Theorem 14, the number of subgroups of rank β is divisible by p. Hence the theorem.

Theorem 17 *If a group of order p^n ($p \geqslant 3$) contains exactly p^β cyclic subgroups of order p^m for a fixed m, $\beta + 1 < m \leqslant n$, then it is of rank β.*

Proof Let γ be the rank of the group considered.

1) If $\gamma < \beta$, then $\gamma < \gamma+1 < \beta+1 < m$. By Theorems 3 and 5, the group would contain either p^γ or no cyclic subgroups of order p^m, according as $m \leqslant n - \gamma$ (then $n \geqslant m + \gamma > 2\gamma$) or $m > n - \gamma$. Hence we get a contradiction.

2) If $\gamma > \beta$, then $\gamma \geqslant \beta + 1$. By Theorem 15, the number of cyclic subgroups of order p^m ($\beta + 1 < m \leqslant n$) would be divisible by $p^{\beta+1}$. Hence we get a contradiction.

Thus the theorem is proved.

References

[1] G. A. Miller. *Proc. London Math. Soc.*, 1904, **2**(2): 142–143.
[2] A. Kulakoff. *Math. Annalen*, 1931, **104**: 778–793.
[3] P. Hall. *Proc. London Math Soc.*, 1933–1934, **36**(2): 29–95.
[4] H. Zassenhaus. *Lehrbuch der Gruppentheorie* I, 1937.
[5] L. K. Hua, H. F. Tuan. *Sic. Rep. Nat. Tsing-Hua Univ.*, 1940, **4**(A): 145–154.
[6] L. K. Hua, H. F. Tuan. *J. Chinese Math. Soc.*, 1940, **2**: 313–319.

On the automorphisms of the symplectic group over any field*

1. Introduction

Let Φ be any field with characteristic $\neq 2$. Unless the contrary is stated, we use small Latin letters to denote elements of the field and capital Latin letters to denote n-rowed matrices over Φ. For matrices which are not $n \times n$, we use $M^{(m)}$ to denote an $m \times m$ matrix. I and 0 denote the n-rowed identity and zero matrices respectively, and $[a_1, \cdots, a_n]$ will denote a diagonal matrix with $a_1, \cdots, a_n$ on its diagonal. We use small Greek letters for automorphisms of the field, i.e.

$$(a+b)^\sigma = a^\sigma + b^\sigma, \quad (ab)^\sigma = a^\sigma b^\sigma$$

for a and b belonging to Φ.

Let
$$\mathfrak{F} = \begin{pmatrix} 0 & I \\ -I & 0 \end{pmatrix}.$$

The group formed by the $2n$-rowed matrices $\mathfrak{T}$ satisfying

$$\mathfrak{T}\mathfrak{F}\mathfrak{T}' = \mathfrak{F}, \tag{1}$$

where $\mathfrak{T}'$ denotes the transposed matrix of $\mathfrak{T}$, is called symplectic and is denoted by $\mathrm{Sp}(\Phi, 2n)$.

Apparently, there are three different types of automorphisms $\mathcal{A}(\mathfrak{T})$ of $\mathrm{Sp}(\Phi, 2n)$:

1) Inner automorphisms. For a symplectic $\mathfrak{P}$, we have

$$\mathcal{A}(\mathfrak{T}) = \mathfrak{P}\mathfrak{T}\mathfrak{P}^{-1}; \tag{2}$$

2) Automophisms induced by those of the field:

$$\mathcal{A}(\mathfrak{T}) = \mathfrak{T}^\sigma; \tag{3}$$

3) Semi-inner automorphisms: We define the group formed by the matrices $\mathfrak{P}$ satisfying

$$\mathfrak{P}\mathfrak{F}\mathfrak{P}' = a\mathfrak{F} \tag{4}$$

---

* Received November 7, 1946, Revised August 15, 1947. *Annals of Mathemattics* Vol. 49, No. 4, October, 1948.

to be the extended symplectic group. Then, for $\mathfrak{P}$ satisfying (4), we have also an automorphism
$$\mathcal{A}(\mathfrak{T}) = \mathfrak{P}^{-1}\mathfrak{T}\mathfrak{P}.$$

Since
$$(b\mathfrak{P})^{-1}\mathfrak{T}(b\mathfrak{P}) = \mathfrak{P}^{-1}\mathfrak{T}\mathfrak{P}$$

and
$$(b\mathfrak{P})\mathfrak{F}(b\mathfrak{P})' = b^2 a \mathfrak{F},$$

we have only to consider those a which represent different cosets of the factor group of the multiplicative group of Φ by the group Φ_2 of square elements of Φ. More precisely, let
$$a, b, \cdots$$
be a representative system of Φ/Φ_2. Then the elements of the extended symplectic group can be expressed as
$$\begin{pmatrix} I & 0 \\ 0 & aI \end{pmatrix} \mathfrak{P},$$
where $\mathfrak{P}$ is symplectic.

It is the purpose of this paper to establish that the group generated by these three types of automorphisms is the largest group of automorphisms of $\mathrm{Sp}(\Phi, 2n)$, provided that the characteristic of Φ is not equal to 2.

In other words, analytically, we have the following

Theorem   *All the automorphisms of* $\mathrm{Sp}(\Phi, 2n)$ *can be expressed as*
$$\mathcal{A}(\mathfrak{T}) = \mathfrak{P}\mathfrak{R}\mathfrak{T}^\sigma \mathfrak{R}^{-1}\mathfrak{P}^{-1}, \tag{5}$$

where σ runs over all automorphisms of the field, $\mathfrak{P}$ runs over all symplectic matrices and
$$\mathfrak{R} = \begin{pmatrix} I & 0 \\ 0 & aI \end{pmatrix}, \tag{6}$$

where a runs over a complete residue system of Φ/Φ_2.

The corresponding problem for the special linear group was solved by O. Schreier and B. L. van der Waerden[①] (see Appendix).

---

① Die Automorphism der projektiven Gruppen. *Hamburg. Univ. Math. Seminar.*, 1928, **6**: 303–322. See Appendix.

2. Involutions

We now consider involutions, that is, matrices $\mathfrak{T}$ of $\mathrm{Sp}(\Phi, 2n)$ such that

$$\mathfrak{T}^2 = \mathfrak{J}, \tag{7}$$

where $\mathfrak{J}$ is the $2n$-rowed identity.

Theorem 1 *Every involution in* $\mathrm{Sp}(\Phi, 2n)$ *is equivalent sympletically to the normal form*

$$\begin{pmatrix} J & 0 \\ 0 & J \end{pmatrix}, \tag{8}$$

where $J = [1, \cdots, 1, -1, \cdots, -1] = \{p, q\}$, say, consists of p ones and q minus ones. Precisely, there exists a symplectic matrix $\mathfrak{P}$ such that

$$\mathfrak{P}\mathfrak{T}\mathfrak{P}^{-1} = \begin{pmatrix} J & 0 \\ 0 & J \end{pmatrix}.$$

The theorem is evident for $n = 1$; in fact, let

$$\mathfrak{T} = \begin{pmatrix} a & b \\ c & d \end{pmatrix}, \quad ad - bc = 1,$$

(7) implies

$$a^2 + bc = 1, \quad (a+d)b = (a+d)c = 0.$$

Consequently, we have

$$\mathfrak{T} = \pm \mathfrak{J}.$$

Now we suppose that the theorem holds for symplectic matrices of order less than $2n$.

From (1) and (7), we have

$$\mathfrak{T}\mathfrak{F} = \mathfrak{F}\mathfrak{T}'^{-1} = \mathfrak{F}\mathfrak{T}' = -(\mathfrak{T}\mathfrak{F})',$$

that is, $\mathfrak{T}\mathfrak{F}$ is skew symmetric, more precisely, $\mathfrak{T}$ has the form

$$\mathfrak{T} = \begin{pmatrix} A & K_1 \\ K_2 & A' \end{pmatrix}, \tag{9}$$

where K_1 and K_2 are skew symmetric. Since $\mathfrak{T}$ is symplectic, we have also that

$$AK_1, \quad K_2A \tag{10}$$

are symmetric and that
$$A^2 + K_1K_2 = I. \tag{11}$$

There is a non-singular matrix Q such that
$$QK_1Q' = \begin{pmatrix} k_1^{(r)} & 0 \\ 0 & 0 \end{pmatrix} = K_1^0,$$

where k_1 is a non-singular skew symmetric r-rowed matrix (in case $r = n$, there are no zeros in the expression). The transformation

$$\begin{pmatrix} Q & 0 \\ 0 & Q'^{-1} \end{pmatrix} \begin{pmatrix} A & K_1 \\ K_2 & A' \end{pmatrix} \begin{pmatrix} Q & 0 \\ 0 & Q'^{-1} \end{pmatrix}^{-1} \tag{12}$$

carries K_1 into K_1^0, where
$$\begin{pmatrix} Q & 0 \\ 0 & Q'^{-1} \end{pmatrix}$$
is clearly symplectic. Thus we may assume at the beginning that
$$K_1 = \begin{pmatrix} k_1^{(r)} & 0 \\ 0 & 0 \end{pmatrix}.$$

From (10), we have
$$A = \begin{pmatrix} a_1^{(r)} & a_2^{(r,n-r)} \\ a_3^{(n-r,r)} & a_4^{(n-r)} \end{pmatrix}, \quad a_3 = 0, \quad a_1k_1 \text{ symmetric}.$$

The symplectic matrix
$$\mathfrak{P} = \begin{pmatrix} I & 0 \\ S & I \end{pmatrix}$$

carries (9) into
$$\mathfrak{P}\mathfrak{T}\mathfrak{P}^{-1} = \begin{pmatrix} I & 0 \\ S & I \end{pmatrix} \begin{pmatrix} A & K_1 \\ K_2 & A' \end{pmatrix} \begin{pmatrix} I & 0 \\ -S & I \end{pmatrix} = \begin{pmatrix} A - K_1S & K_1 \\ * & * \end{pmatrix}. \tag{13}$$

We choose
$$S = \begin{pmatrix} k_1^{-1}a_1 & k_1^{-1}a_2 \\ (k_1^{-1}a_2)' & 0 \end{pmatrix},$$

then the A of $\mathfrak{P}\mathfrak{T}\mathfrak{P}^{-1}$ takes the form

$$A = \begin{pmatrix} 0 & 0 \\ 0 & a_4^{(n-r)} \end{pmatrix},$$

where a_4 is non-singular. Now we assume at the beginning that A has this form. From (10) and (11), we deduce that
$$K_2 = \begin{pmatrix} k_1^{-1} & 0 \\ 0 & k_2 \end{pmatrix}.$$

Then, every involution is equivalent to
$$\mathfrak{T} = \begin{pmatrix} 0 & 0 & k_1 & 0 \\ 0 & a & 0 & 0 \\ k_1^{-1} & 0 & 0 & 0 \\ 0 & k_2 & 0 & a' \end{pmatrix}.$$

This induces two symplectic involutions
$$\mathfrak{T}^{(2r)} = \begin{pmatrix} 0 & k_1 \\ k_1^{-1} & 0 \end{pmatrix}, \quad \mathfrak{T}^{(2n-2r)} = \begin{pmatrix} a & 0 \\ k_2 & a' \end{pmatrix}.$$

By hypothesis of induction, the theorem is true for $r \neq 0$ and $\neq n$. A similar result holds if the rank of k_2 is neither 0 nor n.

If the ranks of k_1 and k_2 are both 0, then $A^2 = I$, it is known that we have Q such that $QAQ^{-1} = J$. The theorem is now evident, since
$$\begin{pmatrix} Q & 0 \\ 0 & Q'^{-1} \end{pmatrix} \begin{pmatrix} A & 0 \\ 0 & A' \end{pmatrix} \begin{pmatrix} Q & 0 \\ 0 & Q'^{-1} \end{pmatrix}^{-1} = \begin{pmatrix} J & 0 \\ 0 & J \end{pmatrix}.$$

If $r = n$, then we have a suitable S such that (13) carries $\mathfrak{T}$ into the form (9) with $A = 0$. That is
$$\mathfrak{T} = \begin{pmatrix} 0 & K \\ K^{-1} & 0 \end{pmatrix}. \tag{14}$$

Note that this case can only happen for even $n, = 2m$ say. We can choose Q such that (12) carries (14) into the same form with
$$K = \begin{pmatrix} 0 & I^{(m)} \\ -I^{(m)} & 0 \end{pmatrix}. \tag{15}$$

If (15) is satisfied, the symplectic matrix
$$\begin{pmatrix} 0 & 0 & I^{(m)} & 0 \\ 0 & I^{(m)} & 0 & 0 \\ -I^{(m)} & 0 & 0 & 0 \\ 0 & 0 & 0 & I^{(m)} \end{pmatrix}$$

carries (14) into

$$\begin{pmatrix} 0 & 0 & I & 0 \\ 0 & I & 0 & 0 \\ -I & 0 & 0 & 0 \\ 0 & 0 & 0 & I \end{pmatrix} \begin{pmatrix} 0 & 0 & 0 & I \\ 0 & 0 & -I & 0 \\ 0 & -I & 0 & 0 \\ I & 0 & 0 & 0 \end{pmatrix} \begin{pmatrix} 0 & 0 & -I & 0 \\ 0 & I & 0 & 0 \\ I & 0 & 0 & 0 \\ 0 & 0 & 0 & I \end{pmatrix} = \begin{pmatrix} 0 & -I & 0 & 0 \\ -I & 0 & 0 & 0 \\ 0 & 0 & 0 & -I \\ 0 & 0 & -I & 0 \end{pmatrix}.$$

This gives $\mathfrak{T}$ with zero K_1 and the theorem follows from the previous case.

Definition An involution equivalent to (8) is called an *involution of signature* $\{p,q\}$.

3. Non equivalence of symplectic groups

Now we consider a set of symplectic matrices $\mathfrak{T}_i$ $(1 \leqslant i \leqslant l)$ satisfying

$$\mathfrak{T}_i \mathfrak{T}_j = \mathfrak{T}_j \mathfrak{T}_i, \quad \mathfrak{T}_i^2 = \mathfrak{J}. \tag{16}$$

Theorem 2 *We have $l \leqslant 2^n$ and the maximum is attained. More precisely, any set of commutative symplectic involutions is equivalent symplectically to a subset of the set of matrices of the form*

$$\begin{pmatrix} J & 0 \\ 0 & J \end{pmatrix},$$

where J is a diagonal matrix with ± 1 on its diagonal.

The proof of the theorem is evident for $n = 1$ (cf. the first paragraph of the proof of Theorem 1). Now we apply induction on n.

Let $\mathfrak{T}_1$ be an involution different from $\pm \mathfrak{J}$. By Theorem 1, we may assume that it has the form

$$\mathfrak{T}_1 = \begin{pmatrix} J & 0 \\ 0 & J \end{pmatrix}, \quad J = \{p, q\}, \quad p > 0, \quad q > 0.$$

These matrices commutative with $\mathfrak{T}_1$ are of the form

$$\mathfrak{T}_i = \begin{pmatrix} a_1^{(p)} & 0 & b_1^{(p)} & 0 \\ 0 & a_2^{(q)} & 0 & b_2^{(q)} \\ c_1^{(p)} & 0 & d_1^{(p)} & 0 \\ 0 & c_2^{(q)} & 0 & d_2^{(q)} \end{pmatrix}_i.$$

Thus we have

$$\begin{pmatrix} a_1 & b_1 \\ c_1 & d_1 \end{pmatrix}_i, \quad \begin{pmatrix} a_2 & b_2 \\ c_2 & d_2 \end{pmatrix}_i,$$

two sets of commutative involutions of $\text{Sp}(\Phi, 2p)$ and $\text{Sp}(\Phi, 2q)$, respectively. By the hypothesis of induction we have Theorem 2.

Since an isomorphic mapping carries involutions into involutions and commutative involutions into commutative involutions, we have

Theorem 3   *For a fixed field Φ, no two $\text{Sp}(\Phi, 2n)$ with different n are isomorphic*

4. Generators of $\text{Sp}(\Phi, 2n)$

Theorem 4   *The group $\text{Sp}(\Phi, 2n)$ is generated by the elements*

$$\begin{pmatrix} A & 0 \\ 0 & A^{-1} \end{pmatrix}, \quad \begin{pmatrix} I & S \\ 0 & I \end{pmatrix}, \quad S = S'$$

and

$$\begin{pmatrix} M & I - M \\ -(I - M) & M \end{pmatrix},$$

where M is a diagonal matrix with zero and one on its diagonal.

Proof   For a given symplectic matrix $\mathfrak{T}$, we can find a suitable[①] M such that

$$\mathfrak{T} = \begin{pmatrix} M & I - M \\ -(I - M) & M \end{pmatrix} = \begin{pmatrix} A & B \\ C & D \end{pmatrix}$$

with non-singular A. Then

$$\begin{pmatrix} A & B \\ C & D \end{pmatrix} \begin{pmatrix} A^{-1} & 0 \\ 0 & A' \end{pmatrix} = \begin{pmatrix} I & S \\ P & Q \end{pmatrix},$$

where $S = S'$. Further, since

$$\begin{pmatrix} I & S \\ P & Q \end{pmatrix} \begin{pmatrix} I & -S \\ 0 & I \end{pmatrix} = \begin{pmatrix} I & 0 \\ R & I \end{pmatrix},$$

where R is symmetric and

$$\begin{pmatrix} M & I - M \\ -(I - M) & M \end{pmatrix}^{-1} \begin{pmatrix} I & 0 \\ R & I \end{pmatrix} \begin{pmatrix} M & I - M \\ -(I - M) & M \end{pmatrix} = \begin{pmatrix} I & -R \\ 0 & I \end{pmatrix}$$

with $M = 0$, we have the theorem.

[①] L. K. Hua. Geometrics of Matrices III. *Trans. Amer. Math. Soc.*, 1947, **61**: 229–255, Theorem 7. Notice that the proof holds for any field.

Theorem 5 *The group Γ generated by all square elements of $\mathrm{Sp}(\Phi, 2n)$ is identical with $\mathrm{Sp}(\Phi, 2n)$.*

Proof Since the field is of characteristic $\neq 2$, we deduce that as S runs over all symmetric matrices, so does $2S$. From

$$\begin{pmatrix} I & S \\ 0 & I \end{pmatrix}^2 = \begin{pmatrix} I & 2S \\ 0 & I \end{pmatrix},$$

we see that Γ contains all elements of the form

$$\mathfrak{P} \begin{pmatrix} I & S \\ 0 & I \end{pmatrix} \mathfrak{P}^{-1}, \tag{$*$}$$

where $\mathfrak{P}$ is any symplectic matrix and S is any symmetric matrix. In particular Γ contains

$$\begin{pmatrix} I & 0 \\ -J & I \end{pmatrix} = \begin{pmatrix} 0 & I \\ -I & 0 \end{pmatrix} \begin{pmatrix} I & S \\ 0 & I \end{pmatrix} \begin{pmatrix} 0 & I \\ -I & 0 \end{pmatrix}^{-1}.$$

For non-singular S, we have

$$\begin{pmatrix} S & 0 \\ 0 & S^{-1} \end{pmatrix} = \begin{pmatrix} I & -S \\ 0 & I \end{pmatrix} \begin{pmatrix} I & 0 \\ S^{-1} - I & I \end{pmatrix} \begin{pmatrix} I & I \\ 0 & I \end{pmatrix} \begin{pmatrix} I & 0 \\ -(I-S) & I \end{pmatrix},$$

thus, Γ contains elements of the form

$$\begin{pmatrix} S & 0 \\ 0 & S^{-1} \end{pmatrix},$$

where $S = S'$.

We shall justify later that any non-singular matrix A can be expressed as product of symmetric matrices, the group contains

$$\begin{pmatrix} A & 0 \\ 0 & A^{-1} \end{pmatrix}.$$

Since

$$\begin{pmatrix} 0 & 1 \\ -1 & 0 \end{pmatrix} = \begin{pmatrix} 1 & 2 \\ 0 & 1 \end{pmatrix} \left(\begin{pmatrix} 1 & 0 \\ 1 & 1 \end{pmatrix} \begin{pmatrix} 1 & 1 \\ 0 & 1 \end{pmatrix} \begin{pmatrix} 1 & 0 \\ 1 & 1 \end{pmatrix}^{-1} \right),$$

the group contains also

$$\begin{pmatrix} M & I - M \\ -(I - M) & M \end{pmatrix}.$$

By Theorem 4, we have the theorem.

Now we are going to justify that every non-singular matrix A is a product of symmetric matrices.

Let J_{ij} ($i \neq j$) denote a matrix with 1 at (i,j) and (j,i) position and with 1 at (k,k) position for $k \neq i$ and j and with zero elsewhere. It is known that the group of non-singular matrices A is generated by J_{ij} and

$$T(d) = \begin{pmatrix} 1 & d & 0 & 0 & \cdots \\ 0 & 1 & 0 & 0 & \cdots \\ 0 & 0 & 1 & 0 & \cdots \\ \vdots & \vdots & \vdots & \vdots & \end{pmatrix}, \quad H(a) = \begin{pmatrix} a & 0 & 0 & \cdots \\ 0 & 1 & 0 & \cdots \\ 0 & 0 & 1 & \cdots \\ \vdots & \vdots & \vdots & \end{pmatrix}.$$

Evidently J_{ij} and $H(a)$ are symmetric and since

$$\begin{pmatrix} 1 & d \\ 0 & 1 \end{pmatrix} = \begin{pmatrix} 0 & 1 \\ 1 & 0 \end{pmatrix} \begin{pmatrix} 0 & 1 \\ 1 & d \end{pmatrix},$$

we have our assertion.

5. Proof of the theorem

1) The theorem is true for $n = 1$[①]. We suppose that the theorem holds for symplectic group of order less than $2n$.

2) Since the normalisator of an involution of signature $\{p,q\}$ is a direct product of two symplectic groups of matrices of orders $2p$ and $2q$, we see that the automorphism $\mathcal{A}(\mathfrak{T})$ carries an involution of signature $\{p,q\}$ into an involution of signature $\{p,q\}$ or $\{q,p\}$. By Theorem we may assume, without loss of generality, that

$$\mathfrak{A} \begin{pmatrix} \{1, n-1\} & 0 \\ 0 & \{1, n-1\} \end{pmatrix} = \begin{pmatrix} \{1, n-1\} & 0 \\ 0 & \{1, n-1\} \end{pmatrix}$$

$$\text{or} = \begin{pmatrix} -\{1, n-1\} & 0 \\ 0 & -\{1, n-1\} \end{pmatrix}.$$

The aggregate of symplectic matrices commutative with

$$\begin{pmatrix} \pm\{1, n-1\} & 0 \\ 0 & \pm\{1, n-1\} \end{pmatrix}$$

---

① Schreier and V. D. Waerden, *ibid*. Note that, in their theorem, they identified two matrices $-\mathfrak{T}$ and $\mathfrak{T}$ as a single element. But we can find the present conclusion easily. See Appendix.

consists of all elements of the form

$$\mathfrak{T} = \begin{pmatrix} a & 0 & b & 0 \\ 0 & A_1 & 0 & B_1 \\ c & 0 & d & 0 \\ 0 & C_1 & 0 & D_1 \end{pmatrix}, \quad A_1 = A_1^{(n-1)}, \quad B_1 = B_1^{(n-1)}, \quad \text{etc.} \tag{17}$$

We shall call $\mathfrak{T}$ the direct product of two matrices

$$\mathfrak{t} = \mathfrak{t}^{(2)} = \begin{pmatrix} a & b \\ c & d \end{pmatrix}, \quad \mathfrak{T}_1 = \mathfrak{T}_1^{(2n-2)} = \begin{pmatrix} A_1 & B_1 \\ C_1 & D_1 \end{pmatrix},$$

which are symplectic. We use the notation $\mathfrak{T} = \mathfrak{t} \times \mathfrak{T}_1$.

Consider the group Λ formed by the elements of the form

$$\begin{pmatrix} a & 0 & b & 0 \\ 0 & \pm I & 0 & 0 \\ c & 0 & d & 0 \\ 0 & 0 & 0 & \pm I \end{pmatrix}.$$

$\mathcal{A}$ carries Λ into

$$\begin{pmatrix} a^* & 0 & b^* & 0 \\ 0 & A_1^* & 0 & B_1^* \\ c^* & 0 & d^* & 0 \\ 0 & C_1^* & 0 & D_1^* \end{pmatrix},$$

where

$$\begin{pmatrix} A_1^* & B_1^* \\ C_1^* & D_1^* \end{pmatrix}$$

for a group $\mathfrak{T}_1^*$. It is clear that $\mathfrak{T}_1^*$ is a normal subgroup of $\mathfrak{T}_1$. But the factor group $\mathrm{Sp}(\Phi, 2n-2)$ over its centrum $(\mathfrak{J}^{(2n-2)}, -\mathfrak{J}^{(2n)-2})$ is simple,[①] and then $\mathfrak{T}_1^*$ is either $\mathfrak{T}_1$ or identity or the centrum. In case $n \neq 2$, the first case can never happen (Theorem 3). The second case also cannot happen, for otherwise, the inverse of $\mathcal{A}$ will cease to carry elements of the form

$$\begin{pmatrix} a & 0 & b & 0 \\ 0 & -I & 0 & 0 \\ c & 0 & d & 0 \\ 0 & 0 & 0 & -I \end{pmatrix}$$

[①] Van der Waerden. Gruppen von Linearen Transforman Theorem. *Ergebnisse der Math.*, 1935, 4(2): 10–11.

into elements of the form (17). Squaring all the elements of Λ, we conclude, by Theorem 5, that $\mathcal{A}$ induces an automorphism on $\mathfrak{t}$. Similarly, $\mathcal{A}$ induces an automorphism on $\mathfrak{T}_1$. In case $n = 2$, there is the additional possibility of interchanging both factors. But the inner automorphism

$$\begin{pmatrix} 0 & 1 & 0 & 0 \\ 1 & 0 & 0 & 0 \\ 0 & 0 & 0 & 1 \\ 0 & 0 & 1 & 0 \end{pmatrix} \mathfrak{T} \begin{pmatrix} 0 & 1 & 0 & 0 \\ 1 & 0 & 0 & 0 \\ 0 & 0 & 0 & 1 \\ 0 & 0 & 1 & 0 \end{pmatrix}$$

carries the factors into the right order. By the hypothesis of induction, we have a two-rowed symplectic matrix $\mathfrak{p}$, a matrix

$$\mathfrak{r} = \begin{pmatrix} 1 & 0 \\ 0 & a_0 \end{pmatrix}$$

and an automorphism α of Φ such that

$$\mathcal{A}(\mathfrak{t}) = \mathfrak{p}\mathfrak{r}\mathfrak{t}^\alpha \mathfrak{r}^{-1} \mathfrak{p}^{-1}$$

and we have a $(2n-2)$-rowed symplectic matrix $\mathfrak{P}_1$, a matrix

$$\mathfrak{R}_1 = \begin{pmatrix} I^{(n-1)} & 0 \\ 0 & b_0 I^{(n-1)} \end{pmatrix}$$

and an automorphism β of Φ such that

$$\mathcal{A}(\mathfrak{T}_1) = \mathfrak{P}_1 \mathfrak{R}_1 \mathfrak{T}^\beta \mathfrak{R}_1^{-1} \mathfrak{P}_1^{-1}.$$

We define

$$\mathfrak{P} = \mathfrak{p} \times \mathfrak{P}_1$$

and

$$\mathfrak{R} = \begin{pmatrix} 1 & 0 \\ 0 & b_0 I \end{pmatrix}.$$

Then the automorphism

$$(\mathfrak{R}^{-1} \mathfrak{P}^{-1} \mathcal{A}(\mathfrak{T}) \mathfrak{P} \mathfrak{R})^{\beta^{-1}}$$

carries $\mathfrak{T}$ given by (17) into

$$\begin{pmatrix} a^\tau & 0 & pb^\tau & 0 \\ 0 & A_1 & 0 & B_1 \\ (1/p)c^\tau & 0 & d^\tau & 0 \\ 0 & C_1 & 0 & D_1 \end{pmatrix},$$

where $\tau = \alpha\beta^{-1}$ and $p = a_0 b_0^{-1}$.

Therefore, without loss of generality, we may assume that

$$\mathcal{A}\begin{pmatrix} a & 0 & b & 0 \\ 0 & A_1 & 0 & B_1 \\ c & 0 & d & 0 \\ 0 & C_1 & 0 & D_1 \end{pmatrix} = \begin{pmatrix} a^\tau & 0 & pb^\tau & 0 \\ 0 & A_1 & 0 & B_1 \\ (1/p)c^\tau & 0 & d^\tau & 0 \\ 0 & C_1 & 0 & D_1 \end{pmatrix}, \quad (18)$$

where p is either 1 or not a square element of Φ.

3) Let Δ be the sub-group formed by matrices of the form

$$\begin{pmatrix} I & [a_1, \cdots, a_n] \\ 0 & I \end{pmatrix}.$$

Then the equation (18) asserts that $\mathcal{A}$ induces a mapping carrying Δ onto itself. Now we consider all the symplectic matrices commutative with every element of Δ. These matrices form a group Σ_2. The elements of Σ_2 are of the form

$$\begin{pmatrix} E & S \\ 0 & E \end{pmatrix}, \quad S = S', \quad E = [\pm 1, \cdots, \pm 1].$$

In fact, from

$$\begin{pmatrix} A & B \\ C & D \end{pmatrix}\begin{pmatrix} I & [a_1, \cdots, a_n] \\ 0 & I \end{pmatrix} = \begin{pmatrix} I & [a_1, \cdots, a_n] \\ 0 & I \end{pmatrix}\begin{pmatrix} A & B \\ C & D \end{pmatrix}, \quad (19)$$

we find immediately

$$[a_1, \cdots, a_n]C = 0,$$

that is $C = 0$. Then $D = A'^{-1}$ and $B = AS$. (19) implies also

$$A[a_1, \cdots, a_n] + B = B + [a_1, \cdots, a_n]D,$$

that is

$$A[a_1, \cdots, a_n]A' = [a_1, \cdots, a_n]$$

for all $[a_1, \cdots, a_n]$. It follows immediately that $A = [\pm 1, \pm 1, \cdots, \pm 1]$.

Therefore $\mathcal{A}$ induces a mapping carrying Σ_2 onto itself. All the square elements form also a group Σ. Then $\mathcal{A}$ induces also an automorphism on Σ. The elements of Σ takes the form

$$\begin{pmatrix} I & S \\ 0 & I \end{pmatrix}$$

and conversely every element of this form belongs to Σ since the characteristic of Φ is not equal to 2. Thus we have

$$\mathcal{A}\begin{pmatrix} I & S \\ 0 & I \end{pmatrix} = \begin{pmatrix} I & S^* \\ 0 & I \end{pmatrix}. \tag{20}$$

This induces a mapping carrying symmetric matrices onto symmetric matrices.

By (18), we have

$$\mathcal{A}\begin{pmatrix} 0 & I \\ -I & 0 \end{pmatrix} = \begin{pmatrix} 0 & D \\ -D^{-1} & 0 \end{pmatrix}, \quad D = [p, 1, \cdots, 1]. \tag{21}$$

Combining (20) and (21), we have

$$\mathcal{A}\begin{pmatrix} I & 0 \\ S & I \end{pmatrix} = \mathcal{A}\left(\begin{pmatrix} 0 & I \\ -I & 0 \end{pmatrix}\begin{pmatrix} I & -S \\ 0 & I \end{pmatrix}\begin{pmatrix} 0 & -I \\ I & 0 \end{pmatrix}\right)$$

$$= \begin{pmatrix} 0 & D \\ -D^{-1} & 0 \end{pmatrix}\begin{pmatrix} I & -S^* \\ 0 & I \end{pmatrix}\begin{pmatrix} 0 & -D \\ D^{-1} & 0 \end{pmatrix}$$

$$= \begin{pmatrix} I & 0 \\ D^{-1}S^*D^{-1} & I \end{pmatrix}.$$

Then we have

$$\mathcal{A}\begin{pmatrix} I & 0 \\ T & I \end{pmatrix} = \begin{pmatrix} I & 0 \\ T^{**} & I \end{pmatrix}, \tag{22}$$

where T and T^{**} are symmetric. The group formed by elements of the form

$$\begin{pmatrix} I & 0 \\ T & I \end{pmatrix}$$

is denoted by Σ_0.

Consider those elements carrying the groups Σ and Σ_0 onto themselves. They form a group with elements of the form

$$\begin{pmatrix} A & 0 \\ 0 & A'^{-1} \end{pmatrix}.$$

Therefore we have

$$\mathcal{A}\begin{pmatrix} A & 0 \\ 0 & A'^{-1} \end{pmatrix} = \begin{pmatrix} A^* & 0 \\ 0 & A^{*\prime -1} \end{pmatrix}, \tag{23}$$

that is $\mathcal{A}$ induces an autormorphism of the general linear group $(A \leftrightarrow A^*)$.

4) In (18), we take

$$A_1 = aI, \quad b = c = 0, \quad B_1 = 0, \quad C_1 = 0,$$

then we have

$$\mathcal{A}\begin{pmatrix} aI^{(n)} & 0 \\ 0 & a^{-1}I^{(n)} \end{pmatrix} = \begin{pmatrix} [a^\tau, a, \cdots, a] & 0 \\ 0 & [a^\tau, a, \cdots, a]^{-1} \end{pmatrix}.$$

If the field contains more than three elements, we can find an element $a \neq \pm 1$. If $a^\tau \neq a$ and a^{-1}, the normalizer of

$$\begin{pmatrix} aI & 0 \\ 0 & a^{-1}I^{(n)} \end{pmatrix} \quad \text{and} \quad \begin{pmatrix} [a^\tau, a, \cdots, a] & 0 \\ 0 & [a^\tau, a, \cdots, a]^{-1} \end{pmatrix}$$

cannot be isomorphic. This is absurd. Therefore $a^\tau = a$ or a^{-1} for all a belonging to Φ. If there is an element a in Φ such that

$$a^\tau = a^{-1},$$

then, if Φ is of characteristic $\neq 3$, we have

$$2a^{-1} = a^{-1} + a^{-1} = a^\tau + a^\tau = (2a)^\tau = \begin{cases} (2a)^{-1}, \\ \text{or } 2a. \end{cases}$$

Both are impossible for $a \neq \pm 1$. If Φ is of characteristic 3, we have

$$1 + a^{-1} = 1 + a^\tau = (1+a)^\tau = \begin{cases} 1 + a, \\ \text{or } (1+a)^{-1}. \end{cases}$$

Both are also impossible for $a \neq \pm 1$. Therefore for $a \neq \pm 1$, we have

$$a^\tau = a.$$

This holds also for $a = \pm 1$. Therefore we have $\tau = 1$, the identity automorphism.

The field having only three elements is the field:

$$-1, 0, 1 \pmod 3.$$

The only automorphism of the field is the identity automorphism. We have also $\tau = 1$. Therefore we have $\tau = 1$ in (18).

5) Another particular case of (18) is

$$\mathcal{A}\begin{pmatrix} I & [b_1,\cdots,b_n] \\ 0 & I \end{pmatrix} = \begin{pmatrix} I & [pb_1,\cdots,b_n] \\ 0 & I \end{pmatrix}. \tag{24}$$

Combining (23) and (24), we have

$$\mathcal{A}\begin{pmatrix} I & A[b_1,\cdots,b_n]A' \\ 0 & I \end{pmatrix} = \mathcal{A}\left(\begin{pmatrix} A & 0 \\ 0 & A'^{-1} \end{pmatrix}\begin{pmatrix} I & [b_1,\cdots,b_n] \\ 0 & I \end{pmatrix}\begin{pmatrix} A & 0 \\ 0 & A'^{-1} \end{pmatrix}^{-1}\right)$$

$$= \begin{pmatrix} I & A^*[pb_1,b_2,\cdots,b_n]A^{*\prime} \\ 0 & I \end{pmatrix}.$$

This asserts that (20) induces a mapping carrying the set of symmetric matrices onto itself and the ranks of S and S^* are equal.

Combining (20) and (24), we have

$$\mathcal{A}\begin{pmatrix} I & S-[b_1,\cdots,b_n] \\ 0 & I \end{pmatrix} = \mathcal{A}\left(\begin{pmatrix} I & S \\ 0 & I \end{pmatrix}\begin{pmatrix} I & [b_1,\cdots,b_n] \\ 0 & I \end{pmatrix}^{-1}\right)$$

$$= \begin{pmatrix} I & S^*-[pb_1,b_2,\cdots,b_n] \\ 0 & I \end{pmatrix}.$$

Consequently, if we have a set of values $b_1,\cdots,b_n$ such that

$$\det(S - [b_1,\cdots,b_n]) = 0,$$

then it satisfies also

$$\det(S^* - [pb_1, b_2, \cdots, b_n]) = 0$$

and vice versa.

Since these equations are linear in each of the variables $b_1,\cdots,b_n$, and the characteristic of Φ is not equal to 2[①], the coefficients are proportional, that is, if we put $S = (s_{ij})$ and $S^* = (s^*_{ij})$,

$$\frac{1}{p} = \frac{s_{11}}{s^*_{11}} = \frac{s_{ii}}{ps^*_{ii}} \qquad \text{for} \quad 1 < i$$

$$= \frac{s_{11}s_{jj} - s^2_{1j}}{s^*_{11}s^*_{jj} - s^{*2}_{1j}} \qquad \text{for} \quad 1 < j$$

---

① Notice that by eliminating b_1, we have an equation which is quadratic in each of $b_2,\cdots,b_n$ and equal to zero for all $b_2,\cdots,b_n$. Thus the coefficients of the equation are identical zero. Consequently we have the assertion.

$$= \frac{s_{ii}s_{jj} - s_{ij}^2}{p(s_{ii}^*s_{jj}^* - s_{ij}^{*2})} \quad \text{for} \quad 1 < i < j$$

$$= \begin{vmatrix} s_{11} & s_{12} & s_{13} \\ s_{12} & s_{22} & s_{23} \\ s_{13} & s_{23} & s_{33} \end{vmatrix} \bigg/ \begin{vmatrix} s_{11}^* & s_{12}^* & s_{13}^* \\ s_{12}^* & s_{22}^* & s_{23}^* \\ s_{13}^* & s_{23}^* & s_{33}^* \end{vmatrix}, \tag{25}$$

that is

$$s_{11}^* = s_{11}p, \quad s_{ii}^* = s_{ii}, \quad 1 < i,$$
$$s_{ij}^2 = s_{ij}^{*2}, \quad 1 < i < j,$$
$$p(s_{11}s_{jj} - s_{1j}^2) = s_{11}^*s_{jj}^* - s_{ij}^{*2} = ps_{11}s_{jj} - s_{ij}^{*2}.$$

Then we have

$$ps_{ij}^2 = s_{ij}^{*2}.$$

This is only possible when p is a square element. Then, we have

$$p = 1.$$

Consequently, we have

$$s_{ii} = s_{ii}^*, \quad s_{ij} = \pm s_{ij}^*.$$

Another special case of (18) is

$$\mathcal{A}\begin{pmatrix} I & \begin{pmatrix} 1 & 0 \\ 0 & B \end{pmatrix} \\ 0 & I \end{pmatrix} = \begin{pmatrix} I & \begin{pmatrix} 1 & 0 \\ 0 & B \end{pmatrix} \\ 0 & I \end{pmatrix}. \tag{26}$$

Using (26) instead of (24), we have, by the same argument, that

$$s_{ij} = s_{ij}^* \quad \text{for} \quad 1 < i < j.$$

Let $G_{ij}(i \neq j)$ denote a matrix with 1 at (i,j) and (j,i)-position with 0 elsewhere. We have already proved that

$$G_{ij}^* = G_{ij} \quad \text{for} \quad 1 < i < j,$$
$$G_{ii}^* = G_{ii}.$$

We have a matrix $A = [-1, 1, \cdots, 1]$ such that

$$AG_{12}A' = -G_{12}.$$

Thus without loss of generality, we may assume that $\mathcal{A}$ satisfies (18) with $p = 1$ and $\tau = 1$ and $G_{12}^* = G_{12}$.

If $G_{13}^* = -G_{13}$, by (25), we have

$$0 = \begin{vmatrix} 1 & 1 & 1 \\ 1 & 0 & 1 \\ 1 & 1 & 1 \end{vmatrix} = \begin{vmatrix} 1 & 1 & -1 \\ 1 & 0 & 1 \\ -1 & 1 & 1 \end{vmatrix} = 4,$$

this is impossible. Therefore
$$G_{1i}^* = G_{1i}.$$

Since
$$[a, 1, \cdots, 1] G_{12} [a, 1, \cdots, 1]' = a G_{12},$$

the additive property of
$$S^* = S$$

the module formed by symmetric matrices implies for all symmetric S. Therefore

$$\mathcal{A} \begin{pmatrix} I & S \\ 0 & I \end{pmatrix} = \begin{pmatrix} I & S \\ 0 & I \end{pmatrix}. \tag{27}$$

6) Combining (23) and (27), we have

$$ASA' = A^* S A^{*\prime}$$

for all symmetric matrices S. Then we have

$$A^* = \pm A.$$

Those A satisfying $A^* = A$ form a subgroup Γ. The index of Γ in the full linear group is either 1 or 2. Now we are going to prove that it cannot be two, therefore Γ is the full linear group. By (18), all these A of the form

$$\begin{pmatrix} a_1 & 0 \\ 0 & A_1^{(n-1)} \end{pmatrix} \tag{28}$$

belong to Γ. For $n \geqslant 3$ the full linear group is generated by the elements (28) and J_{12}, where J_{ij} denotes a matrix with 1 at (i,j) and (j,i) position and with 1 at (k,k) position for $k \neq i, k \neq j$ and with 0 elsewhere. If the index is two then

$$J_{12}^* = -J_{12}.$$

By (28), we have

$$J_{ij}^* = J_{ij} \quad \text{for } i > 1, j > 1. \tag{29}$$

Then, from
$$J_{13} = J_{23}J_{12}J_{23}, \quad J_{23} = J_{13}J_{12}J_{13},$$
we deduce that
$$J_{13}^* = -J_{13}, \quad J_{23}^* = -J_{23}.$$
The last equation contradicts (29). Therefore
$$A = A^*$$
for all non-singular A, that is
$$\mathcal{A}\begin{pmatrix} A & 0 \\ 0 & A'^{-1} \end{pmatrix} = \begin{pmatrix} A & 0 \\ 0 & A'^{-1} \end{pmatrix}. \tag{30}$$

In case $n = 2$, by (18), we have that all those A of the form
$$\begin{pmatrix} a_1 & 0 \\ 0 & a_2 \end{pmatrix}$$
belong to Γ. Further we have
$$\begin{pmatrix} 1 & s \\ 0 & 1 \end{pmatrix}^* = \pm \begin{pmatrix} 1 & s \\ 0 & 1 \end{pmatrix}.$$
Squaring both sides, we have
$$\begin{pmatrix} 1 & 2s \\ 0 & 1 \end{pmatrix}^* = \begin{pmatrix} 1 & 2s \\ 0 & 1 \end{pmatrix}$$
for all s, then we have
$$\begin{pmatrix} 1 & s \\ 0 & 1 \end{pmatrix}^* = \begin{pmatrix} 1 & s \\ 0 & 1 \end{pmatrix}.$$
Similarly, we have
$$\begin{pmatrix} 1 & 0 \\ t & 1 \end{pmatrix}^* = \begin{pmatrix} 1 & 0 \\ t & 1 \end{pmatrix}.$$
Since
$$\begin{pmatrix} 0 & 1 \\ -1 & 0 \end{pmatrix} = \begin{pmatrix} 1 & 1 \\ 0 & 1 \end{pmatrix}\begin{pmatrix} 1 & 0 \\ -1 & 1 \end{pmatrix}\begin{pmatrix} 1 & 1 \\ 0 & 1 \end{pmatrix},$$
we have
$$\begin{pmatrix} 0 & 1 \\ 1 & 0 \end{pmatrix}^* = \begin{pmatrix} 0 & 1 \\ 1 & 0 \end{pmatrix}.$$

Evidently the full linear group is generated by elements of the form

$$\begin{pmatrix} a_1 & 0 \\ 0 & a_2 \end{pmatrix}, \quad \begin{pmatrix} 1 & s \\ 0 & 1 \end{pmatrix}, \quad \begin{pmatrix} 0 & 1 \\ 1 & 0 \end{pmatrix}.$$

The assertion is also true for $n = 2$.

7) From (18), we have

$$\mathcal{A} \begin{pmatrix} M & I - M \\ -(I - M) & M \end{pmatrix} = \begin{pmatrix} M & I - M \\ -(I - M) & M \end{pmatrix}. \tag{31}$$

Combining (27), (30) and (31) and using Theorem 4 we have the theorem.

Appendix: automorphisms of unimodular group

Since the proof of O. Schreier and B. v. d. Waerden contains an unjustifiable point[①], for completeness we include the following appendix.

In the appendix, we do not assume that the characteristic p of the field is not equal to 2 and we only consider two-rowed matrices. The theorem to be proved is the following.

Theorem *Every automorphism of the unimodular group of order two is of the form*

$$PT^\sigma P^{-1},$$

where P is non-singular and σ is an automorphism of the field.

Definition For $p \neq 0$, a unimodular matrix A is called *parabolic*, if $A^{2p} = I$. For $p \neq 0$, a unimodular matrix A is called *parabolic*, if there are infinitely many unimodular matrices similar to A and commutative with A.

Lemma 1 *A necessary and sufficient condition for a matrix A to be parabolic is that the characteristic equation has a double root ± 1, and consequently, every parabolic element is similar to*

$$\pm \begin{pmatrix} 1 & s \\ 0 & 1 \end{pmatrix}$$

---

① Loc. cit, p. 313. Let S be a linear fractional transformation on Φ. Let $\mathfrak{N}_S$ be the normalizer of S and $\overline{\mathfrak{N}}_S$ be the normalizer of $\mathfrak{N}_S$. They assert that

$$\overline{\mathfrak{N}}_S : \mathfrak{N}_S = \begin{cases} 1, & \text{if } S \text{ is parabolic,} \\ 2, & \text{if } S \text{ is elliptic and hyperbolic.} \end{cases}$$

However this is not true for S being elliptic. Since the fixed elements of an elliptic transformation may not be in the field, sometimes there does not exist a transformation in Φ to permit them.

under the unimodular group.

Proof 1) $p \neq 0$. In an extended field, the characteristic roots of A are α and $1/\alpha$. From $A^{2p} = I$, we deduce
$$\alpha^{2p} = 1.$$
Consequently $\alpha^2 = 1$ and $\alpha = \alpha^{-1} = \pm 1$.

2) $p = 0$. Suppose $\alpha \neq \pm 1$, we have in an extended field a matrix P such that
$$PAP^{-1} = \begin{pmatrix} \alpha & 0 \\ 0 & \alpha^{-1} \end{pmatrix}.$$
If B is commutative with A, we have
$$PBP^{-1} = \begin{pmatrix} \beta & 0 \\ 0 & \beta^{-1} \end{pmatrix}.$$
Since A and B are similar, we have either $\alpha = \beta$ or $\alpha = \beta^{-1}$. That is, we have only finite number of B commutative with and similar to A.

3) If the matrix A has a double characteristic root, we have a matrix P in the field such that
$$PAP^{-1} = \pm \begin{pmatrix} 1 & s \\ 0 & 1 \end{pmatrix}.$$
Evidently, all the matrices of the form
$$\pm \begin{pmatrix} 1 & sa^2 \\ 0 & 1 \end{pmatrix} = \pm \begin{pmatrix} a & 0 \\ 0 & a^{-1} \end{pmatrix} \begin{pmatrix} 1 & s \\ 0 & 1 \end{pmatrix} \begin{pmatrix} a & 0 \\ 0 & a^{-1} \end{pmatrix}^{-1}$$
are similar to and commutative with $\pm \begin{pmatrix} 1 & s \\ 0 & 1 \end{pmatrix}$.

Lemma 2 Any two non-commutative parabolic elements can be carried simultaneously into
$$\pm \begin{pmatrix} 1 & s \\ 0 & 1 \end{pmatrix}, \quad \pm \begin{pmatrix} 1 & 0 \\ t & 1 \end{pmatrix}$$
by a unimodular transformation.

Proof By Lemma 1, we may assume that they take the form
$$\pm \begin{pmatrix} 1 & s \\ 0 & 1 \end{pmatrix}, \quad \begin{pmatrix} a & b \\ c & d \end{pmatrix}, \quad a+d = \pm 2, \quad ad-bc = 1, \quad c \neq 0.$$

We can always find a unique t in the field such that
$$-ct^2 + (d-a)t + b = 0.$$
Then
$$\begin{pmatrix} 1 & t \\ 0 & 1 \end{pmatrix} \begin{pmatrix} a & b \\ c & d \end{pmatrix} \begin{pmatrix} 1 & t \\ 0 & 1 \end{pmatrix}^{-1}$$
carries the second matrix into our required form.

Proof of the theorem By Lemma 2, we may assume that after subjecting the given automorphism to an inner automorphism we have s_0, s_0^*, t_0, t_0^*, none of them being zero and a choice of $\pm$ such that
$$\mathcal{A}\left(\pm \begin{pmatrix} 1 & s_0 \\ 0 & 1 \end{pmatrix}\right) = \pm \begin{pmatrix} 1 & s_0^* \\ 0 & 1 \end{pmatrix}$$
and
$$\mathcal{A}\left(\pm \begin{pmatrix} 1 & 0 \\ t_0 & 1 \end{pmatrix}\right) = \pm \begin{pmatrix} 1 & 0 \\ t_0^* & 1 \end{pmatrix}.$$
The normalisator of $\pm \begin{pmatrix} 1 & s_0 \\ 0 & 1 \end{pmatrix}$ is the group formed by elements of the form
$$\pm \begin{pmatrix} 1 & s \\ 0 & 1 \end{pmatrix}.$$
Squaring, we have
$$\begin{pmatrix} 1 & 2s \\ 0 & 1 \end{pmatrix}.$$
Thus, for $p \neq 2$, $\mathcal{A}$ carries the group Γ_1
$$\begin{pmatrix} 1 & s \\ 0 & 1 \end{pmatrix}$$
onto itself. For $p = 2$, $\pm \begin{pmatrix} 1 & s \\ 0 & 1 \end{pmatrix} = \begin{pmatrix} 1 & s \\ 0 & 1 \end{pmatrix}$, the statement is also true. Similarly $\mathcal{A}$ carries the group Γ_2 formed by matrices
$$\begin{pmatrix} 1 & 0 \\ t & 1 \end{pmatrix}$$
onto itself.

Consider those elements which carry Γ_1 and Γ_2 onto themselves, they take the form
$$\begin{pmatrix} a & 0 \\ 0 & a^{-1} \end{pmatrix}$$
and constitute a group Σ.

Further consider those elements which permute the groups Γ_1 and Γ_2, they take the form
$$\begin{pmatrix} 0 & d \\ -d^{-1} & 0 \end{pmatrix}.$$

Since
$$\begin{pmatrix} d^{-1} & 0 \\ 0 & d \end{pmatrix} P \begin{pmatrix} d^{-1} & 0 \\ 0 & d \end{pmatrix}^{-1}$$
keeps Γ_1, Γ_2 and Σ unaltered and
$$\begin{pmatrix} d^{-1} & 0 \\ 0 & d \end{pmatrix} \begin{pmatrix} 0 & d \\ -d^{-1} & 0 \end{pmatrix} \begin{pmatrix} d^{-1} & 0 \\ 0 & d \end{pmatrix}^{-1} = \begin{pmatrix} 0 & 1 \\ -1 & 0 \end{pmatrix},$$
therefore, we may assume that after another inner automorphism,
$$\mathcal{A}\begin{pmatrix} 0 & 1 \\ -1 & 0 \end{pmatrix} = \begin{pmatrix} 0 & 1 \\ -1 & 0 \end{pmatrix}$$
and
$$\mathcal{A}\begin{pmatrix} 1 & s \\ 0 & 1 \end{pmatrix} = \begin{pmatrix} 1 & \sigma(s) \\ 0 & 1 \end{pmatrix}, \quad \sigma(s_1 + s_2) = \sigma(s_1) + \sigma(s_2)$$
and
$$\mathcal{A}\begin{pmatrix} a & 0 \\ 0 & a^{-1} \end{pmatrix} = \begin{pmatrix} \rho(a) & 0 \\ 0 & \rho(a)^{-1} \end{pmatrix}, \quad \rho(a_1 a_2) = \rho(a_1)\rho(a_2).$$

Consequently
$$\mathcal{A}\begin{pmatrix} 1 & 0 \\ t & 1 \end{pmatrix} = \mathcal{A}\left(\begin{pmatrix} 0 & 1 \\ -1 & 0 \end{pmatrix} \begin{pmatrix} 1 & -t \\ 0 & 1 \end{pmatrix} \begin{pmatrix} 0 & 1 \\ -1 & 0 \end{pmatrix}^{-1}\right) = \begin{pmatrix} 1 & 0 \\ \sigma(t) & 1 \end{pmatrix}.$$

Since
$$\left(\begin{pmatrix} 1 & a \\ 0 & 1 \end{pmatrix} \begin{pmatrix} 0 & 1 \\ -1 & 0 \end{pmatrix}\right)^3 = I$$
if and only if $a = 1$, we have
$$\sigma(1) = 1.$$

Since
$$\begin{pmatrix} x & 0 \\ 0 & x^{-1} \end{pmatrix} = \begin{pmatrix} 1 & -x \\ 0 & 1 \end{pmatrix} \begin{pmatrix} 1 & 0 \\ x^{-1}-1 & 1 \end{pmatrix} \begin{pmatrix} 1 & 1 \\ 0 & 1 \end{pmatrix} \begin{pmatrix} 1 & 0 \\ -(1-x) & 1 \end{pmatrix},$$
we have
$$\begin{pmatrix} \rho(x) & 0 \\ 0 & \rho(x)^{-1} \end{pmatrix} = \begin{pmatrix} 1 & -\sigma(x) \\ 0 & 1 \end{pmatrix} \begin{pmatrix} 1 & 0 \\ \sigma(x^{-1})-1 & 1 \end{pmatrix} \begin{pmatrix} 1 & 1 \\ 0 & 1 \end{pmatrix} \begin{pmatrix} 1 & 0 \\ -1+\sigma(x) & 1 \end{pmatrix}$$
$$= \begin{pmatrix} * & 1-\sigma(x)\sigma(x^{-1}) \\ * & \sigma(x^{-1}) \end{pmatrix}.$$

We deduce
$$1 - \sigma(x)\sigma(x^{-1}) = 0, \quad \rho(x)^{-1} = \sigma(x^{-1}),$$
that is, for $x \neq 0$,
$$\sigma(x) = \rho(x).$$

This asserts that $\sigma(x)$ is an autmorphism of the field.

On the existence of solutions of certatin equations in a finite field*

In another paper[①] one of the authors stated that he had arrived at limits, both inferior and superior, for the number of solutions of the equation

$$c_1 x_1^{a_1} + c_2 x_2^{a_2} + \cdots + c_s x_s^{a_s} + c_{s+1} = 0 \tag{1}$$

in the x's where a's are integers such that $0 < a < p^n - 1$; $s \geqslant 2$ for $c_{s+1} \neq 0$ and $s > 2$ for $c_{s+1} = 0$, the c's being given elements of a finite field of order p^n, p prime, which will be designated by $F(p^n)$; and

$$c_1 \cdots c_s x_1 \cdots x_s \neq 0. \tag{2}$$

As a consequence of this result, one can obtain

Theorem I *The equation (1) with the restriction (2) always has at least k solutions in the x's for k any given positive integer provided p^n exceeds a certain limit.*

In this paper we shall give two quite different approaches to establish this theorem. The first is closely related to the one previously mentioned and the other argument, although subject to the limitation $s > 2$ when $c_{s+1} \neq 0$, is far simpler and is based on a method[②] which was introduced by one of the authors in the study of generalized Gaussian sums over a finite field.

The limit given here can be sharpened, and the proof of this will be published later.

Elsewhere[③] it was shown that the exact number of solutions of (1) may be

---

* Reprinted from *Proceedings of the National Academy of Sciences*, 1948, **34**(6): 258–263. By L. K. Hua and H. S. Vandiver.

① These Proceedings, 1947, **33**: 236–242. In this paper reference to previous results was made, but an important paper by Mordell. *Mathematische Zeitschrift*, 1933, **37**: 207, which bore more directly on the contents, was, unfortunately, not mentioned. Other relevant references are Pellet, *Bull. Math. Soc. France*, 1886, **15**, 80–93; Dickson, *Crelle*, 1909, **135**: 181–188; Hurwitz, *Crelle*, 1909, **136**: 272–292; Mitchell, *Ann Math.*, II, 1917, **18**: 120; Davenport, *Jour. London Math. Soc.*, 1931, **6**: 49–54; Schur, I., *Jahresber. Deutsch. Math. Verein.*, 1916, **25**: 114.

② Cf. a paper by Hua, Loo-Keng. "On a double exponential sum", soon to appear in the *Science Reports of Tsing Hua University*. An abstract is given in the *Science Record of the Acad. Sinica*, **1**, Nos. 1–2.

③ These Proceedings, 1946, **32**: 47–52.

determined directly if we know the exact number of solutions of $\theta^{m\alpha}, \theta^{m\beta}$ of the equation
$$\theta^{i+m\alpha} + \theta^{j+m\beta} + 1 = 0 \tag{3}$$
for each i and j; θ being a primitive root of $F(p^n)$; $(p^n - 1, a_i) = d_i$, $i = 1, 2, \cdots, s$; m is the L.C.M of $d_1, d_2, \cdots, d_s$. Further in this argument some or all of the a's may equal $p^n - 1$. To prove this we employed formulae (5), (12) and (15) of the paper just mentioned and all of these contain positive terms only.

We may adapt this method, however, to finding limits from the number of solutions of (1) by first finding limits for the number of solutions of (3) and then employing (5), (12) and (15) of the previous paper[①].

In the latter the formulae
$$A_{00} = (c-1)^2 + c(m-1); \quad A_{h0} = A_{0k} = c^2 - c,$$
$$A_{hk} = c^2, \quad h \not\equiv k; \quad A_{hh} = c^2 - c,$$
$$h \not\equiv 0 \pmod{m}, \quad k \not\equiv 0 \pmod{m}, \quad p^n = 1 + mc \tag{4}$$
are derived; where
$$A_{hk} = \sum_{i,j}^{0 \text{ to } m-1} [i,j][i+h, j+k]$$
and $[i,j]$ is the number of different pairs r and t in the set $0, 1, \cdots, c-1$ which satisfy the equation
$$1 + \theta^{i+rm} = \theta^{j+tm}, \quad p^n - 1 = cm. \tag{5}$$
If $\{i,j\}$ denotes the number of solutions of (3) then we note that
$$[i,j] = \{i, j+\varepsilon\}, \tag{6}$$
where $\varepsilon = \text{ind}(-1)$, with $x = \theta^{\text{ind } x}$. We have
$$\sum_{i,j}^{0 \text{ to } m-1} (m^2[i,j] - p^n)^2 = \sum_{i,j}^{0 \text{ to } m-1} (m^4[i,j]^2 - 2m^2 p^n [i,j]) + p^{2n} m^2,$$
which gives, after employing (4) on the right as well as the known relations
$$[i,j] = [j+\varepsilon, i+\varepsilon] = [-j+\varepsilon, i-j] = [i-j+\varepsilon, -j]$$
$$= [-i, j-i] = [j-i+\varepsilon, -i+\varepsilon, -i], \tag{7}$$
$$\left(\frac{p^n}{m^2} + \frac{D}{\sqrt{d}}\right) \geqslant [i,j] \geqslant \left(\frac{p^n}{m^2} - \frac{D}{\sqrt{d}}\right), \tag{8}$$

---
① These Proceedings, 1946, **32**: 47–52.

where
$$D = \frac{1}{m}\sqrt{p^n(m^2 - 3m + 2) + 3m + 1}.$$

Here d is the number of different pairs in (7) in the sense that $[h, k]$ is different from $[h_1, k_1]$ if $h \not\equiv h_1$, or $k \not\equiv k_1 \pmod{m}$.

The use of (8) with (6) of this paper and (5), (12) and (15) of the former paper yield superior and inferior limits for the number of solutions of (1), and we obtain a proof of Theorem I.

For the second proof of Theorem I (except for $s = 2$ with $c_{s+1} \neq 0$), we proceed as follows:

Any element α of $F(p^n)$ may be written uniquely in the form, if θ is a primitive root in $F(p^n)$,
$$\alpha = d_0 + d_1\theta + \cdots + d_{n-1}\theta^{n-1} = h(\theta), \tag{9}$$

where the d's are in the included field $F(p)$. Now θ satisfies an irreducible equation $f(x) = 0$ of degree n with coefficients in $F(p)$ and whose roots are $\theta_1, \theta_2, \cdots, \theta_n$, with $\theta_1 = \theta$. Set (here tr is short for trace)
$$\sum_{i=1}^{n} h(\theta_i) = \text{tr } h(\theta).$$

We shall now prove

Lemma 1 If $\zeta = e^{2i\pi/p}$, then, for $n < p$,
$$\sum_{\alpha \in F(p^n)} \zeta^{\text{tr}(\alpha t)} = \begin{cases} 0, & \text{for } t \neq 0, \\ p^n, & \text{for } t = 0. \end{cases} \tag{10}$$

This statement is obvious for $t = 0$. For $t \neq 0$, αt ranges over all the elements of $F(p^n)$. Hence it is sufficient to prove that
$$\sum_{\alpha \in K} \zeta^{\text{tr}(\alpha t)} = \sum_{d_0=0}^{p-1} \cdots \sum_{d_{n-1}=0}^{p-1} \zeta^T = \prod_{i=0}^{n-1} \left(\sum_{d_i=0}^{p-1} \zeta^{d_i \text{tr}(\theta^i)} \right) = 0, \tag{11}$$

where
$$T = \text{tr}(d_0 + d_1\theta + \cdots + d_{n-1}\theta^{n-1}).$$

There is one of the i's such that p does not divide $\text{tr}(\theta^i)$. For, if the contrary is true then $\text{tr}(\theta^i) = 0$ for all $0 \leqslant i \leqslant n - 1$. Since $\theta_1 = \theta$ satisfies $(x) = 0$ of degree n as do also $\theta_2, \cdots, \theta_n$, then Newton's relations between the elementary symmetric functions of these roots and the sum of the nth powers of said roots give easily $\theta^n = c$, with c

in $F(p)$. This yields $\theta^{n(p-1)} = 1$, contradicting the definition of θ as a generator of $F(p^n)$ for $n > 1$, since $p^n - 1 = (p-1)(p^{n-1} + \cdots + 1) > n(p-1)$. This proves (10).

We now prove

Lemma 2 Write K for $F(p^n)$. If

$$\psi(t, a, \zeta) = \sum_{\substack{x \in K \\ x \neq 0}} \zeta^{\operatorname{tr}(tx^a)},$$

then for $t \neq 0$,

$$|\psi(t, a, \zeta)| \leqslant (p^n - 1, a) p^{n/2}. \tag{12}$$

Now

$$\frac{1}{p^n} \sum_{t \in K} (\psi(t, a, \zeta) \psi(t, a, \zeta^{-1})) = \sum_{t \in K} \sum_{\substack{x, y \in K \\ x, y \neq 0}} \zeta^{\operatorname{tr}(tx^a - ty^a)}. \tag{13}$$

Using Lemma 1, the right-hand member reduces to $(p^n - 1)(p^n - 1, a)$, since this is the number of different solutions in x and $y \neq 0$ of

$$x^a = y^a$$

in $F(p^n)$. Since $|\psi(t, a, \zeta)| = |\psi(t, a, \zeta^{-1})|$, then (12) gives, if we set $\psi(t, a, \zeta) = \psi(t, a)$,

$$\frac{1}{p^n} \sum_{t \in K} |\psi(t, a)|^2 = (p^n - 1)(p^n - 1, a). \tag{14}$$

Since, if $t \neq 0$, the number of solutions of $ty^a = q$ is $\leqslant (p^n - 1, a)$, then

$$(p^n - 1)|\psi(t, a)|^2 = \sum_{\substack{y \in K \\ y \neq 0}} |\psi(ty^a, a)|^2$$

$$\leqslant (p^n - 1, a) \sum_{q \in K} |\psi(q, a)|^2$$

$$\leqslant (p^n - 1, a)^2 p^n (p^n - 1),$$

and by (14), we then have Lemma 2. Then for the proof of Theorem I, we have

$$N = \frac{1}{p^n} \sum_{t \in K} \psi(c_1 t, a_1) \cdots \psi(c_s t, a_s) \zeta^{\operatorname{tr}(c_{s+1} t)}$$

$$= \frac{(p^n - 1)^s}{p^n} + \frac{1}{p^n} \sum_{\substack{t \in K \\ t \neq 0}} \psi(c_1 t, a_1) \cdots \psi(c_s t, a_s) \zeta^{\operatorname{tr}(c_{s+1} t)}, \tag{15}$$

where N is the number of solutions of (1), and now we count all the solutions under the restriction (2) (in connection with (5) we did not do this exactly, that is, all the solutions of (5) would be $m^2[i,j]$). Then

$$\left|N - \frac{(p^n-1)^s}{p^n}\right| \leqslant \frac{1}{p^n} \sum_{\substack{t \in K \\ t \neq 0}} |\psi(c_1 t, a_1)| \cdots |\psi(c_s t, a_s)|,$$

and by Lemma 2, we have

$$\left|N - \frac{(p^n-1)^s}{p^n}\right| \leqslant \prod_{i=1}^{s}(p^n-1, a_i) \cdot p^{ns/2},$$

whence

$$\frac{(p^n-1)^s}{p^n} - Ap^{ns/2} \leqslant N \leqslant \frac{(p^n-1)^s}{p^n} + Ap^{ns/2},$$

$$A = \prod_{i=1}^{s}(p^n-1, a_i). \tag{16}$$

The quantity $A > 0$ never exceeds the fixed value $\prod_{i=1}^{s} a_i$. Set $p^n - 1 = h$. Then the left-hand member of (16) becomes, if $r = (s+2)/2$,

$$\frac{h^r \left(h^{(s-r)} - A\left(1 + \frac{1}{h}\right)^r\right)}{p^n}, \tag{17}$$

which for $s > 2$ and p^n sufficiently large is obviously $> k$ for k any integer. This proves Theorem I, for $s > 2$. From (16) we may also write

$$N = \frac{(p^n-1)^s}{p^n} + 0\,(p^{ns/2}). \tag{18}$$

We may now apply Theorem I to congruences with respect to an ideal prime modulus in any commutative ring R with a unity element. We may consider the congruence

$$\alpha_1 x_1^{a_1} + \alpha_2 x_2^{a_2} + \cdots + \alpha_s x_s^{a_s} + \alpha_{s+1} \equiv 0 \pmod{\mathfrak{p}}, \tag{19}$$

where now the α's are fixed elements in R, that is, we may fix the a's and α's and consider the above congruence for various values of $\mathfrak{p}$. This is a bit different from the situation in Theorem I where the domain of the coefficients changes with each $\mathfrak{p}$. But if the number of incongruent residues (norm) in R of the ideal $\mathfrak{p}$ is finite then the residue classes form a field, since a finite integral domain is a field. Hence we may apply Theorem I and obtain

Theorem II *The congruence* (19) *always has at least k solutions in the x's for k any positive integer provided*

$$\alpha_1 \alpha_2 \cdots \alpha_s x_1 x_2 \cdots x_s \not\equiv 0 \pmod{\mathfrak{p}}, \tag{20}$$

$s \geqslant 2$ *for* $c_{s+1} \not\equiv 0 \pmod{\mathfrak{p}}$ *and* $s > 2$ *for* $c_{s+1} \equiv 0 \pmod{\mathfrak{p}}$; $\mathfrak{p}$ *is a prime ideal of finite norm in a commutative ring R with a unity element, the a's are fixed positive integers, the α's are fixed elements in R, and the norm of $\mathfrak{p}$ is sufficiently large.*

If R is the ring of integers in an algebraic field then every ideal has a finite norm so that Theorem II holds for any algebraic ring of this type with $\mathfrak{p}$ any prime ideal in the ring. If R is the ring of rational integers, then $n = 1$ and R is the system of residue classes modulo p, and we have in particular the

Corollary *If in* (19) *the x's are rational integers and* $\mathfrak{p} = (p)$ *with p a prime rational integer then this congruence always has k solutions for p sufficiently large with the other conditions in Theorem II holding, k any integer.*

This corollary was given by Mordell when $\alpha_{s+1} \not\equiv 0$, and we include *all*[①] solutions of (19). We considered only solutions prime to $\mathfrak{p}$ (primitive solutions) in our work, following the conditions (2) and (20).

---

① Mordell's method of proof is different from either of those used in the present paper.

In some ways there is quite a distinction between finding the primitive solutions of and equation in a finite field and finding all solutions. The congruence

$$x_1^2 + x_2^2 + \cdots + x_s^2 \equiv 0 \pmod{2}$$

has no solutions in integers prime to 2, if s is odd, but evidently has solutions for some of the x's even. Again the congruence $x^7 + y^7 + 1 \equiv 0 \pmod{491}$ has no solutions in integers x and y prime to 491. But the congruence $x^7 + 1 \equiv 0 \pmod{491}$ obviously has solutions.

Characters over certain types of rings with applications to the theory of equations in a finite field*

Characters, modulo m, have been defined (for example by Landau[1]) and have been employed extensively in Group Theory and Number Theory, particularly in some of the most important parts of Analytic Number Theory. However, we may define them by means of any ring with a unity element and which contains only a finite number of elements which are not zero divisors, instead of the special ring consisting of the residue classes modulo m, and a number of extensions of known results concerning said special case are immediate. We will set up this concept and then specialize it for the finite field using results which are then applied to the theory of certain equations in a finite field which were considered in another paper[2] by the authors. This type of equation is

$$c_1 x_1^{a_1} + c_2 x_2^{a_2} + \cdots + c_s x_s^{a_s} + c_{s+1} = 0, \qquad (1)$$

where the a's are integers such that $0 < a < p^n - 1$; $s \geqslant 2$ for $c_{s+1} \neq 0$ and $s > 2$ for $c_{s+1} = 0$, the c's being given elements of a finite field of order p^n, p prime, which will be designated by $F(p^n)$ and

$$c_1 \cdots c_s x_1 \cdots x_s \neq 0. \qquad (2)$$

In the present paper we obtain a new explicit expression for the number of solutions of (1) in terms of generalized Lagrange resolvents and (see the τ number defined in Theorem I) and using this we find better limits for the number of solutions of (1) than those derived in E.

---

  * Reprinted from *Proceedings of the National Academy of Sciences*, 1949, **35**(2): 94–99. By L. K. Hua and H.S. Vandiver.

  [1] *Vorlesungen über Zahlentheorie*. Leipzig, Hirzel, 1927, Bd. I, 83–87. Cf. also Dickson. *Modern Elementary Theory of Numbers*. University of Chicago Press, 1939: 272–276.

  [2] These Proceedings, 1948, **34**: 258–263. This article will be referred to as E. Characters defined by a general finite field have been considered by a number of writers. Cf. the paper referred to in our footnote [2] in page 190 of this book and the references given therein as well as Davenport. *Acta Mathematica*, 1939, **71**: 99–121.

Let R be a ring with a unity element which has the property of containing only a finite number of elements which are not divisors of zero. A set of such elements forms a group. For, since it does not contain zero or any divisors of zero, the cancellation holds and a finite semi-group having this property is known to be a group. Or, otherwise expressed, these elements are units in R.

Let an arithmetical function $\chi(a)$ be defined for any element a of R so that

I. $\chi(a)$ is a complex number.

II. $\chi(1) \neq 0$.

III. $\chi(a) = 0$ when a is zero or any divisor of zero.

IV. If $a_1 = a_2$ in R, then $\chi(a_1) = \chi(a_2)$.

V. $\chi(a_1 a_2) = \chi(a_1)\chi(a_2)$.

The arguments employed by Landau for the special ring formed by the residue classes modulo m by which be obtains his Satz 127 to Satz 133 inclusive may be easily extended to yield the following:

For each character χ we have

$$\chi(1) = 1. \tag{3}$$

There is a character designated by χ_0 such that

$$\chi_0(a) = 1, \tag{3a}$$

for each a in R. Let $\phi(R) = h$ be the number of distinct units in R. We shall call h the *indicator*[①] of R. Then since the units u form a group, $u^h = 1$ in R, and by I and V,

$$\chi(a) = e^{2i\pi b/h}, \tag{4}$$

for $0 \leqslant b < h$.

We also obtain

Lemma I

$$\sum_{a \in R} \chi(a) = \begin{cases} h, & \text{for } \chi = \chi_0, \\ 0, & \text{for } \chi \neq \chi_0. \end{cases} \tag{5}$$

Suppose that we have, if $h \equiv 0 \pmod{k}$, a character such that $(\chi(a))^k = 1$ for each unit a in R, this will be called a special character and designated by χ_k. In particular χ_0 is a special character for any k.

---

① This term is employed for the reason that the special R consisting of the residue classes modulo m, our units correspond to the integers less than m and prime to it.

Now let R be a finite field of order p^n, p prime, and designate the same by $F(p^n)$. Let θ be a primitive root of this field, and let β be a primitive (p^n-1)th root of unity. Then by (4)
$$\theta = \beta^b, \quad 0 \leqslant b < p^n - 1,$$
and since all the elements, $\neq 0$, of $F(p^n)$ are given by powers of θ then the whole character, by V, is given by powers of β^b. Now consider χ_k. Then
$$\chi_k(\theta) = \beta^d \tag{6}$$
and by definition of χ_k, $(\chi_k(\theta))^k = \beta^{dk} = 1$ which gives $d \equiv 0 \pmod{(p^n-1)/k}$. It follows from this that there are exactly k special characters, and in particular there are h distinct characters defined by R. We now have

Lemma II
$$\sum_{\chi_k} \chi_k(a) = \begin{cases} k, & \text{for } a \text{ a } k\text{th power,} \\ 0, & \text{otherwise.} \end{cases} \tag{7}$$

This follows easily from (6) and the remarks following it.

We also have[①]

Lemma III If $\zeta = e^{2i\pi/p}$, then in $F(p^n)$,
$$\sum_{a \in k} \zeta^{\text{tr}(at)} = \begin{cases} 0, & \text{for } t \neq 0, \\ p^n, & \text{for } t = 0. \end{cases} \tag{8}$$

We now proceed to[②]

Theorem I The number of solutions N of (1) with the restriction (2) is
$$N = \frac{(p^n-1)^s}{p^n} + \frac{T}{p^n},$$
where
$$T = \sum_{\substack{a \in K \\ a \neq 0}} \prod_{i=1}^{s} \sum_{\chi_{k_i}} \chi_{k_i}(a^{-1}c_i^{-1}) \tau(\chi_{k_i}) \zeta^{\text{tr}(ac_{s+1})},$$

[①] In this paper we are using the symbols 1 and 0 both for the unity and zero elements in the complex field and the unity and zero elements in R. This should not cause confusion if we keep in mind that $\chi(a)$ is always in the complex field while a is always in R.

[②] This was proved for $n < p$ in E, and the argument employed there may be extended to the case $n \geqslant p$. However, the result was previously known in general, in fact it follows from the results in Stickelberger. *Math. Annalen*, 1890, **37**: 321, §1, paragraph 5, since any finite field may be represented by a complete set of residue classes with respect to a prime ideal modulus in an algebraic field.

$$\tau(\chi_{k_i}) = \sum_{b \in K} \chi_{k_i}(b) \zeta^{\operatorname{tr}(b)}$$

and $k_i = (a_i, p^n - 1)$. Also K stands for $F(p^n)$.

To prove this we note that

$$N = \frac{1}{p^n} \sum_{a \in K} \sum_{i=1}^{s} \sum_{\substack{x_i \in K \\ x_i \neq 0}} \zeta^{\operatorname{tr}(aG(x))},$$

$$G(x) = c_1 x_2^{k_1} + c_2 x_3^{k_2} + \cdots + c_s x_s^{k_s} + c_{s+1}. \tag{9}$$

First, it is known that the number of solutions of (1) and the number of solutions of $G(x) = 0$ are the same if $(p^n - 1, a_i) = k_i$. Also when we carry out the summation in (9a) for the x_i's all the terms are unity when we have a set of x's which satisfy $G(x) = 0$ with the restriction (2) and when we sum with respect to a we obtain p^n according to Lemma III, but when we have a set of x_i's such that (1) is not satisfied then by the same lemma the sum with respect to a gives zero (the relation (15) in E, p. 261, is a transformation of (9)). In the right-hand member, if we separate the zero and non-zero values of a, we obtain from (9),

$$N = \frac{(p^n - 1)^s}{p^n} + \frac{T}{p^n}, \tag{10}$$

where

$$T = \frac{1}{p^n} \sum_{\substack{a \in K \\ a \neq 0}} \sum_{i=1}^{s} \sum_{\substack{x_i \in K \\ x_i \neq 0}} \zeta^{\operatorname{tr}(aG(x))}.$$

Now employing Lemma II,

$$\sum_{\substack{x \in K \\ x \neq 0}} \zeta^{\operatorname{tr}(cx^k)} = \sum_{y \in K} \sum_{\chi_k} \chi_k(c^{-1}y) \zeta^{\operatorname{tr}(y)}$$

$$= \sum_{\chi_k} \chi_k(c^{-1}) \tau(\chi_k)$$

and then (10) gives

$$T = \sum_{\substack{a \in K \\ a \neq 0}} \prod_{i=1}^{s} \sum_{\chi_{k_i}} \chi_{k_i}(c_i^{-1} a^{-1}) \tau(\chi_{k_i}) \zeta^{\operatorname{tr}(ac_{s+1})}. \tag{11}$$

We now employ (10) and (11) to find limits for N. We see from (4) that for a fixed a

$$\prod_{i=1}^{j} \chi_{u_i}(a) = \chi_v(a), \tag{11a}$$

for some v, Also it is konwn[①] that

$$|\tau(\chi)| = p^{n/2}, \quad \chi \neq \chi_0,$$
$$|\tau(\chi_0)| = 1. \tag{11b}$$

If we write by (11a),

$$\prod_{i=1}^{s} \chi_{k_i}(a^{-1}) = \chi(a),$$

$$\sum_{a \in K} \chi(ac_{s+1}^{-1})\zeta^{\text{tr}(a)} = \chi(c_{s+1}^{-1}) \sum_{a \in K} \chi(a)\zeta^{\text{tr}(a)} = \chi(c_{s+1}^{-1})\tau(\chi), \quad c_{s+1} \neq 0,$$

and employ (11) and (11b) noting that there are k characters χ_k, and that χ is a χ_k,

$$|T| \leq \prod_{i=1}^{s} \sum_{\chi_{k_i}} |\chi_{k_i}(c_i^{-1})| |\tau(\chi)| |\tau(\chi_{k_i})|,$$

$$|T| \leq \prod_{i=1}^{s} (1 + (k_i - 1)p^{n/2}) p^{n/2}. \tag{12}$$

Let us now suppose that $c_{s+1} = 0$. Then we have $\zeta^{\text{tr}(ac_{s+1})} = 1$ in (11) and

$$\sum_{\substack{a \in K \\ a \neq 0}} \chi(c_i^{-1} a^{-1}) \leq p^n - 1$$

and then

$$|T| \leq (p^n - 1) \prod_{i=1}^{s} (1 + (k_i - 1)p^{n/2}). \tag{13}$$

Hence (10) gives for $c_{s+1} \neq 0$,

$$\left| N - \frac{(p^n - 1)^s}{p^n} \right| \leq \prod_{i=1}^{s} (1 + (k_i - 1)p^{n/2}) p^{-n/2} \tag{14}$$

and for $c_{s+1} = 0$,

$$\left| N - \frac{(p^n - 1)^s}{p^n} \right| \leq \frac{(p^n - 1)}{p^n} \prod_{i=1}^{s} (1 + (k_i - 1)p^{n/2}). \tag{15}$$

Whence we have

[①] Davenport and Hasse. *J. f. Math.* (*Crelle*), 1935, **172**: 151–174 found, for the case $s = 2$ a formula related to our (10). See their relation (6.2), p. 173, and the value of N given at the top of p. 174. Their result will be discussed in a future paper by the authors.

Theorem II *If N is the number of solutions of (1) under the condition (2), then for $c_{s+1} \neq 0, (p^n - 1, a_i) = k_i$,*

$$\frac{(p^n - 1)^s}{p^n} - D \leqslant N \leqslant \frac{(p^n - 1)^s}{p^n} + D \tag{16}$$

and for $c_{s+1} = 0$,

$$\frac{(p^n - 1)^s}{p^n} - D_1 \leqslant N \leqslant \frac{(p^n - 1)^s}{p^n} + D_1, \tag{17}$$

where

$$D = p^{-n/2} \prod_{i=1}^{s} (1 + (k_i - 1)p^{n/2})$$

and

$$D_1 = \frac{p^n - 1}{p^n} \prod_{i=1}^{s} (1 + (k_i - 1)p^{n/2}).$$

From the above theorem we infer immediately that

$$N = \frac{(p^n - 1)^s}{p^n} + 0 \; (p^{\frac{n(s-v)}{2}}), \tag{18}$$

where $v = 1$ or 0 according as $c_{s+1} \neq 0$ or $c_{s+1} = 0$.

If we consider (1) without the restriction $x_1, x_2, \cdots, x_s \neq 0$, we may find limits for the number of solutions by considering one or more of the c's in (1) as being zero in turn and employ for each of the resulting equations the limits (16) and (17) and adding the results we obtain inferior and superior limits of somewhat the same type as (16) and (17) and in particular we infer that if N_1 is said number then

$$N_1 = p^{n(s-1)} + 0 \; (p^{n(s-v)/2}), \tag{19}$$

v having the same meaning as in (18).

On the automorphisms of a sfield*

Let K be a sfield. A mapping $(a \to a^\sigma)$ of K onto itself is called a semiautomorphism if it satisfies

$$(a+b)^\sigma = a^\sigma + b^\sigma, \tag{1}$$

$$(aba)^\sigma = a^\sigma b^\sigma a^\sigma \tag{2}$$

and

$$1^\sigma = 1. \tag{3}$$

The well-known examples of semi-automorphisms are automorphisms, which satisfy $(ab)^\sigma = a^\sigma b^\sigma$, and anti-automorphisms, which satisfy $(ab)^\sigma = b^\sigma a^\sigma$.

It is a known problem about the existence of semi-automrphism other than automorphisms and anti-automorphisms. It is the aim of the note to settle this problem, namely:

Theorem 1 *Every semi-automorphism is either an automorphism or an anti-automorphism.*

To prove the theorem we need several simple consequences of (1), (2) and (3). Putting $b = a^{-1}$ in (2), we have

$$(a^{-1})^\sigma = (a^\sigma)^{-1} \tag{4}$$

and replacing a by $a+b$ and b by 1 in (2), we have, by (3),

$$(ab)^\sigma + (ba)^\sigma = a^\sigma b^\sigma + b^\sigma a^\sigma. \tag{5}$$

Applying (2) twice, we obtain, by (4),

$$(ba)^\sigma = (ab(ab)^{-1}ba)^\sigma = a^\sigma b^\sigma (ab)^{\sigma-1} b^\sigma a^\sigma. \tag{6}$$

Substituting (6) into (5), we have

$$(ab)^\sigma + a^\sigma b^\sigma (ab)^{\sigma-1} b^\sigma a^\sigma = a^\sigma b^\sigma + b^\sigma a^\sigma,$$

---

* Reprinted from *Proceedings of the National Academy of Sciences*, 1949, **35**(7): 386–389.

which is equivalent to
$$((ab)^\sigma - a^\sigma b^\sigma)(1 - (ab)^{\sigma-1} b^\sigma a^\sigma) = 0. \tag{7}$$

We deduce immediately
$$(ab)^\sigma = \begin{cases} a^\sigma b^\sigma, & \text{or} \\ b^\sigma a^\sigma. \end{cases} \tag{8}$$

Suppose that we have a pair of elements a and b such that
$$(ab)^\sigma = b^\sigma a^\sigma \neq a^\sigma b^\sigma. \tag{9}$$

Then, for any c, we have
$$(ac)^\sigma = c^\sigma a^\sigma. \tag{10}$$

In fact, otherwise, we would have, by (8),
$$(ac)^\sigma = a^\sigma c^\sigma \neq c^\sigma a^\sigma,$$

and, by (9) and (1)
$$a^\sigma c^\sigma + b^\sigma a^\sigma = (ac)^\sigma + (ab)^\sigma = (a(b+c))^\sigma = \begin{cases} a^\sigma(b^\sigma + c^\sigma), & \text{or} \\ (b^\sigma + c^\sigma)a^\sigma. \end{cases}$$

Both conclusions are impossible. Similarly, we prove that, for any d,
$$(db)^\sigma = b^\sigma d^\sigma. \tag{11}$$

Now we are going to prove that
$$(dc)^\sigma = c^\sigma d^\sigma.$$

Suppose the contrary, that is, by (8),
$$(dc)^\sigma = d^\sigma c^\sigma \ (\neq c^\sigma d^\sigma). \tag{12}$$

Similar to the previous argument, we can establish that
$$(ac)^\sigma = a^\sigma c^\sigma \tag{13}$$

and
$$(db)^\sigma = d^\sigma b^\sigma. \tag{14}$$

Now we consider the elements:

$$b^\sigma a^\sigma + (ac)^\sigma + (db)^\sigma + d^\sigma c^\sigma = ((a+d)(b+c))^\sigma$$

$$= \begin{cases} (a^\sigma + d^\sigma)(b^\sigma + c^\sigma), & \text{or} \\ (b^\sigma + c^\sigma)(a^\sigma + d^\sigma). \end{cases}$$

The first conclusion contradicts (9), by (10) and (11), and the second conclusion contradicts (12), by (13) and (14).

Therefore, if there is a pair of elements a, b such that $(ab)^\sigma = b^\sigma a^\sigma (\neq a^\sigma b^\sigma)$, then, for any pair of elements c and d, we have $(cd)^\sigma = d^\sigma c^\sigma$. This proves our theorem.

By means of the previous theorem, we also settle a problem in the study of projective geometry on a line over a sfield. Namely:

Theorem 2 *Any one to one mapping carrying the projective line over a sfield of characteristic $\neq 2$ onto itself and keeping harmonic relation invariant is a semi-linear transformation induced by an automorphism or an anti-automorphism.*

This theorem was established for the quaternion algebra (Ancochea[①]) and later for division algebra (Ancochea [②]), characteristic $\neq 2$ and he left the general problem open.

As an easy consequence of the following theorem on geometry of matrices (the proof will be given elsewhere), we can extend Theorem 1 to any semisimple ring with a descending chain condition.

Theorem 3 *Two $n \times m$ matrices Z and W are said to coherent, if the rank of their difference $Z - W$ is one. Suppose $1 < n \leqslant m$. Any one to one mapping carrying $n \times m$ matrices into $n \times m$ matrices and leaving the coherence relation invariant is of the form*

$$Z_1 = PZ^\sigma Q + R, \tag{15}$$

where $P \ (= P^{(n)})$ and $Q \ (= Q^{(m)})$ are non-singular and $R = R^{(n,m)}$, and σ is an automorphism of the sfield. In case $n = m$, in addition to (15), we have also

$$Z_1 = PZ'^\sigma Q + R, \tag{16}$$

where σ is an anti-automorphism of the sfield and Z' is the transposed matrix of Z.

[①] Ancochea G. *J. Math.*, 1942, **184**: 192–198.
[②] Ancochea G. *Annals Math.*, 1947, **48**: 147–153.

By entirely different methods, Ancochea and Kaplansky [1] treated the problem of semi-automorphisms under some restrictions. The former established the corresponding theorem for a simple algebra over a field of characteristic $\neq 2$, and the later extended it to semi-simple algebra over any field. Both of their methods rooted in the structure theory of linear algebras, therefore neither of them can be extended to the general case.

---

[1] Kaplansky I. *Duke Math. J.*, 1947, **14**: 521–525.

On the number of solutions of some trinomial equations in a finite field*

In the present paper we shall use the results in two previous papers[①] by the authors to obtain (Theorem I) the exact number, explicitly in terms of p, n, k_1 and k_2, of solutions of the equation, if $0 < a_i < p^n - 1$, $(p^n - 1, a_i) = k_i$, $i = 1, 2$,

$$c_1 x_1^{a_1} + c_2 x_2^{a_2} + c_3 = 0, \tag{1}$$

in x_1 and x_2, non-zero elements of a finite field $F(p^n)$ of order p^n, p an odd prime with c_1, c_2, c_3 given elements of $F(p^n)$, $c_1 c_2 c_3 \neq 0$, in certain special cases. Secondly, we shall find limits (Theorem II) for the number of solutions of (1) which are better than those given in C for the solutions of this particular equation, and which have the unusual property that they agree with the exact values found for the number of solutions in the special cases treated in Theorem I. We shall employ the notation used in H (defined in (9b) just below relation (14) of that paper).

Let α be a primitive $(ds_1 s_2)$-th root of unity, $m = ds_1 s_2$ where $ds_1 = k_1, ds_2 = k_2$ where $(k_1, k_2) = d$, whence $(s_1, s_2) = 1$. Let χ_{k_1} be a special k_1-character and χ_{k_2} a special k_2-character in $F(p^n)$ and write, if K stands for $F(p^n)$,

$$\tau(\chi_{k_1}^{\mu_1}) = \sum_{a \in K} \alpha^{\mu_1 s_2 \operatorname{ind}(a)} \zeta^{\operatorname{tr}(a)}$$

and

$$\tau(\chi_{k_2}^{\mu_2}) = \sum_{a \in K} \alpha^{\mu_2 s_1 \operatorname{ind}(a)} \zeta^{\operatorname{tr}(a)},$$

where $\operatorname{tr}(a)$ is the trace of a in $F(p^n)$ with respect to $F(p)$, and $g^{\operatorname{ind}(a)}$ is defined by

$$g^{\operatorname{ind}(a)} = a$$

---

* Reprinted from *Proceedings of the National Academy of Sciences*, 1949, **35**(8): 477–481. By Loo-Keng Hua and Vandiver H S.

① These Proceedings, 1949, **35**: 94–99, this will be referred to as C; 1949, **35**: 451–457, this will be referred to as H.

in $F(p^n)$, g being a primitive root of $F(p^n)$ and we define ind (0) as zero. Hence, as is known, provided $\chi_{k_1}^{\mu_1}\chi_{k_2}^{\mu_2} \neq 1$,

$$\frac{\tau(\chi_{k_1}^{\mu_1})\tau(\chi_{k_2}^{\mu_2})}{\tau(\chi_{k_1}^{\mu_1}\chi_{k_2}^{\mu_2})} = \sum_a \alpha^{\mu_1 s_2 \operatorname{ind}(a) + \mu_2 s_1 \operatorname{ind}(1-a)}.$$

Also, setting $(-a_1)$ for (a) we obtain

$$\sum_{a \in K} \alpha^{\mu_1 s_2 \operatorname{ind}(a) + \mu_2 s_1 \operatorname{ind}(1-a)}$$
$$= \sum_{a_1 \in K} \alpha^{\mu_1 s_2 \operatorname{ind}(-a_1) + \mu_2 s_1 \operatorname{ind}(1+a_1)}$$
$$= \sum_{a_1 \in K} \alpha^{\mu_1 s_2 \operatorname{ind}(-1)} \cdot \alpha^{\mu_1 s_2 \operatorname{ind}(a_1) + \mu_2 s_1 \operatorname{ind}(1+a_1)}.$$

Let $n = 2n_1$, then $p^{2n_1} - 1 \equiv 0 \pmod{4}$ and if $(\mu_1 s_2(p^{2n_1}-1))/2 = e$ so that $\alpha^{\mu_1 s_2 \operatorname{ind}(-1)} = (-1)^e = 1$,

$$\frac{\tau(\chi_{k_1}^{\mu_1})\tau(\chi_{k_2}^{\mu_2})}{\tau(\chi_{k_1}^{\mu_1}\chi_{k_2}^{\mu_2})} = \sum_{a \in K} \alpha^{\mu_1 s_2 \operatorname{ind}(a) + \mu_2 s_1 \operatorname{ind}(a+1)} = \psi(\alpha). \tag{2}$$

Now Mitchell[1] showed that if $\mu_1 s_2 \not\equiv 0 \pmod{m}$, $\mu_2 s_1 \not\equiv 0 \pmod{m}$ and $\mu s_2 + \mu' s_1 \not\equiv 0 \pmod{m}$ then

$$\psi(\alpha) = (-1)^{r-1} p^{n_1}, \tag{3}$$

where p belongs to the exponent $2t$, modulo m and $n_1 = rt$ with $p^t \equiv -1 \pmod{m}$. We now employ (18) of H, which may be written as follows:

$$N = p^n - 2 - \sum_{\mu_1=0}^{k_1-1} \chi_{k_1}^{\mu_1}\left(-\frac{c_3}{c_1}\right) - \sum_{\mu_2=0}^{k_2-1} \chi_{k_2}^{\mu_2}\left(-\frac{c_3}{c_2}\right) - \sum_{v=1}^{d-1} \chi_{k_1}^{s_1 v}\left(-\frac{c_2}{c_1}\right)$$
$$+ \sum_{\mu_1,\mu_2} \chi_{k_1}^{\mu_1}\left(-\frac{c_3}{c_1}\right)\chi_{k_2}^{\mu_2}\left(-\frac{c_3}{c_2}\right)\frac{\tau(\chi_{k_1}^{\mu_1})\tau(\chi_{k_2}^{\mu_2})}{\tau(\chi_{k_1}^{\mu_1}\chi_{k_2}^{\mu_2})}, \tag{4}$$

where in the last term $\chi_{k_1}^{\mu_1} \neq 1$, $\chi_{k_2}^{\mu_2} \neq 1$, $\chi_1^{\mu_1}\chi_2^{\mu_2} \neq 1$, and divide the discussion into seven cases in order to express s in terms of integers only. Because of (3) and $\chi(\alpha) = -1$ for $\mu_1 = 0, \mu_2 \neq 0$, also for $\mu_1 \neq 0, \mu_2 = 0$, this is always possible.

The cases are:

I. $c_3 = c_2 = c_1$. II. $c_3 = c_1, c_2 \neq c_1$ with ind $c_2 \equiv$ ind $c_1 \pmod{d}$. III. $c_3 = c_1$, ind $c_2 \not\equiv$ ind $c_1 \pmod{d}$. IV. $c_3 = c_2, c_1 \neq c_2$, ind $c_2 \equiv$ ind $c_1 \pmod{d}$. V. $c_3 = c_2$,

---

[1] Ann. Math, II, 1917, **18**: 120.

$c_2 \not\equiv c_1 \pmod{d}$. VI. $c_3 \neq c_2$, $c_3 \neq c_1$, ind $c_2 \equiv$ ind $c_1 \pmod{d}$. VII. $c_3 \neq c_1, c_3 \neq c_2$, $c_1 \not\equiv c_2 \pmod{d}$. Write

$$A = \sum_{\mu_1,\mu_2} \chi_{k_1}^{\mu_1}\left(-\frac{c_3}{c_1}\right) \chi_{k_2}^{\mu_2}\left(-\frac{c_3}{c_2}\right) \frac{\tau(\chi_{k_1}^{\mu_1})\tau(\chi_{k_2}^{\mu_2})}{\tau(\chi_{k_1}^{\mu_1}\chi_{k_2}^{\mu_2})}, \tag{5}$$

where $\chi_{k_1}^{\mu_1} \neq 1, \chi_{k_2}^{\mu_2} \neq 1, \chi_{k_1}^{\mu_1}\chi_{k_2}^{\mu_2} \neq 1$.

In case I, we find

$$\sum_{\mu_1=0}^{k_1-1} \chi_{k_1}^{\mu_1}\left(-\frac{c_3}{c_1}\right) = k_1,$$

$$\sum_{\mu_2=0}^{k_2-1} \chi_{k_2}^{\mu_2}\left(-\frac{c_3}{c_2}\right) = k_2,$$

$$\sum_{v=1}^{d-1} \chi_{k_1}^{s_1 v}\left(-\frac{c_2}{c_1}\right) = d-1.$$

To reduce A in this case, we take all the terms in which $\chi_{k_1}^{\mu_1} \neq 1, \chi_{k_2}^{\mu_2} \neq 1$ and then subtract all the terms from the result in which $\chi_{k_1}^{\mu_1}\chi_{k_2}^{\mu_2} = 1$ with $\chi_{k_1}^{\mu_1} \neq 1, \chi_{k_2}^{\mu_2} \neq 1$ and employ (3). The terms of the first type give

$$(k_1-1)(k_2-1)p^{n_1}(-1)^{r-1}.$$

As in the proof of (18) of s there are $(d-1)$ terms of the second type just mentioned. Hence (4) reduces to

$$p^n + 1 - k_1 - k_2 - d + ((k_1-1)(k_2-1) - (d-1))(-1)^{r-1}p^{n_1}. \tag{6}$$

For case II, we find the same value for the first summation in (4) but for the second, we have the value 0 since $c_3 \neq c_2$. In reducing A for this case and separating the summation in the same way it was separated in case I, we have for the first part $(k_1-1)p^{n_1}(-1)^r$ and for the second part $(d-1)p^{n_1}(-1)^r$. Hence, we have for case II the value, if N_i denotes the number of solutions in case i, $i = $ I, II, $\cdots$, VII,

$$N_2 = p^n + 1 - d - k_1 + (k_1 + d - 2)p^{n_1}(-1)^r. \tag{7}$$

Similarly for cases III, IV, etc., we have

$$N_3 = p^n + 1 - k_1 + (k_1 - 2)p^{n_1}(-1)^r, \tag{8}$$

$$N_4 = p^n + 1 - d - k_2 + (k_2 + d - 2)p^{n_1}(-1)^r, \tag{9}$$

$$N_5 = p^n + 1 - k_2 + (k_2 - 2)p^{n_1}(-1)^r, \tag{10}$$

$$N_6 = p^n + 1 - d + p^{n_1}(-1)^r(d-2), \tag{11}$$

$$N_7 = p^n + 1 + (-1)^{r-1} 2p^{n_1}. \tag{12}$$

Theorem I *If N_i is the number of solutions of the equation*

$$c_1 x_1^{a_1} + c_2 x_2^{a_2} + c_3 = 0$$

in x_1 and x_2, non-zero elements of a finite field $F(p^n)$ of order p^n, p an odd prime, with c_1, c_2, c_3 given non-zero elements of $F(p^n)$ with N_i, $i = 1, 2, \cdots, 7$, the number of solutions in the corresponding seven cases numbered I, II, $\cdots$, VII, given just above the relation (5); $0 < a_i < p^n - 1$, $(a_i, p^n - 1) = k_i$, $i = 1, 2$; $m = ds_1s_2$, $ds_1 = k_1$, $ds_2 = k_2$, $(k_1, k_2) = d$; $n = 2n_1$, p belongs to the exponent $2t$ modulo m, $n_1 = rt$ with $p^t \equiv -1 \pmod{m}$, then the values of N_i are given by the relations (6)–(12), inclusive.

Mithcell obtained these results for the special case when $k_1 = k_2$ by a more complicated method.

We now use (4) to find closer limits for N in the special case when $s = 2$, than those found in C for equation (1) of that paper. Set

$$M = p^n \sum_{\mu_1=0}^{k_1-1} \chi_{k_1}^{\mu_1}\left(-\frac{c_3}{c_1}\right) - \sum_{\mu_2=0}^{k_2-1} \chi_{k_2}^{\mu_2}\left(-\frac{c_3}{c_2}\right) - \sum_{v=1}^{d-1} \chi_{k_1}^{vs_1}\left(-\frac{c_2}{c_1}\right).$$

Now every term on the right of (4) is real so that we may write $N \leqslant M + |A|$ and $N \geqslant M - |A|$ and using the known result

$$|\tau(\chi)| = p^{n/2}, \quad \chi \neq \chi_0, \tag{13}$$

we have, if

$$D = |((k_1-1)(k_2-1) - (d-1))p^{n/2}|, \tag{13a}$$

$$N \leqslant M + D, \tag{14}$$

$$N \geqslant M - D. \tag{15}$$

To obtain limits in terms of rational integers only, divide the discussion into the seven cases I—VII defined in the proof of Theorem I. Then we can evaluate M exactly as in the determination of N in Theorem I in each of the seven cases[①].

[①] As a check on the accuracy of our work, the present problem was discussed in a mathematical seminar, using a different method of proof. During the resulting discussion, Mr. Olin B. Faircloth made a suggestion, which, when used in connection with the method employed in this paper yielded a better superior limit for N, than that which we originally had.

Write N_i, $i = 1, 2, \cdots, 7$ for the value of N corresponding to each of the seven cases referred to in Theorem I, respectively. Then from (14) and (15) we have the

Theorem II If
$$c_1 x_1^{a_1} + c_2 x_2^{a_2} + c_3 = 0,$$
where c_1, c_2, c_3 are given elements in a finite field $F(p^n)$ of order p^n, p prime, $c_1 c_2 c_3 \neq 0$, then we have the following limits for N_i, the number of solutions in the x's, neither zero, of the above equation,
$$M_i - D \leqslant N_i \leqslant M_i + D, \tag{16}$$
where $M_1 = p^n + 1 - k_1 - k_2 - d$; $M_2 = p^n + 1 - d - k_1$; $M_3 = p^n + 1 - k_1$; $M_4 = p^n + 1 - d - k_2$; $M_5 = p^n + 1 - k_2$; $M_6 = p^n + 1 - d$; $M_7 = p^n + 1$; $k_1 = (p^n - 1, a_1)$, $k_2 = (p^n - 1, a_2)$, $d = (k_1, k_2)$ and D is defined in (13a), M_i; $i = 1, 2, \cdots, 7$, is the value of M corresponding to each of the seven cases, I, II, $\cdots$, VII, of Theorem I, respectively.

For the case when $n = 1$ and $k_1 = k_2$ is an odd prime, Hurwitz[①] by a different method proved the above theorem using rational integers only. His argument is much longer than ours. For the particular cases treated in Theorem I the *exact value for N agrees with the limits found in Theorem* II, *when* $c_1 = c_2 = c_3$, *with the superior limit when r is odd and the inferior limit when r is even*. For the case $n = 1, k_1 = k_2$ an odd prime, this fact was noted by Mitchell, who also stated that Hurwitz's method could be extended so as to obtain limits for the case when n is general, which had a similar property.

---

① *Crelle*, 1909, **136**: 272–292.

On the nature of the solutions of certain equations in a finite field*

In two recent papers[①] the authors considered the equation

$$c_1 x_1^{a_1} + c_2 x_2^{a_2} + \cdots + c_s x_s^{a} + c_{s+1} = 0 \qquad (1)$$

in the x's, where the a's are integers such that $0 < a < p^n - 1$; $s \geqslant 2$ for $c_{s+1} \neq 0$, and $s > 2$ for $c_{s+1} = 0$ the c's being given elements of a finite field of order p^n, p an odd prime, which will be designated by $F(p^n)$; and

$$c_1 c_2 \cdots c_s x_1 x_2 \cdots x_s \neq 0 \qquad (2)$$

in $F(p^n)$. Limits were found for the number of solutions of (1) and it was proved that for p^n sufficiently large (1) always had solutions under the restriction (2). In the statement above concerning (1) we made the provision that $s \geqslant 2$ for $c_{s+1} \neq 0$. If we consider the case where $s = 1$, (1) reduces to the equation

$$c_1 x_1^{a_1} + c_2 = 0, \qquad (3)$$

with $c_1 c_2 \neq 0$. Since any finite field can be represented by means of the residue classes with respect to some prime ideal in an algebraic field, then examination of the solutions of (3) is included in the theory of the congruence

$$x^n \equiv \alpha \pmod{\mathfrak{p}}, \qquad (4)$$

where α is an integer in an algebraic field K, $\mathfrak{p}$ is a prime ideal in that field and $\alpha \not\equiv 0 \pmod{\mathfrak{p}}$. The study of the last relation led to the theory of the class field and the laws of reciprocity for nth powers[②].

We wish to point out the sharp distinction between this problem and the problem of finding the solutions or the number of solutions of the equation (1) and when $s \geqslant 2$ for $c_{s+1} \neq 0$.

---

* Reprinted from *Proceedings of the National Academy of Sciences*, 1949, **35** (8): 481-487 (By Hua L K and Vandiver H S.)

① These Proceedings, 1948, **34**: 258-263, this will be referred to as E, 1949, **35**: 94-99, this will be referred to as C.

② Hasse H. *Ber, Deut Math Verein*, 1926, **35**: 1-55; 1927, **36**: 233-311; *Erganzungsband*, 1930 **VI**.

In addition to this, we shall also obtain an expression (Theorem I) for the number of solutions of (1) when $c_{s+1} \neq 0$ and $s = 2$, which leads to the proof of Theorem I and which we shall also have occasion to make use of in another paper. Also we obtain Theorem II, which gives the exact number of solutions of (1) when the k's are prime each to each, $c_{s+1} = 0$ and $(a_i, p^n - 1) = k_i$, $i=1,2,\cdots,s$.

We consider equation (4) and confine ourselves to the case where K is an algebraic field which contains a primitive lth root of unity with l prime. Then

$$x^l - \alpha \equiv 0 \pmod{\mathfrak{p}} \tag{5}$$

if and only if the lth power character of α, modulo $\mathfrak{p}$ is unity, or

$$\left(\frac{\alpha}{\mathfrak{p}}\right)_l = 1.$$

However, by a known theorems[1]

$$\left(\frac{\alpha}{\mathfrak{p}}\right)_l \neq 1$$

for an infinity of distinct prime ideals $\mathfrak{p}$ provided α does not equal an lth power of an integer in K. Now the last statement is contrary to what we have for $s > 1$ in (1) with $c_{s+1} \neq 0$. For, as we have shown in E (p.262) the congruence

$$\alpha_1 x_1^{a_1} + \alpha_2 x_2^{a_2} + \cdots + \alpha_s x_s^{a_s} + \alpha_{s+1} \equiv 0 \pmod{\mathfrak{p}}, \tag{6}$$

with

$$\alpha_1 \alpha_2 \cdots \alpha_s x_1 x_2 \cdots x_s \not\equiv 0 \pmod{\mathfrak{p}}$$

for $s \geqslant 2$ with $c_{s+1} \neq 0 \pmod{\mathfrak{p}}$ always has solutions if the norm of $\mathfrak{p}$ exceeds a certain limit, in fact (6) always has at least k solutions in the x's for k any positive integer if the norm of p exceeds a certain limit. In particular when we consider the case where there are no solutions it is known[2] that $x^7 + y^7 + 1 \equiv 0 \pmod{p}$ has no solutions in integers prime to p for just four primes p, namely 29, 71, 113, 491, but has solutions for all the other primes p. As for $x^5 + y^5 + 1 \equiv 0 \pmod{p}$ it has no solutions for $p = 11, 41, 71, 101$. Also $x^3 + y^3 + 1 \equiv 0 \pmod{p}$ has no solutions only for $p = 7$ and 13. It is also knoun[3] that for $m < 10l$, and $p = 1 + ml$ and $h(l)$ is the number of primes p for which

$$x^l + y^l + 1 \equiv 0 \pmod{p} \tag{7}$$

---

[1] Takagi T. *J, Coll Sci, Imp. Univ. Tokyo*, 1920, **42**: 16.
[2] Dickson L E. *Messenger Math.*, 1908, **38**: 14-33.
[3] These results are due to Beeger N G W H, unpublished.

has no solutions for $xy \not\equiv 0 \pmod{p}$ then $h(11) = 3$, $h(13)=4$, $h(17)=4$, $h(19)=6$. In view of these facts it may happen that $h(l)$ increases as l increases. Note also from the cases 3, 5 and 7 that there is a range of values for p in which the corresponding congruences may or may not have solutions.

From the above considerations we may say that associated with any form of the type given by the left-hand member of (4) there is an integer $e(k)$ such that there are exactly $e(k)$ values of $\mathfrak{p}$ such that (4) has k solutions. Even in the case where k is zero, the class of forms satisfying the condition may be rather general. For example,[①] it was shown that if c is a prime and m is an integer such that $p = 1 + mc$ with p prime then
$$a_1 x_1^m + a_2 x_2^m + \cdots + a_s x_s^m \equiv 0 \pmod{p}$$
has only the solutions $x_1 \equiv x_2 \equiv \cdots \equiv x_s \equiv 0 \pmod{p}$ provided $s \leqslant c - 2$ the sum of no n of the a's is $\equiv 0 \pmod{p}$, $0 < n \leqslant s$, and
$$(|a_1| + |a_2| + \cdots + |a_s|)^{\phi(c)} < p,$$
where $\phi(c)$ is the indicator of c. Select $a_1 = a_2 = \cdots = a_{s-1} = 1$ and $a_s = -s$, and also an m and p with $p = 1 + mc$ with c fixed and satisfying the other conditions just mentioned then it is evident that
$$x_1^m + x_2^m + \cdots + x_{s-1}^m - s x_s^m \equiv 0 \pmod{p}$$
has no primitive solutions (incidentally, in view of Theorem III of the article just referred to, the equation
$$x_1^m + x_2^m + \cdots + x_{s-1}^m - s x_s^m = 0$$
has no solutions in rational integers unless each x is zero, if there is a prime p with $p = 1 + mc$ and $(2s-1)^{\varphi(c)} < p$).

We now recur to the subject of the number of solutions of equation (1), treated in E and C. The case $s = 2$ with $c_3 \neq 0$, has received considerable attention in the literature.[②] Let us examine
$$c_1 x_1^{a_1} + c_2 x_2^{a_2} + c_3 = 0, \tag{8}$$
with
$$x_1 x_2 c_1 c_2 c_3 \neq 0 \tag{8a}$$

---

① Vandiver, H, S. these Proceedings, 1946, **32**: 101-106.
② Cf. footnote 1 of E.

Theorem I of C, we see will give, if N is the number of solutions of (8) in x_1 and x_2 under the conditions (8a),

$$N = \frac{(p^n - 1)^2}{p^n} + \frac{T}{p^n}, \tag{9}$$

where we may write T in the form, if $k_i = (a_i, p-1)$, μ_1 ranges over $0, 1, \cdots, k_1 - 1$; μ_2 over $0, 1, \cdots, k_2 - 1$,

$$\sum_{\substack{a \in K \\ a \neq 0}} \sum_{\mu_1, \mu_2} \tau(\chi_{k_1}^{\mu_1}) \tau(\chi_{k_2}^{\mu_2}) \chi_{k_1}^{\mu_1}(c_1^{-1}) \chi_{k_2}^{\mu_2}(c_2^{-1}) \chi_{k_1}^{\mu_1}(a^{-1}) \chi_{k_2}^{\mu_2}(a^{-1}) \zeta^{\mathrm{tr}(ac_3)}, \tag{9a}$$

where χ_{k_1} is a special character defined by k_1 or as we shall say, a k_1-character, and similarly χ_{k_2} is a special character defined by k_2 and where, if we write K for $F(p^n)$,

$$\tau(\chi, \zeta) = \tau(\chi) = \sum_{b \in K} \chi(b) \zeta^{\mathrm{tr}(b)}; \quad \zeta = e^{2i\pi/p}, \tag{9b}$$

with $\chi(0) = 0$. This is obtained from the fact that if α is a primitive $(p^n - 1)$th root of unity then $\alpha^{\mu \ \mathrm{ind} \ a}$, where a ranges over all elements of $F(p^n)$, $\neq 0$, gives a character as defined in C where ind a is given by

$$g^{\mathrm{ind} \ a} = a,$$

and where g is a primitive root in $F(p^n)$, so we obtain all characters by taking $\mu = 0, 1, \cdots, p^n - 2$. We have from C the relations, is χ_0 is the principal character,

$$\left. \begin{array}{l} |\tau(\chi)| = p^{n/2}; \ \chi \neq \chi_0, \\ \tau(\chi_0) = -1. \end{array} \right\} \tag{10}$$

To reduce (9a) we divide the discussion into four cases.

First Case. $\chi_{k_1}^{\mu_1} = 1, \chi_{k_2}^{\mu_2} = 1$. In this event $\chi_{k_1}^{\mu_1}(a^{-1}) = \chi_{k_2}^{\mu_2}(a^{-1}) = 1$, $\chi_{k_1}^{\mu_1}(c_1^{-1}) = \chi_{k_2}^{\mu_2}(c_2^{-1}) = 1$, which reduces (9a), using (10), to

$$\sum_a \zeta^{\mathrm{tr}(ac_3)} = -1. \tag{11}$$

Second Case. $\chi_{k_1}^{\mu_1} = 1; \chi_{k_2}^{\mu_2} \neq 1$. This gives from (9a) and (10),

$$-\sum_a \tau(\chi_{k_2}^{\mu_2}) \chi_{k_2}^{\mu_2}(c_2^{-1} a^{-1}) \zeta^{\mathrm{tr}(ac_3)} = \sum_a -\tau(\chi_{k_2}) \chi_{k_2}(c_2^{-1}) (\chi_{k_2}(a))^{-1} \zeta^{\mathrm{tr}(ac_3)},$$

and setting $ac_3 = -a_1$, using $c_3 \neq 0$, this reduces to

$$-\sum_{a_1} \tau(\chi_{k_2}^{\mu_2}) \chi_{k_2}^{\mu_2}(-c_2^{-1} c_3)(\chi_{k_2}^{\mu_2}(a_1))^{-1} \zeta^{-\mathrm{tr} a_1} = -\sum \tau(\chi_{k_2}) \overline{\tau(\chi_{k_2})} \chi_{k_2}\left(-\frac{c_3}{c_2}\right),$$

and using the known relation

$$\tau(\chi_{k_2}^{\mu_2})\overline{\tau(\chi_{k_2}^{\mu_2})} = p^n, \tag{11a}$$

this is seen to equal

$$-p^n \sum_{\mu_2} \chi_{k_2}^{\mu_2}\left(-\frac{c_3}{c_2}\right). \tag{12}$$

Also, if $\chi_{k_1}^{\mu_1} \neq 1$ and $\chi_{k_2}^{\mu_2} = 1$ we find in a similar way that the result is

$$-p^n \sum_{\mu_1} \chi_{k_1}^{\mu_1}\left(-\frac{c_3}{c_1}\right). \tag{13}$$

Third Case. $\chi_{k_1}^{\mu_1}\chi_{k_2}^{\mu_2} = 1$ with $\chi_{k_1}^{\mu_1} \neq 1, \chi_{k_2}^{\mu_2} \neq 1$, (9a) reduces to

$$\sum_a \sum_{\mu_1} \chi_{k_1}^{\mu_1}(c_1^{-1})\chi_{k_1}^{-\mu_1}(c_2^{-1})\tau(\chi_{k_1}^{\mu_1})\tau(\chi_{k_1}^{-\mu_1},\zeta)\zeta^{\operatorname{tr}(ac_3)}.$$

Now

$$\tau(\chi_{k_1}^{-\mu_1},\zeta) = \tau(\chi_{k_1}^{-\mu_1},\zeta^{-1})\chi_{k_1}(-1) = \overline{\tau(\chi_{k_1}^{\mu_1})}\chi_{k_1}^{\mu_1}(-1).$$

Hence for the admissible values of μ_1, (9a) reduces to

$$-\sum_{\mu_1} \chi_{k_1}^{\mu_1}\left(-\frac{c_2}{c_1}\right)p^n. \tag{14}$$

Now we must determine the possible values of μ_1 from $\chi_{k_1}^{\mu_1}\chi_{k_2}^{\mu_2} = 1$. Let α be a primitive (ds_1s_2)th root of unity where $ds_1 = k_1, ds_2 = k_2$ and where $(k_1, k_2) = d$, whence $(s_1, s_2) = 1$. Then $\chi_{k_1}^{\mu_1}\chi_{k_2}^{\mu_2} = 1$ gives integers μ_1 and μ_2 such that $\alpha^{s_1\mu_1 + s_2\mu_2} = 1$ or $s_2\mu_1 + s_1\mu_2 \equiv 0 \pmod{(ds_1s_2)}$ so that $\mu_1 = v_1s_1, \mu_2 = v_2s_2$. Hence the possible values of μ_1 are $s_1v_1; v_1 = 1, 2, \cdots, d-1$, as the case $v_1 = 0$ was excepted. Hence (9a) reduces to, for this case,

$$-\sum_{v_1=1}^{d-1} \chi_{k_1}^{s_1v_1}\left(-\frac{c_2}{c_1}\right)p^n. \tag{15}$$

Fourth Case. Here $\chi_{k_1}^{\mu_1} \neq 1$, $\chi_{k_2}^{\mu_2} \neq 1$, $\chi_{k_1}^{\mu_1}\chi_{k_2}^{\mu_2} \neq 1$. Consider such terms in (9a). They may be written in the form

$$\sum_{\mu_1,\mu_2}\sum_{a\in K} \tau(\chi_{k_1}^{\mu_1})\tau(\chi_{k_2}^{\mu_2})\chi_{k_1}^{\mu_1}\chi_{k_2}^{\mu_2}(a^{-1})\zeta^{\operatorname{tr}(ac_3)}\chi_{k_1}^{\mu_1}(c_1^{-1})\chi_{k_2}^{\mu_2}(c_2^{-1}). \tag{16}$$

Write $a_1 = ac_3$, then $a^{-1} = a_1^{-1}c_3$ and using $c_3 \neq 0$,

$$\sum_{\substack{a_1 \in K \\ a_1 \neq 0}} \chi_{k_1}^{\mu_1}\chi_{k_2}^{\mu_2}(a^{-1}) = \sum_{a_1 \in K} \chi_{k_1}^{\mu_1}\chi_{k_2}^{\mu_2}(c_3)(\chi_{k_1}^{\mu_1}\chi_{k_2}^{\mu_2}(a))^{-1}\zeta^{\text{tr}(a_1)}$$

$$= \sum_{a_1 \in K} \chi_{k_1}^{\mu_1}\chi_{k_2}^{\mu_2}(c_3)(\chi_{k_1}^{\mu_1}\chi_{k_2}^{\mu_2}(a_1))^{-1}\chi_{k_1}^{\mu_1}\chi_{k_2}^{\mu_2}(-1)\zeta^{-\text{tr}(a_1)}.$$

Hence using (11a), we find from (9a) that the terms in case 4 in T may be written as

$$p^n \sum_{\mu_1,\mu_2} \chi_{k_1}^{\mu_1}\left(-\frac{c_3}{c_1}\right)\chi_{k_2}^{\mu_2}\left(-\frac{c_3}{c_2}\right)\frac{\tau(\chi_{k_1}^{\mu_1})(\chi_{k_2}^{\mu_2})}{\tau(\chi_{k_1}^{\mu_1}\chi_{k_2}^{\mu_2})}, \tag{17}$$

$\chi_{k_1}^{\mu_1} \neq 1$, $\chi_{k_2}^{\mu_2} \neq 1$, $\chi_{k_1}^{\mu_1}\chi_{k_2}^{\mu_2} \neq 1$. Using (9), (9a), (11), (12), (13), (14), (15) and (17), we find

$$N = p^n - \sum_{\mu_1=0}^{k_1-1} \chi_{k_1}^{\mu_1}\left(-\frac{c_3}{c_1}\right) - \sum_{\mu_2=0}^{k_2-1}\chi_{k_2}^{\mu_2}\left(-\frac{c_3}{c_2}\right) - \sum_{v=1}^{d-1}\chi_{k_1}^{s_1 v}\left(-\frac{c_2}{c_1}\right)$$

$$+ \sum_{\mu_1,\mu_2}\chi_{k_1}^{\mu_1}\left(-\frac{c_3}{c_1}\right)\chi_{k_2}^{\mu_2}\left(-\frac{c_3}{c_2}\right)\frac{\tau(\chi_{k_1}^{\mu_1})\tau(\chi_{k_2}^{\mu_2})}{\tau(\chi_{k_1}^{\mu_1}\chi_{k_2}^{\mu_2})}, \tag{18}$$

where, in the last term, $\chi_{k_1}^{\mu_1} \neq 1$, $\chi_{k_2}^{\mu_2} \neq 1$, $\chi_{k_1}^{\mu_1}\chi_{k_2}^{\mu_2} \neq 1$.

Whence we have the (cf. Davenport-Hasse[①]).

Theorem I *If N is the number of solutions in x_1 and x_2 of*

$$c_1 x_1^{a_1} + c_2 x_2^{a_2} + c_3 = 0,$$

with c_1, c_2, c_3, x_1, x_2 in a finite field $F(p^n)$ of order p^n with p an odd prime, $c_1 c_2 c_3 x_1 x_2 \neq 0$ in $F(p^n)$; $(a_1, p^n - 1) = k_1$, $0 < k_1 < p^n - 1$, $(a_2, p^n - 1,) = k_2$, $0 < k_2 < p^n - 1$, x_{k_1} is a special k_1-character, χ_{k_2} is a special k_2-character, $d = (k_1, k_2)$, $ds_1 = k_1$, $\tau(\chi)$ is defined as in (9b), then the relation (18) holds.

[①] The proof of this theorem we gave as a direct application of our general theorem I of C to the special case $s = 2$. Another proof may be obtained by using the result of Davenport and Hasse, *Jour für Math* (Crelle) **172**, p.174 relation (6.5) which gives the number of solutions of $ax^m + by^n = c$, $abc \neq 0$ in $F(p^n)$, where x and y are nor restricted, as they were in our work, to non-zero values in $F(p^n)$. To do this we consider the possible solutions involving zero and subtract the number of them from the right-hand member of the relation (6.5) just referred to. We note that the number of solutions of $ax^m = c$ and $by^n = c$ are, respectively,

$$\sum_{\mu=0}^{m-1}\chi^n\left(\frac{c}{a}\right); \quad \sum_{\nu=0}^{n-1}\psi\left(\frac{c}{b}\right),$$

which when taken in connection with (6.5) yields a formula equivalent to our relation (18).

We shall now obtain N explicitly in terms of p, n and s, for special values of k_1 when $c_{s+1} = 0$. By the remark immediately following (9b), we may write

$$\chi_{k_i}^{\mu_i}(a^{-1}) = \alpha_i^{\mu_i \text{ind}(a^{-1})} \tag{19}$$

for any i in the set $0, 1, \cdots, k_i - 1$, and where α_i is a primitive k_ith root of unity. Set in Theorem I of C, $c_{s+1} = 0$, then for this case we have

$$T = \sum_{\mu_i=0}^{k_i-1} \sum_{\substack{a \in K \\ a \neq 0}} (\alpha_1^{\mu_1} \alpha_2^{\mu_2} \cdots \alpha_s^{\mu_s})^{\text{ind}(a^{-1})} \cdot \prod_{i=1}^{s} \alpha_i^{\mu_i \text{ind}(a^{-1})} \tau(\alpha_i^{\mu_i}). \tag{20}$$

Assume that the k's are prime each to each and carry out the summation with respect to a we obtain

Theorem II *If N_s is the number of solutions of (1) in x's under the conditions (2) and the other conditions stated in connection with (1), and also $k_1, k_2, \cdots, k_s$ are prime each to each with $c_{s+1} = 0$, then*

$$N_s = \frac{p^n - 1}{p^n}((p^n - 1)^{s-1} + (-1)^s). \tag{21}$$

This result may be proved in a simpler way as we shall now show.

Consider the equation

$$c_1 x_1^{k_1} + \cdots + c_s x_s^{k_s} = 0, \quad (k_i, k_j) = 1, \quad i \neq j \tag{22}$$

and $A = k_1 k_2 \cdots k_s$ dividing $p^n - 1$. Since $(k_1, k_2, \cdots, k_s) = 1$, we have integers λ and μ such that

$$\lambda k_1 + \mu k_2 \cdots k_s = 1, \quad (\lambda, p^n - 1) = 1.$$

Putting

$$x_1 = y_1^\lambda, \quad x_2 = y_1^{-\mu A/k_1 k_2} y_2, \quad x_3 = y_1^{-\mu A/k_1 k_3} y_3, \cdots.$$

Then we have

$$c_1 y_1^{\lambda k_1} + y_1^{-\mu A/k_1}(c_2 y_2^{k_2} + \cdots + c_s y_s^{k_s}) = 0,$$

i.e.,

$$c_1 y_1 + c_2 y_2^{k_2} + \cdots + c_s y_s^{k_s} = 0.$$

Since, for given elements of $F(p^n)$, $y_2, \cdots, y_s$ such that

$$c_2 y_2^{k_2} + \cdots + c_s y_s^{k_s} \neq 0,$$

we have a unique y_1, therefore, we have

$$N_s = (p^n - 1)^{s-1} - N_{s-1}.$$

Consequently

$$\begin{aligned} N_s &= (p^n - 1)^{s-1} - (p^n - 1)^{s-2} + \cdots + (-1)^s(p^n - 1) \\ &= \frac{p^n - 1}{p^n}((p^n - 1)^{s-1} + (-1)^s). \end{aligned}$$

We also have, since the residue classes with respect to a rational prime modulus p form a finite field of order p, the

Corollary I *If N is the number of incongruent solutions in the y's of the congruence, with p an odd prime,*

$$d_1 y_1^{a_1} + d_2 y_2^{a_2} + \cdots + d_s y_s^{a_s} \equiv 0 \; (\mathrm{mod}\; p),$$

where $d_1 d_2 \cdots d_s \not\equiv 0 \;(\mathrm{mod}\; p)$, $y_1 y_2 \cdots y_s \not\equiv 0 \;(\mathrm{mod}\; p)$; $k_1, k_2, \cdots, k_s$ prime each to each with $(a_i, p - 1) = k_i$; $0 < k_i < p - 1$, $i = 1, 2, \cdots, s$, then

$$N = \frac{p-1}{p}((p-1)^{s-1} + (-1)^s).$$

It may also be noted that the procedure used in the second proof of Theorem II gives us a method for determining the solutions of (22).

Some properties of a sfield*

All the results of this paper are initiated from the almost trivial identity that if $ab \neq ba$, then

$$a = (b^{-1} - (a-1)^{-1}b^{-1}(a-1))(a^{-1}b^{-1}a - (a-1)^{-1}b^{-1}(a-1))^{-1}. \qquad (1)$$

Among these results, there are two interesting ones which are the perfect form of two theorems due to H. Cartan[①] and J. Dieudonné[②] respectively.

Theorem 1 *Every sfield is generated by a non-central element and its conjugates.*

Let K be the sfield and L be the conjugate set. The identity (1) asserts that if there is an element b of L such that $ab \neq ba$, then a belongs to the sfield K_1 generated by L. That is, $K - K_1$ and K_1 are commutative elementwise.

Suppose our theorem is false; we have an element a of $K - K_1$ and two elements b and b' of K_1 such that $bb' \neq b'b$. Since a and ab belong to $K - K_1$, we have $(ab)b' = b'(ab) = ab'b$, which contradicts $bb' \neq b'b$. The theorem follows.

An immediate consequence of Theorem 1 is the following interesting result:

Theorem 2 *Every proper normal subsfield of a sfield is contained in the center.*

H. Cartan proved the theorem under the assumption that the rank of the sfield over its center is finite. His proof is far more complicated than the present one.

More precise results can be obtained by specializing the subgroup.

Theorem 3 *Every sfield, which is not a field, has no proper subsfield containing all its commutators.*

It is enough to prove the existence of a commutator which does not belong to the center. Suppose $ab \neq ba$. From a modification of the identity (1):

$$a = (1 - (a-1)^{-1}b^{-1}(a-1)b)(a^{-1}b^{-1}ab - (a-1)^{-1}b^{-1}(a-1)b)^{-1}, \qquad (2)$$

we deduce that at least one of $a^{-1}b^{-1}ab$ and $(a-1)^{-1}b^{-1}(a-1)b$ does belong to the center.

As corollaries of Theorem 3, we have

---

* Reprinted from *Proceedings of the National Academy of Sciences*, 1949, **35** (9): 533-537.
① Cartan H. *Ann. école normale superieure*, 1947, **64**: 59–77, Theorem 4.
② Dieudonné. J, *Bull. Soc. math. France*, 1943, **71**: 27 – 45.

Theorem 4 *If the center of a sfield contains all the commutators of the sfield, then the sfield is commutative. An element of a sfield commutative with all the commutators of the sfield belongs to the center.*

Moreover, as a simple consequence of

$$a^{-1}c^{-1}ac = a^{-2}(ac^{-1})^2 c^2 \qquad (3)$$

and Theorem 3, we have

Theorem 5 *Every sfield, which is not a field, has no proper subsfield containing all its square elements.*

Theorem 6 *If the center of a sfield contains all the square elements of the sfield, then the sfield is commutative. An element of a sfield commutative with all the square elements of the sfield belongs to the center.*

By means of an identity in the theory of finite differences:

$$\Delta^{k-1} x^k = k! \left(x + \frac{1}{2}(k-1) \right),$$

where $\Delta f(x) = f(x+1) - f(x)$ and $\Delta^i f(x) = \Delta(\Delta^{i-1} f(x))$, we have

Theorem 7 *Every sfield of characteristic $\geqslant k$, which is not commutative, has no proper subsfield containing all its k-th power elements.*

Let us denote $a^{-1}b^{-1}ab$ by (a,b). An r-commutator ($r \geqslant 2$) is defined inductively by the relation

$$(a_1, \cdots, a_r) = (a_1, (a_2, \cdots, a_r)). \qquad (4)$$

Let C_r be the set of all r-commutators of the sfield.

Theorem 8 *Let $r \geqslant 2$. Every sfield, which is not commutative, has no proper subsfield containing all its r-commutators. An element of a sfield which permutes with all the r-commutators of the sfield belongs to the center.*

Let A_r and B_r denote the two propositions of our theorem. B_r follows from A_r immediately. Theorems 3 and 5 assert that both A_2 and B_2 are true. We shall prove the theorem by induction, we assume that A_{r-1} and B_{r-1} are both true. Let a be an element of the sfield, which does not belong to the center. By B_{r-1}, we have an element belonging to C_{r-1} such that $ab \neq ba$. The identity (2) asserts that a belongs to the sfield generated by C_r. Therefore A_r is true. The theorem is proved.

The theorem asserts that the lower central series of the multiplicative group of a sfield can never be ended at the identity. It is an almost trivial fact that the upper central series can never go up. It is a comparatively difficult question about the derived series. Elsewhere I shall prove

Theorem 9 *Let $r \geqslant 1$. Every sfield is generated by all the elements of the r-th derived group.*

An application of Theorem 1 is to establish the simplicity of the projective special linear group $\mathrm{PSL}_n(K)$① of dimension n over the sfield K.

Theorem 10 *The projective special linear group $\mathrm{PSL}_n(K)$ is simple except when $n = 2$ and K has 2 or 3 elements.*

In case K is a field of characteristic $\neq 2$ or K is a complete field, the theorem was proved by Burnside-Jordan-Dickson (see Dickson, ② p. 85 and Van der Waerden, ③ p.7). For the incomplete field, it was proved by Iwasawa ④. For a sfield, Dieudonné proved the theorem for $n > 2$, and for $n = 2$ ⑤ except in the case when the center of K has 2, 3 or 5 elements. Since both exceptional cases in Theorem 10 are not simple, it is the final answer to the problem about the simplicity of $\mathrm{PSL}_n(K)$.

Owing to the previous known results, we assume, from now on, that $n = 2$ and the sfield is not commutative. It is the same thing to prove that

Theorem 11 *Every normal subgroup of $\mathrm{SL}_2(K)$ is contained in the center, except when K has 2 or 3 elements.*

Let A be an element of $\mathrm{SL}_2(K)$ which does not belong to the center. We are going to prove that A and its conjugates generate $\mathrm{SL}_2(K)$. We shall prove that there is a matrix Q belonging to $\mathrm{SL}_2(K)$ such that QAQ^{-1} has a non-zero element at $(1,2)$ position. In fact, let $A = \begin{pmatrix} \alpha & \beta \\ \gamma & \delta \end{pmatrix}$. If $\beta \neq 0$, $Q = I$ satisfies our requirement, and if $\gamma \neq 0$, then $Q = \begin{pmatrix} 0 & 1 \\ -1 & 0 \end{pmatrix}$ does. If $\beta = \gamma = 0$, we have an element x such that $\alpha x \neq x\delta$, since A does not belong to the center. Then

$$\begin{pmatrix} 1 & x \\ 0 & 1 \end{pmatrix} \begin{pmatrix} \alpha & 0 \\ 0 & \delta \end{pmatrix} \begin{pmatrix} 1 & x \\ 0 & 1 \end{pmatrix}^{-1} = \begin{pmatrix} * & -\alpha x + x\delta \\ * & * \end{pmatrix}.$$

We may assume that the subgroup generated by A and its conjugates contains an

---

① For the definition and properties of $\mathrm{PSL}_n(K)$, see Dieudonné.④
② Dickson L E. *Linear Groups.* Leipzig, 1901.
③ Van der Waerden B L. *Gruppen von linearen Transformationen.* Berlin, 1935: 7.
④ Iwasawa M. *Proc. Imp. Acad. Japan,* 1941, **17**: 57.
⑤ The author had some difficulty in understanding Dieudonné's proof. In fact, all the parabolic elements of $\mathrm{PSL}_2(K)$ do not form a single conjugate set in $\mathrm{PSL}_2(K)$.

element with $\beta \neq 0$. Since

$$\begin{pmatrix} 1 & 0 \\ \beta^{-1}\alpha & 1 \end{pmatrix} \begin{pmatrix} \alpha & \beta \\ \gamma & \delta \end{pmatrix} \begin{pmatrix} 1 & 0 \\ \beta^{-1}\alpha & 1 \end{pmatrix}^{-1} = \begin{pmatrix} 0 & * \\ * & * \end{pmatrix},$$

we have assumed that the group contains an element of the form $A = \begin{pmatrix} 0 & \beta \\ \gamma & \delta \end{pmatrix}$.
We take

$$B = \begin{pmatrix} -\gamma^{-1}\delta\beta^{-1} & \gamma^{-1} \\ -\gamma\kappa & 0 \end{pmatrix},$$

where κ belongs to the commutator subgroup of K, so that B belongs to $\mathrm{SL}_2(K)$. We have

$$A(BA^{-1}B^{-1}) = AB(BA)^{-1} = \begin{pmatrix} -\beta\gamma\kappa & 0 \\ * & -(\gamma\kappa\beta)^{-1} \end{pmatrix}.$$

Now we shall prove that we can choose κ such that $-\beta\gamma\kappa$ does not belong to the center. For otherwise $(-\beta\gamma\kappa_1)^{-1}(-\beta\gamma\kappa_2) = \kappa_1^{-1}\kappa_2$ belongs to the center for any two commutators κ_1 and κ_2, that is, the commutators of κ belong to the center of K. By Theorem 4, K is a field which contradicts our supposition. Therefore the subgroup contains an element of the form $C = ABA^{-1}B^{-1} = \begin{pmatrix} a & 0 \\ c & d \end{pmatrix}$, where a does not belong to the center. We have an element λ of K such that $d\lambda \neq \lambda a$, then the subgroup contains

$$\begin{pmatrix} a & 0 \\ c & d \end{pmatrix} \begin{pmatrix} 1 & 0 \\ \lambda & 1 \end{pmatrix} \begin{pmatrix} a & 0 \\ c & d \end{pmatrix}^{-1} \begin{pmatrix} 1 & 0 \\ -\lambda & 1 \end{pmatrix} = \begin{pmatrix} 1 & 0 \\ d\lambda a^{-1} - \lambda & 1 \end{pmatrix}.$$

We may suppose that our subgroup contains $\begin{pmatrix} 1 & 0 \\ c & 1 \end{pmatrix}$, $c \neq 0$.

We are going to prove that the subgroup contains $\begin{pmatrix} 1 & 0 \\ 1 & 1 \end{pmatrix}$. In fact, if $c = \pm 1$, our assertion is trivial. Let us suppose $c \neq 0, \neq \pm 1$. The field K_1, obtained from the center of K by adjunction of c, contains more than 3 elements. $\mathrm{SL}_2(K_1)$ is generated by any one element, not in the center, and its conjugates; in particular, $\begin{pmatrix} 1 & 0 \\ c & 1 \end{pmatrix}$ and its conjugates generate $\mathrm{SL}_2(K_1)$. Thus $\begin{pmatrix} 1 & 0 \\ 1 & 1 \end{pmatrix}$ is in the subgroup under consideration. Consequently the subgroup contains

$$\begin{pmatrix} \lambda^{-1} & 0 \\ 0 & \lambda \end{pmatrix} \begin{pmatrix} 1 & 0 \\ 1 & 1 \end{pmatrix} \begin{pmatrix} \lambda^{-1} & 0 \\ 0 & \lambda \end{pmatrix}^{-1} = \begin{pmatrix} 1 & 0 \\ \lambda^2 & 1 \end{pmatrix}$$

for all λ.

If the characteristic of the sfield is not equal to 2, then the subgroup contains all the elements of the form

$$\begin{pmatrix} 1 & 0 \\ (\lambda+1)^2 & 1 \end{pmatrix} \begin{pmatrix} 1 & 0 \\ \lambda^2 & 1 \end{pmatrix}^{-1} \begin{pmatrix} 1 & 0 \\ 1 & 1 \end{pmatrix}^{-1} = \begin{pmatrix} 1 & 0 \\ 2\lambda & 1 \end{pmatrix}.$$

Therefore we have the theorem.

If the characteristic of the sfield equals 2, since

$$\begin{pmatrix} 1 & (\lambda+1)^{-2} \\ 0 & 1 \end{pmatrix} \begin{pmatrix} 1 & 0 \\ \lambda^2 & 1 \end{pmatrix} \begin{pmatrix} 1 & 1 \\ 0 & 1 \end{pmatrix} \begin{pmatrix} 1 & 0 \\ (1+\lambda^{-1})^{-2} & 1 \end{pmatrix} = \begin{pmatrix} (\lambda+1)^{-2} & 0 \\ 0 & (\lambda+1)^2 \end{pmatrix},$$

the subgroup contains elements of the form

$$\begin{pmatrix} a^2 & 0 \\ 0 & a^{-2} \end{pmatrix}$$

for all a. Let a be any element of the sfield. From

$$\begin{pmatrix} 1 & a \\ 0 & 1 \end{pmatrix} \begin{pmatrix} a^2 & 0 \\ 0 & a^{-2} \end{pmatrix} \begin{pmatrix} 1 & \lambda^2 \\ 0 & 1 \end{pmatrix} \begin{pmatrix} 1 & a \\ 0 & 1 \end{pmatrix}^{-1} \begin{pmatrix} a^{-2} & 0 \\ 0 & a^2 \end{pmatrix}$$
$$= \begin{pmatrix} 1 & a^2\lambda^2 a^2 + a^5 + a \\ 0 & 1 \end{pmatrix} = \begin{pmatrix} a^2 & 0 \\ 0 & a^{-2} \end{pmatrix} \begin{pmatrix} 1 & \lambda^2 \\ 0 & 1 \end{pmatrix} \begin{pmatrix} a^2 & 0 \\ 0 & a^{-2} \end{pmatrix}^{-1}$$
$$\times \begin{pmatrix} (a+1)^2 & 0 \\ 0 & (a+1)^{-2} \end{pmatrix} \begin{pmatrix} 1 & a \\ 0 & 1 \end{pmatrix} \begin{pmatrix} (a+1)^2 & 0 \\ 0 & (a+1)^{-2} \end{pmatrix}^{-1},$$

we deduce that $\begin{pmatrix} 1 & a \\ 0 & 1 \end{pmatrix}$ belongs to the subgroup. Since $\begin{pmatrix} 1 & a \\ 0 & 1 \end{pmatrix}$ and its conjugates generate $\mathrm{SL}_2(K)$ as a runs over all elements of K, the theorem follows.

The previous method can be used to establish the easier

Theorem 12 *Every normal subgroup of the general linear group* $\mathrm{GL}_n(K)$, *which is not contained in the center, contains* $\mathrm{SL}_n(K)$, *except the cases when* $n = 2$ *and K has 2 or 3 elements.*

On the generators of the symplectic modular group*

Introduction

Let n be a positive integer. Throughout this paper, unless the contrary is stated, we shall use capital Latin letters to denote n-rowed matrices and capital German letters to denote $2n$-rowed matrices. Furthermore, an r-rowed matrix R will be denoted by $R^{(r)}$. Let

$$\mathfrak{F} = \begin{pmatrix} 0 & I \\ -I & 0 \end{pmatrix},$$

where I and 0 denote the identity and zero matrices respectively. Let Γ be the group of all matrices $\mathfrak{M}$ with rational integral elements which satisy

$$\mathfrak{M}\mathfrak{F}\mathfrak{M}' = \mathfrak{F}, \tag{1}$$

where $\mathfrak{M}'$ denotes the transpose of $\mathfrak{M}$. Let Γ_0 be the factor group of Γ over its centrum; Γ_0 is called the symplectic modular group. It can be thought of as being obtained from Γ by identifying the elements $\mathfrak{M}$ and $-\mathfrak{M}$. In applications to modular functions of the nth degree[1] and to the projective geometry of matrices[2] it is customary to identify $\mathfrak{M}$ and $-\mathfrak{M}$ as a single transformation. For this reason we have considered Γ_0 rather than Γ; it might be pointed out, however, that the generators of Γ_0 obtained in this paper happen to be a set of generators of Γ.

It is the aim of this paper to find the generators of the symplectic modular group. It will be proved here that this group is generated by two or four independent elements, according as $n = 1$ or $n > 1$. The method used here can be extended so as to give a set of generators for matrices with elements in any Euclidean ring. In particular, we give the details for the generalized Picard group at the end of this paper.

---

*Presented to the Society, February 28, 1948; received by the editors March 6,1948. (By Hua L K and Reiner I). Reprinted from *Transactions of the American Mathematical Society*, (1949, **65**(3): 415-426.)

[1] Siegel C L. *Math. Ann*, 1939, **116**: 617-657.
[2] Hua L K. *Trans. Amer. Math. Soc.*, 1945, **57**: 441-490.

Problems of this type have been considered previously. Poincaré[①] stated without proof that every matrix $\mathfrak{M}$ with integral elements for which $\mathfrak{M}\mathfrak{G}\mathfrak{M}' = \mathfrak{G}$, where $\mathfrak{G}$ is the direct sum of n two-rowed skew-symmetric matrices, is expressible as a product of elementary matrices of two simple types. Brahana[②] proved this and extended the result to the case where $\mathfrak{G}$ is any skew-symmetric matrix by showing in this case that every such matrix $\mathfrak{M}$ is expressible as a product of matrices taken from some finite set of matrices. From the results given in the present paper, a much stronger form of Brahana's result can be easily deduced.

1. If we set
$$\mathfrak{M} = \begin{pmatrix} A & B \\ C & D \end{pmatrix}, \tag{2}$$

(1) is equivalent to
$$AB' = BA', \quad CD' = DC', \quad AD' - BC' = I. \tag{3}$$

By taking inverses of both sides of (1) and using $\mathfrak{F}^{-1} = -\mathfrak{F}$, we can deduce that $\mathfrak{M}'\mathfrak{F}\mathfrak{M} = \mathfrak{F}$, so that
$$A'C = C'A, \quad B'D = D'B, \quad A'D - C'B = I. \tag{4}$$

We shall begin by showing in §3 that Γ_0 is generated by the following types of elements:

(I) *Translations:*
$$\mathfrak{T} = \begin{pmatrix} I & S \\ 0 & I \end{pmatrix},$$

where S is symmetric.

(II) *Rotations:*
$$\mathfrak{R} = \begin{pmatrix} U & 0 \\ 0 & U'^{-1} \end{pmatrix},$$

where U is unimodular, that is, abs $U=1$ (where abs U denotes the absolute value of the determinant of U).

(III) *Semi-involutions:*
$$\mathfrak{S} = \begin{pmatrix} J & I-J \\ J-I & J \end{pmatrix},$$

---

① Poincaré H. *Rend. Circ. Mat. Palermo*, 1904, **18**: 45-110.
② Brahana H R. *Ann. of Math*, 1923, **24** (2): 265-270.

where J is a diagonal matrix whose diagonal elements are 0's and 1's, so that $J^2 = J$ and $(I - J)^2 = I - J$.

It is easily verified that matrices of types I, II and III satisfy (1).

2. In this section we prove two lemmas on matrices.

Lemma 1 *Let m be a nonzero integer, and let T be an n-rowed symmetric matrix at least one of whose elements is not divisible by m. There exists a symmetric matrix S with integral elements such that*

$$0 < \text{abs}(T - mS) < |m|^n. \tag{5}$$

Proof The lemma is evident for $n = 1$. Consider next $n = 2$; let $T = (t_{ij})$, $S = (S_{ij})$. Then

$$\text{abs}(T - mS) = |(t_{11} - ms_{11})(t_{22} - ms_{22}) - (t_{12} - ms_{12})^2|. \tag{6}$$

If m divides both t_{11} and t_{22}, it cannot divide t_{12}; we can then choose S so that $t_{11} - ms_{11} = t_{22} - ms_{22} = 0$ and $0 < |t_{12} - ms_{12}| < |m|$. Suppose on the other hand that m does not divide one of the diagonal elements, say t_{11}. Fix s_{12} arbitrarily, and choose s_{11} so that $0 < |t_{11} - ms_{11}| < |m|$. Since (6) can be written as

$$\text{abs}(T - mS) = |-m(t_{11} - ms_{11})s_{22} + \cdots|,$$

where $\cdots$ represents terms not involving s_{22}, we can choose an integer s_{22} by the Euclidean algorithm so that

$$0 < \text{abs}(T - mS) \leqslant |m(t_{11} - ms_{11})| < |m|^2.$$

Suppose now that the result has been established for $n = r - 1$ with $r \geqslant 3$; we shall deduce it for $n = r$. Let $T = T^{(r)}$ and let some element t_{ij} of T be not divisible by m. Since $r \geqslant 3$, there exists a diagonal element t_{kk} of T which is not in the same row or column as t_{ij}. Let T_1 be the symmetric $(r - 1)$-rowed matrix obtained from T by omitting the kth row and kth column; let S_1 be similarly related to S. By the induction hypothesis, we may choose S_1 symmetric so that

$$0 < \text{abs}(T_1 - mS_1) < |m|^{r-1}. \tag{7}$$

However, we have

$$\text{abs}(T - mS) = |(t_{kk} - ms_{kk})\det(T_1 - mS_1) + \cdots|, \tag{8}$$

where $\cdots$ represents terms not involving s_{kk}. Choose s_{lk} arbitrarily for $l = 1, 2, \cdots, k-1, k+1, \cdots, r$. Then by the Euclidean algorithm we can choose s_{kk} so that

$$0 < \mathrm{abs}(T - mS) \leqslant |m|\mathrm{abs}(T_1 - mS_1) < |m|^r.$$

This completes the proof of the lemma.

Lemma 2 *Let A and B satisfy $AB' = BA'$ and let $\det A \neq 0$. There exists a symmetric matrix S such that either*

$$B - AS = 0, \tag{9}$$

or

$$0 < \mathrm{abs}(B - AS) < \mathrm{abs}A. \tag{10}$$

Proof From $AB' = BA'$ and $\det A \neq 0$, we may deduce that A^*B is symmetric, where A^* denotes the adjoint of A. We apply Lemma 1 with $T = A^*B$ and $m = \det A$. Either every element of A^*B is divisible by m, in which case there exists a symmetric matrix S with $A^*B = mS$, or else there exist symmetric matrices R and S such that $A^*B = mS + R$ with $0 < \mathrm{abs}\, R < |m|^n$. In virtue of the relation $A^*A = mI$, these alternatives become: either $B = AS$ (in which case (9) holds), or $B - AS = AR/m$; however,

$$\mathrm{abs}\frac{AR}{m} = \frac{(\mathrm{abs}A)(\mathrm{abs}R)}{|m|^n} = \frac{\mathrm{abs}R}{|m|^{n-1}},$$

so that

$$0 < \mathrm{abs}(B - AS) < |m| = \mathrm{abs}A.$$

3. We are now ready to show that Γ_0 is generated by matrices of types I, II and III. Let $\mathfrak{M}$ given by (2) be an arbitrary element of Γ_0. It suffices to prove that by multiplying $\mathfrak{M}$ by matrices of types I, II and III, one obtains a product of matrices of those types.

(3) implies that not both A and B are 0. Since

$$\begin{pmatrix} A & B \\ C & D \end{pmatrix} \begin{pmatrix} 0 & I \\ -I & 0 \end{pmatrix} = \begin{pmatrix} -B & A \\ * & * \end{pmatrix},$$

we may assume that A has rank $r > 0$. Furthermore,

$$\begin{pmatrix} U & 0 \\ 0 & U'^{-1} \end{pmatrix} \begin{pmatrix} A & B \\ C & D \end{pmatrix} \begin{pmatrix} V & 0 \\ 0 & V'^{-1} \end{pmatrix} = \begin{pmatrix} UAV & * \\ * & * \end{pmatrix},$$

so that A may be taken to be of the form

$$A = \begin{pmatrix} A_1 & 0 \\ A_2 & 0 \end{pmatrix}, \tag{11}$$

where A_1 is an r-rowed nonsingular matrix. We similarly decompose B as

$$B = \begin{pmatrix} B_1^{(r)} & * \\ * & * \end{pmatrix}.$$

From (3) it is easily seen that $A_1 B_1' = B_1 A_1'$. By Lemma 2, there exists a symmetric matrix S_1 with either $A_1 S_1 + B_1 = 0$ or $0 <$abs $R_1 <$abs A_1, where $R_1 = A_1 S_1 + B_1$. Define

$$S = \begin{pmatrix} S_1^{(r)} & 0 \\ 0 & 0 \end{pmatrix}.$$

Then

$$\begin{pmatrix} A & B \\ C & D \end{pmatrix} \begin{pmatrix} I & S \\ 0 & I \end{pmatrix} = \begin{pmatrix} A & AS + B \\ * & * \end{pmatrix}, \tag{12}$$

so that A remains unaltered while B_1 of B is replaced by 0 or R_1. If the second alternative occurs, we proceed as follows: let

$$J = \begin{pmatrix} 0 & 0 \\ 0 & I^{(n-r)} \end{pmatrix}. \tag{13}$$

Then

$$\begin{pmatrix} A & B \\ C & D \end{pmatrix} \begin{pmatrix} J & I - J \\ J - I & J \end{pmatrix} = \begin{pmatrix} \overline{A} & * \\ * & * \end{pmatrix}, \tag{14}$$

where

$$\overline{A} = AJ - B(I - J) = \begin{pmatrix} -R_1 & 0 \\ * & 0 \end{pmatrix}.$$

We now repeat the process as before, and so on. Since there are only finitely many positive integers less than abs A_1, this process eventually terminates. Thus, by multiplying $\mathfrak{M}$ by matrices of types I, II and III one arrives at a matrix

$$\begin{pmatrix} A_0 & B_0 \\ * & * \end{pmatrix}$$

with

$$A_0 = \begin{pmatrix} R & 0 \\ * & 0 \end{pmatrix}, \quad B_0 = \begin{pmatrix} 0 & * \\ * & * \end{pmatrix}$$

and det $R \neq 0$. One readily deduces from $A_0 B_0' = B_0 A_0'$ that B_0 must be of the form

$$B_0 = \begin{pmatrix} 0 & * \\ 0 & * \end{pmatrix}.$$

But then

$$\begin{pmatrix} 0 & I \\ -I & 0 \end{pmatrix} \begin{pmatrix} A_0 & B_0 \\ C_0 & D_0 \end{pmatrix} \begin{pmatrix} J & I-J \\ J-I & J \end{pmatrix} = \begin{pmatrix} A^+ & B^+ \\ 0 & D^+ \end{pmatrix},$$

where J is given by (13). Finally we notice that for a matrix

$$\begin{pmatrix} A & B \\ 0 & D \end{pmatrix}$$

of Γ_0, we must have $A = U$ unimodular, $D = U'^{-1}$, and thence from (3), $B = SU'^{-1}$ with symmetric S. Therefore

$$\begin{pmatrix} A & B \\ 0 & D \end{pmatrix} = \begin{pmatrix} I & S \\ 0 & I \end{pmatrix} \begin{pmatrix} U & 0 \\ 0 & U'^{-1} \end{pmatrix}.$$

This completes the proof that Γ_0 is generated by the matrices of types I, II and III.

4. The set of matrices of types I, II and III which generate Γ_0 are certainly not independent generators. Let us reduce the number of generators as much as possible. Since

$$\begin{pmatrix} I & S_1 \\ 0 & I \end{pmatrix} \begin{pmatrix} I & S_2 \\ 0 & I \end{pmatrix} = \begin{pmatrix} I & S_1 + S_2 \\ 0 & I \end{pmatrix},$$

the subgroup formed by matrices of type I is generated by those type I matrices whose S's are given by

$$S_0 = \begin{pmatrix} 1 & 0 & \cdots & 0 \\ 0 & 0 & \cdots & 0 \\ \vdots & \vdots & & \vdots \\ 0 & 0 & \cdots & 0 \end{pmatrix}, \quad S_1 = \begin{pmatrix} 0 & 1 & 0 & \cdots & 0 \\ 1 & 0 & 0 & \cdots & 0 \\ 0 & 0 & 0 & \cdots & 0 \\ \vdots & \vdots & \vdots & & \vdots \\ 0 & 0 & 0 & \cdots & 0 \end{pmatrix} \tag{15}$$

and all matrices obtained from these by interchanging any two rows and the corresponding columns. Next we note that

$$\begin{pmatrix} U & 0 \\ 0 & U'^{-1} \end{pmatrix} \begin{pmatrix} I & S \\ 0 & I \end{pmatrix} \begin{pmatrix} U^{-1} & 0 \\ 0 & U' \end{pmatrix} = \begin{pmatrix} I & USU' \\ 0 & I \end{pmatrix},$$

so that the group generated by matrices of types I and II is the same as that generated by all type II matrices and the two translations whose S's are given by (15). However, we have

$$\begin{pmatrix} 1 & -1 \\ 0 & 1 \end{pmatrix} \begin{pmatrix} 1 & 0 \\ 0 & -1 \end{pmatrix} \begin{pmatrix} 1 & 0 \\ -1 & 1 \end{pmatrix} = \begin{pmatrix} 0 & 1 \\ 1 & -1 \end{pmatrix} = \begin{pmatrix} 0 & 1 \\ 1 & 0 \end{pmatrix} - \begin{pmatrix} 0 & 0 \\ 0 & 1 \end{pmatrix}.$$

Hence the translation with S_1 is obtainable from that with S_0 and the matrices of type II. Therefore Γ_0 is generated by the matrix

$$\mathfrak{T}_0 = \begin{pmatrix} I & S_0 \\ 0 & I \end{pmatrix} \tag{16}$$

with S_0 given by (15), and all matrices of types II and III.

Since

$$\begin{pmatrix} U & 0 \\ 0 & U'^{-1} \end{pmatrix} \begin{pmatrix} V & 0 \\ 0 & V'^{-1} \end{pmatrix} = \begin{pmatrix} UV & 0 \\ 0 & (UV)'^{-1} \end{pmatrix},$$

in order to find the generators of the subgroup of rotations we have merely to find the generators of the group of unimodular matrices. These are given by the following theorem.

Theorem 1 Let $n \geqslant 2$. Every unimodular matrix with rational integral elements is a product of the matrices U_1, U_2, U_3 and their inverses, where

$$U_1 = \begin{pmatrix} 0 & 0 & \cdots & 0 & 1 \\ 1 & 0 & \cdots & 0 & 0 \\ \vdots & \vdots & & \vdots & \vdots \\ 0 & 0 & \cdots & 0 & 0 \\ 0 & 0 & \cdots & 1 & 0 \end{pmatrix}, \quad U_2 = \begin{pmatrix} 1 & 1 & \cdots & 0 & 0 \\ 0 & 1 & \cdots & 0 & 0 \\ \vdots & \vdots & & \vdots & \vdots \\ 0 & 0 & \cdots & 1 & 0 \\ 0 & 0 & \cdots & 0 & 1 \end{pmatrix},$$

$$U_3 = \begin{pmatrix} -1 & 0 & \cdots & 0 & 0 \\ 0 & 1 & \cdots & 0 & 0 \\ \vdots & \vdots & & \vdots & \vdots \\ 0 & 0 & \cdots & 1 & 0 \\ 0 & 0 & \cdots & 0 & 1 \end{pmatrix}. \tag{17}$$

Proof It is known[1] that every unimodular matrix is a product of U_1, U_2, U_3,

---

[1] See for example, MacDuffiee C C. *The theory of matrices*. Berlin, 1933: 34, Theorem 22.5.

and
$$U_4 = \begin{pmatrix} 0 & 1 & \cdots & 0 & 0 \\ 1 & 0 & \cdots & 0 & 0 \\ \vdots & \vdots & & \vdots & \vdots \\ 0 & 0 & \cdots & 1 & 0 \\ 0 & 0 & \cdots & 0 & 1 \end{pmatrix},$$

and their inverses. It is sufficient to show that U_4 is expressible as a product of U_1, U_2, U_3 and their inverses. We define $T = U_2 U_1$ for the remainder of this proof, and let $\mathfrak{r} = (r_1, \cdots, r_n)'$ be a column vector. Then

$$T\mathfrak{r} = \begin{pmatrix} r_n + r_1 \\ r_1 \\ \vdots \\ r_{n-1} \end{pmatrix},$$

$$T^2\mathfrak{r} = \begin{pmatrix} r_{n-1} + r_n + r_1 \\ r_n + r_1 \\ r_1 \\ \vdots \\ r_{n-2} \end{pmatrix}, \cdots, T^{n-1}\mathfrak{r} = \begin{pmatrix} r_2 + r_3 + \cdots + r_n + r_1 \\ r_3 + \cdots + r_n + r_1 \\ \vdots \\ r_n + r_1 \\ r_1 \end{pmatrix}.$$

Therefore

$$U_1^{-1} T^{n-1} \mathfrak{r} = \begin{pmatrix} r_3 + \cdots + r_n + r_1 \\ \vdots \\ r_n + r_1 \\ r_1 \\ r_2 + r_3 + \cdots + r_n + r_1 \end{pmatrix},$$

so that

$$(T^{-1})^{n-2} U_1^{-1} T^{n-1} \mathfrak{r} = \begin{pmatrix} r_1 \\ r_2 + \cdots + r_n + r_1 \\ r_3 \\ \vdots \\ r_n \end{pmatrix},$$

$$U_1(T^{-1})^{n-2}U_1^{-1}T^{n-1}\mathfrak{r} = \begin{pmatrix} r_n \\ r_1 \\ r_2 + \cdots + r_n + r_1 \\ r_3 \\ \vdots \\ r_{n-1} \end{pmatrix}.$$

From this we see that

$$T^{n-3}U_1(T^{-1})^{n-2}U_1^{-1}T^{n-1}\mathfrak{r} = \begin{pmatrix} r_3 + r_4 + \cdots + r_n \\ r_4 + \cdots + r_n \\ \vdots \\ r_n \\ r_1 \\ r_2 + \cdots + r_n + r_1 \end{pmatrix}$$

and

$$(T^{-1})^{n-2}U_1 T^{n-3} U_1 (T^{-1})^{n-2} U_1^{-1} T^{n-1} \mathfrak{r} = \begin{pmatrix} r_n \\ r_1 \\ r_2 + r_1 \\ \vdots \\ r_{n-1} \end{pmatrix}.$$

Define

$$U^\dagger = U_1^{-1}(T^{-1})^{n-2} U_1 T^{n-3} U_1 (T^{-1})^{n-2} U_1^{-1} T^{n-1}.$$

Then

$$U^\dagger = \begin{pmatrix} 1 & 0 \\ 1 & 1 \end{pmatrix} \dotplus I^{(n-2)},$$

where $\dotplus$ denotes the direct sum of matrices. But from

$$U_3 = \begin{pmatrix} -1 & 0 \\ 0 & 1 \end{pmatrix} \dotplus I^{(n-2)} \quad \text{and} \quad U_2^{-1} = \begin{pmatrix} 1 & -1 \\ 0 & 1 \end{pmatrix} \dotplus I^{(n-2)},$$

we deduce

$$U_3 U^\dagger U_2^{-1} U^\dagger = \begin{pmatrix} -1 & 0 \\ 0 & 1 \end{pmatrix} \begin{pmatrix} 1 & 0 \\ 1 & 1 \end{pmatrix} \begin{pmatrix} 1 & -1 \\ 0 & 1 \end{pmatrix} \begin{pmatrix} 1 & 0 \\ 1 & 1 \end{pmatrix} \dotplus I^{(n-2)}$$

$$= \begin{pmatrix} 0 & 1 \\ 1 & 0 \end{pmatrix} \dotplus I^{(n-2)} = U_4.$$

This completes the proof of the theorem.

Corollary Let $n \geqslant 2$. Every unimodular matrix with rational integral elements of determinant $+1$ is a product of powers of U_2 and

$$U_5 = \begin{pmatrix} 0 & \cdots & 0 & (-1)^{n-1} \\ 1 & \cdots & 0 & 0 \\ \vdots & & \vdots & \vdots \\ 0 & \cdots & 0 & 0 \\ 0 & \cdots & 1 & 0 \end{pmatrix}.$$

By Theorem 1 we see now that Γ_0 is generated by $\mathfrak{T}_0$ and the set of all semi-involutions and the three rotations defined by

$$\mathfrak{R}_i = \begin{pmatrix} U_i & 0 \\ 0 & U_i'^{-1} \end{pmatrix}, \quad i = 1, 2, 3. \tag{18}$$

We finally consider type III matrices. Let J_r be the diagonal matrix obtained from the identity matrix by replacing the rth 1 by 0. In that case, if $r \neq s$, we have

$$\begin{pmatrix} J_r & I - J_r \\ J_r - I & J_r \end{pmatrix} \begin{pmatrix} J_s & I - J_s \\ J_s - I & J_s \end{pmatrix} = \begin{pmatrix} J_{rs} & I - J_{rs} \\ J_{rs} - I & J_{rs} \end{pmatrix},$$

where J_{rs} is obtained from the identity matrix by replacing the rth and sth ones by 0's. Therefore, in order to obtain all type III matrices, we need only those semi-involutions

$$\begin{pmatrix} J_r & I - J_r \\ J_r - I & J_r \end{pmatrix}, \quad r = 1, 2, \cdots, n, \tag{19}$$

with J_r defined above. Now, let U be that unimodular matrix obtained from I by interchanging the 1st and rth rows; then we have

$$\begin{pmatrix} U & 0 \\ 0 & U'^{-1} \end{pmatrix} \begin{pmatrix} J_r & I - J_r \\ J_r - I & J_r \end{pmatrix} \begin{pmatrix} U^{-1} & 0 \\ 0 & U' \end{pmatrix} = \begin{pmatrix} J_1 & I - J_1 \\ J_1 - I & J_1 \end{pmatrix}.$$

Therefore Γ_0 is generated by the matrices $\mathfrak{T}_0$, $\mathfrak{R}_i$ ($i = 1, 2, 3$) and the matrix

$$\mathfrak{S}_0 = \begin{pmatrix} J_1 & I - J_1 \\ J_1 - I & J_1 \end{pmatrix}, \tag{20}$$

with J, previously defined. But

$$\mathfrak{S}_0^2 = \mathfrak{R}_3,$$

so that $\mathfrak{R}_3$ may be dropped from the list of generators. Therefore we have the following theorem.

Theorem 2 Γ_0 *is generated by the four matrices* $\mathfrak{T}_0, \mathfrak{R}_1, \mathfrak{R}_2$ *and* $\mathfrak{S}_0$ *given by* (15), (18) *and* (20), *for* $n > 1$. *For* $n = 1$, Γ_0 *is generated by* $\mathfrak{T}_0$ *and* $\mathfrak{S}_0$.

5. In this section we shall prove the independence of the generators given in Theorem 2. For $n = 1$, this is trivial because $\mathfrak{S}_0$ is of finite order while $\mathfrak{T}_0$ is not. Hereafter we suppose that $n > 1$.

(1) *Independence of* $\mathfrak{T}_0$. We consider the transformation

$$(X_1, Y_1) = (X, Y)\mathfrak{M}; \tag{21}$$

if XY' is symmetric, it is easily verified that $X_1 Y_1'$ is also symmetric. We shall show that if the diagonal elements of XY' are even, those of $X_1 Y_1'$ will also be even if $\mathfrak{M}$ is $\mathfrak{R}_1, \mathfrak{R}_2$, or $\mathfrak{S}_0$, while if $\mathfrak{M} = \mathfrak{T}_0$ it is possible to choose X and Y so that some diagonal element of $X_1 Y_1'$ is odd. This will show that $\mathfrak{T}_0$ is not expressible as a product of $\mathfrak{R}_1$, $\mathfrak{R}_2$ and $\mathfrak{S}_0$ and their inverses.

Assume now that the diagonal elements of XY' are even. From (21) one readily deduces that if $\mathfrak{M}$ is a rotation, $X_1 Y_1' = XY'$, so that the diagonal elements of $X_1 Y_1'$ are also even. If secondly $\mathfrak{M}$ is a semi-involution, we have

$$X_1 = XJ + Y(I - J), \quad Y_1 = -X(I - J) + YJ,$$

so that

$$X_1 Y_1' = XJY' - Y(I - J)X' = XJY' + YJX' - YX'.$$

Since XJY' is the transpose of YJX', it is again clear that the diagonal elements of $X_1 Y_1'$ are even. Finally, Suppose $\mathfrak{M} = \mathfrak{T}_0$. Then we obtain

$$X_1 Y_1' = X(XS_0 + Y)' = XY' + XS_0 X'$$

and for $X = I$ the first diagonal element of $X_1 Y_1'$ is odd. This completes the proof of the independence of $\mathfrak{T}_0$. We may remark in passing, however, that $\mathfrak{T}_0^2$ is expressible as a product of the powers of $\mathfrak{R}_1, \mathfrak{R}_2$ and $\mathfrak{S}_0$.

(2) *Independence of* $\mathfrak{R}_1$. Let $\mathfrak{r} = (r_1, \cdots, r_n, s_1, \cdots, s_n)'$ be a column vector with $2n$ components. It is clear that the second component r_2 is unaffected when $\mathfrak{r}$ is multiplied on the left by any of the matrices $\mathfrak{T}_0, \mathfrak{R}_2$, and $\mathfrak{S}_0$ and their inverse. Under multiplication on the left by $\mathfrak{R}_1$, however, r_2 is replaced by r_1. Hence $\mathfrak{R}_1$ cannot be expressed as a product of $\mathfrak{T}_0, \mathfrak{R}_2$ and $\mathfrak{S}_0$ and their inverses.

(3) *Independence of* $\mathfrak{R}_2$. Multiplying $\mathfrak{r}$ on the left by $\mathfrak{R}_1$ or $\mathfrak{S}_0$ or their inverses permutes components of $\mathfrak{r}$; under any such permutation, however, any r_i and its corresponding s_i remain n components apart. Since the effect of multiplying on the left by $\mathfrak{T}_0$ is to replace r_1 by $r_1 + s_1$, it is clear that by multiplying $\mathfrak{r}$ on the left by a product of $\mathfrak{R}_1$, $\mathfrak{S}_0$ and $\mathfrak{T}_0$ and their inverses, r_1 may be replaced by a linear combination of r_1 and s_1 and its position may be changed. It is however impossible to replace r_1 by $r_1 + r_2$ in this way, and this is exactly the effect of multiplication of $\mathfrak{r}$ on the left by $\mathfrak{R}_2$. This proves the independence of $\mathfrak{R}_2$.

(4) *Independence of* $\mathfrak{S}_0$. We note that

$$\begin{pmatrix} * & * \\ 0 & * \end{pmatrix} \begin{pmatrix} * & * \\ 0 & * \end{pmatrix} = \begin{pmatrix} * & * \\ 0 & * \end{pmatrix}.$$

Since $\mathfrak{T}_0$, $\mathfrak{R}_1$ and $\mathfrak{R}_2$ and their inverses are all of the form

$$\begin{pmatrix} * & * \\ 0 & * \end{pmatrix}$$

and $\mathfrak{S}_0$ is not of this form, it is clear that $\mathfrak{S}_0$ is not expressible as a product of $\mathfrak{T}_0$, $\mathfrak{R}_1$ and $\mathfrak{R}_2$ and their inverse.

6. Our previous method can be extended to any Euclidean ring; in particular, for the ring formed by the Gaussian integers, we have the following result:

Theorem 3 *Let Γ' be the group of matrices $\mathfrak{M}$ with Gaussian integers as elements which satisfy* (1). *Let Γ'_0 be obtained from Γ' by identifying the four elements $\pm \mathfrak{M}$ and $\pm i\mathfrak{M}$. Then for $n > 1$, Γ'_0 is generated by the matrices $\mathfrak{T}_0$, $\mathfrak{R}_1$, $\mathfrak{R}_2$ and $\mathfrak{S}_0$ defined previously, and the matrix*

$$\mathfrak{T}_1 = \begin{pmatrix} I & S_1 \\ 0 & I \end{pmatrix}, \quad \text{where } S_1 = iS_0. \tag{22}$$

For $n = 1$, Γ'_0 is generated by $\mathfrak{T}_0$, $\mathfrak{T}_1$ and $\mathfrak{S}_0$.

The independence of the generators is shown as follows (with suitable modification when $n = 1$):

(1) *Independence of* $\mathfrak{T}_0$. We use the method of §5, (1). Let XY' be a symmetric matrix with Gaussian integers as elements, such that the real part of each diagonal element is even. This property is preserved when (X, Y) is subjected to the transformations $\mathfrak{T}_1$, $\mathfrak{R}_1$, $\mathfrak{R}_2$ and $\mathfrak{S}_0$ according to (21), but not for the transformation $\mathfrak{T}_0$.

(2) *Independence of* $\mathfrak{T}_1$. This is clear since $\mathfrak{T}_1$ is the only generator which is not real.

(3) The independence of $\mathfrak{R}_1$, $\mathfrak{R}_2$ and $\mathfrak{S}_0$ follows exactly as before.

Tsing Hua University,
Peiping, China.
Institute For Advanved Study,
Princeton, N. J.

Geometry of symmetric matrices over any field with characteristic other than two*

1. Introduction

It is the aim of this paper to establish a generalization of von Staudt's theorem for the geometry of symmetric matrices over any field with characteristic $\neq 2$. Before stating the theorem explicitly, we explain some notations which will be used throughout the paper.

Let Φ be a field with characteristic $\neq 2$. We use capital Latin letters to denote n-rowed matrices over Φ, and I and 0 to denote the n-rowed identity and zero matrices respectively. For matrices which are not $n \times n$, we use $M^{(l,m)}$ to denote an $l \times m$ matrix and $M^{(m)} = M^{(m,m)}$. M' denotes the transposed matrix of M. We use, also, small Greek letters σ and τ as superscripts to denote automorphisms of the field Φ, that is, for any two elements a and b of Φ, we have

$$(a+b)^\sigma = a^\sigma + b^\sigma, \quad (ab)^\sigma = a^\sigma b^\sigma.$$

Let

$$\mathfrak{F} = \begin{pmatrix} 0 & I \\ -I & 0 \end{pmatrix}, \quad \mathfrak{J} = \begin{pmatrix} I & 0 \\ 0 & I \end{pmatrix}.$$

A pair of matrices (X, Y) is called nonsingular and symmetric, if (X, Y) is of rank n and

$$(X, Y)\mathfrak{F}(X, Y)' = 0. \tag{1}$$

We shall say that two nonsingular symmetric pairs (X, Y) and (X_1, Y_1) belong to the same class, if and only if there exists a nonsingular Q such that

$$(X_1, Y_1) = Q(X, Y) \tag{2}$$

holds. A class of nonsingular symmetric pairs of matrices will be called a point $P = \{(X, Y)\}$. The totality of these points forms *a projective space of symmetric matrices*.

---

* Reprinted from *Annals of Mathematics*, 1949, **50**(1): 8-31.
Received February 6, 1947, Revised October 7, 1947.

The rank of
$$\langle P, P_1 \rangle = (X,Y)\mathfrak{F}(X_1,Y_1)'$$
is called the *arithmetic distance* between the two points P and P_1 represented by (X,Y) and (X_1,Y_1) respectively. It is denoted by $r(P,P_1)$. Evidently this notion is independent of the particular choice of the elements from the same class. It is also obvious that two points coincide if and only if their arithmetic distance is zero.

Our problem is to find all the transformations carrying the projective space of symmetric matrices onto itself and keeping arithmetic distance invariant.

Evidently, the symplectic transformations
$$(X_1,Y_1) = Q(X,Y)\mathfrak{T}, \qquad \mathfrak{T}\mathfrak{F}\mathfrak{T}' = \mathfrak{F}, \tag{3}$$
satisfy our requirement. But this does not tell the whole truth; in fact, the transformations
$$(X_1,Y_1) = Q(X^\sigma, Y^\sigma) \tag{4}$$
also satisfy our requirement.

In order to find the full truth, we have to extend the notion of the symplectic group. We consider the transformation
$$(X_1,Y_1) = Q(X,Y)\mathfrak{T}, \qquad \mathfrak{T}\mathfrak{F}\mathfrak{T}' = a\mathfrak{F}, \tag{5}$$
where a is an element $\neq 0$. Further, since
$$(b\mathfrak{T})\mathfrak{F}(b\mathfrak{T})' = b^2 a\mathfrak{F}$$
and
$$(X_1,Y_1) = Qb^{-1}(X,Y)b\mathfrak{T},$$
only those a which represent different classes of the factor group of the multiplicative group of Φ modulo the multiplicative group of all square elements of Φ are essentially taken into consideration. The transformations (5) are called extended symplectic transformations, and they form a group to be called the extended symplectic group. It is apparent that the factor group of the extended symplectic group over the symplectic group is isomorphic to the factor group Φ/Φ_2.

It will be proved in this paper that, for $n > 1$,

Every transformation carrying the projective space of symmetric matrices of order $n > 1$ onto itself and keeping arithmetic distance invariant is of the form
$$(X_1,Y_1) = Q(X_1^\sigma, Y_1^\sigma)\mathfrak{T}, \tag{6}$$

where Q is a nonsingular $n \times n$ matrix, σ is an automorphism of Φ, $\mathfrak{T}$ is a $2n \times 2n$ matrix satisfying
$$\mathfrak{T}\mathfrak{F}\mathfrak{T}' = a\mathfrak{F}$$
and a runs over a complete residue system of Φ/Φ_2.

In nonhomogeneous coordinates ($Z = Y^{-1}X$), (6) takes the form
$$Z_1 = a(AZ^\sigma + B)(CZ^\sigma + D)^{-1}, \tag{7}$$
where $AB' = BA'$, $CD' = DC'$, $AD' - BC' = I$.

If, in particular, Φ is the real field, then it has no automorphism other than identity, and $+1$ and -1 are the representatives of the factor group Φ/Φ_2. Therefore we deduce that any transformation carrying the projective space of real symmetric matrices of order $n > 1$ onto itself and keeping arithmetic distance invariant is of the form
$$Z_1 = (AZ + B)(CZ + D)^{-1},$$
where $AB' = BA'$, $CD' = DC'$, $AD' - BC' = \pm I$.

A comparison with the theorem stated in I[①] reveals that we have eliminated the conditions on "continuity", "sense" and "harmonic separation". This is the final goal of our algebraic treatment.

Notice that, for $n = 1$, the theorem is no longer true. An additional condition concerning harmonic separation is needed. For a field of characteristic 2, the theorem is not true for $n = 1$ even with the condition on harmonic separation.

Before proceeding to the text, it may be worthwhile to mention several connections of the theorem with various fields of mathematics.

Algebraists may have some interest in the following theorem which is suggested by the "affine geometry" of symmetric matrices. Consider only the "finite" points (X, Y); that is, pairs for which Y is nonsingular. They can be represented in nonhomogeneous coordinates $Z (= Y^{-1}X)$. The arithmetic distance between two points Z and Z_1 is simply the rank of the difference $Z - Z_1$. Then we have

Let Σ be the set of all symmetric matrices of order $n \geqslant 2$ over the field Φ with characteristic $\neq 2$. Any mapping of Σ onto itself leaving the rank of the difference of two matrices invariant is of the form

---

①*Geometrics of Matrices. Trans. Amer. Math. Soc.*, 1945, **57**: 441-481. In particular, p.459. The papers under the same title published in *Trans. Amer. Math. Soc.*, 1945, **57**: 441-481; 1947, **61**: 193-228, 229-255 will be referred as I, II and III respectively. Compare also C. R. (*Doklady*) *Acad. Sci. URSS* (N.S.), 1946, **53**: 95-97.

$$Z_1 = aAZ^\sigma A' + S, \qquad S = S', \tag{8}$$

where $a \in \Phi$ and σ is an automorphism of Φ.

This theorem assures the algebraists that their study of symmetric matrices under the congruence group is really a very general one. We shall proceed to prove this theorem by purely algebraic methods, and then we shall deduce our fundamental theorem of the projective geometry of symmetric matrices from it.

From the geometrical point of view, the case $n = 2$ is very interesting. Let (x_1, x_2, x_3, x_4) and (y_1, y_2, y_3, y_4) be two distinct points of ordinary projective 3-space. We use

$$(X, Y) = \begin{pmatrix} x_1 & x_2 & x_3 & x_4 \\ y_1 & y_2 & y_3 & y_4 \end{pmatrix}$$

to denotet the line passing through both points. Two pairs of matrices (X, Y) and (X_1, Y_1) represent the same line if and only if we have a nonsingular matrix Q such that

$$(X, Y) = Q(X_1, Y_1).$$

The lines (X, Y) which satisfy

$$(X, T)\mathfrak{F}(X, Y)' = 0$$

form a line complex. Thus a point of the projective space of 2-rowed symmetric matrices denotes a line on the complex. For two lines (X, Y) and (X_1, Y_1) the rank of $(X, Y)\mathfrak{F}(X_1, Y_1)'$ is either 0 or 1 or 2; its significance is that the lines coincide, intersect, or do not intersect, respectively. Therefore the present statement in its most specific form gives us a rational treatment of the line geometry on a line complex. It seems also to be new. Notice that the line geometry on a line complex is identical with Lie's geometry of circles and that Laguerre's geometry of circles is the "affine geometry" of 2-rowed symmetric matrices. This indicates still other applications of both the homogeneous and nonhomogeneous forms of our theorem.

2. Subspaces

In §2~§6, we consider only the finite points of the projective space of symmetric matrices. Thus there will be no confusion if we use a symmetric matrix to denote a point. Throughout §2~§4, we assume that $n = 2$.

Let S and T be two symmetric matrices with arithmetic distance 1, that is, $S-T$ is of rank 1. The set of symmetric matrices P satisfying

$$r(P, S) \leqslant 1, \qquad r(P, T) \leqslant 1$$

forms a *subspace spanned by S and T* which will be denoted by $\Pi(S,T)$.

We have a transformation of the form (8) carrying S and T into

$$0, \quad \begin{pmatrix} x_0 & 0 \\ 0 & 0 \end{pmatrix}, \quad x_0 \neq 0,$$

respectively. Thus $\Pi(S,T)$ is equivalent under (8) to $\Pi\left(0, \begin{pmatrix} x_0 & 0 \\ 0 & 0 \end{pmatrix}\right)$. Let

$$P = \begin{pmatrix} x & y \\ y & z \end{pmatrix}$$

be a point of $\Pi\left(0, \begin{pmatrix} x_0 & 0 \\ 0 & 0 \end{pmatrix}\right)$, then

$$r(0,P) \leqslant 1, \quad r\left(\begin{pmatrix} x_0 & 0 \\ 0 & 0 \end{pmatrix}, P\right) \leqslant 1$$

implies

$$xz - y^2 = (x - x_0)z - y^2 = 0.$$

Consequently we have $z = 0$ and $y = 0$, that is, $\Pi\left(0, \begin{pmatrix} x_0 & 0 \\ 0 & 0 \end{pmatrix}\right)$ consists of all the matrices of the form

$$\begin{pmatrix} x & 0 \\ 0 & 0 \end{pmatrix}. \tag{9}$$

Therefore every subspace can be carried by a transformation of the form (8) into $\Pi\left(0, \begin{pmatrix} 1 & 0 \\ 0 & 0 \end{pmatrix}\right)$ which has the explicit form given by (9).

Two subspaces are said to be *complemented* if they have one and only one point in common.

Theorem 1 *Every pair of complemented subspaces can be carried simultaneously into*

$$\Pi\left(0, \begin{pmatrix} 1 & 0 \\ 0 & 0 \end{pmatrix}\right), \quad \Pi\left(0, \begin{pmatrix} 0 & 0 \\ 0 & 1 \end{pmatrix}\right)$$

by a transformation of the form (8).

Proof We may suppose that one of the subspaces is $\Pi\left(0, \begin{pmatrix} 1 & 0 \\ 0 & 0 \end{pmatrix}\right)$ and that the common point is 0. Let U be a point in the other subspace different from 0. Then, since $r(0, U) = 1$, we have

$$U = \varepsilon \begin{pmatrix} a^2 & ab \\ ab & b^2 \end{pmatrix}, \quad b \neq 0.$$

The transformation

$$Z_1 = \begin{pmatrix} 1 & -ab^{-1} \\ 0 & 1 \end{pmatrix} Z \begin{pmatrix} 1 & -ab^{-1} \\ 0 & 1 \end{pmatrix}'$$

carries $\Pi\left(0, \begin{pmatrix} 1 & 0 \\ 0 & 0 \end{pmatrix}\right)$ onto itself and carries U into

$$\begin{pmatrix} 0 & 0 \\ 0 & u \end{pmatrix}, \quad u \neq 0.$$

The space spanned by 0 and $\begin{pmatrix} 0 & 0 \\ 0 & u \end{pmatrix}$ is $\Pi\left(0, \begin{pmatrix} 0 & 0 \\ 0 & 1 \end{pmatrix}\right)$, which proves the theorem.

Consider a point Z, outside a pair of complemented subspaces Π_1 and Π_2, whose arithmetic distance from the common point of Π_1 and Π_2 is 2. Suppose that Z is so chosen that there is a point P_1 of Π_1 satisfying

$$r(Z, P_1) = 1$$

and that for each point Q of $\Pi(P_1, Z)$ there exists a point P_2 of Π_2 such that $r(P_2, Q) = 1$. As Q runs over all elements of $\Pi(P_1, Z)$, the set of all points in the subspaces $\Pi(P_2, Q)$ form what we shall call a *reducible subspace* Σ. Π_1 and Π_2 will be called the *components* of Σ.

Theorem 2 *Every reducible subspace can be carried by a transformation of (8) into the set of matrices*

$$\begin{pmatrix} x & 0 \\ 0 & z \end{pmatrix}, \tag{10}$$

where x and z run over all elements of the field.

To prove the theorem, we may assume that

$$\Pi_1 = \Pi\left(0, \begin{pmatrix} 1 & 0 \\ 0 & 0 \end{pmatrix}\right), \quad \Pi_2 = \Pi\left(0, \begin{pmatrix} 0 & 0 \\ 0 & 1 \end{pmatrix}\right).$$

Then
$$Z = \begin{pmatrix} a & b \\ b & c \end{pmatrix}, \qquad ac - b^2 \neq 0.$$

To ensure the existence of P_1, we have to assume that $c \neq 0$ in which case
$$P_1 = \begin{pmatrix} a - b^2/c & 0 \\ 0 & 0 \end{pmatrix}.$$

Then $\Pi(P_1, Z)$ is the set of points:
$$Q = (1-t)\begin{pmatrix} a - b^2/c & 0 \\ 0 & 0 \end{pmatrix} + t\begin{pmatrix} a & b \\ b & c \end{pmatrix},$$

where t runs over all elements of Φ.

The existence of P_2 in Π_2 asserts that
$$(1-t)\left(a - \frac{b^2}{c}\right) + t + a \neq 0,$$

i.e.,
$$a - \frac{b^2}{c} + t\frac{b^2}{c} \neq 0$$

for all t. Consequently $b = 0$, $ac \neq 0$ and
$$P_2 = \begin{pmatrix} 0 & 0 \\ 0 & tc \end{pmatrix}.$$

Now Q takes the form
$$\begin{pmatrix} a & 0 \\ 0 & tc \end{pmatrix}$$

and the space spanned by Q and P_2 consists of the points
$$\begin{pmatrix} sa & 0 \\ 0 & tc \end{pmatrix}.$$

This proves the theorem.

3. Construction of involutions

We begin by defining an involution. Let Σ be a reducible subspace and Z a point not on Σ. Let Δ be the set of elements X on Σ for which
$$r(X, Z) \leqslant 1.$$

Then there is one and only one symmetric matrix $\overline{Z}$, different from Z, such that
$$r(X, \overline{Z}) \leqslant 1$$
for all X belonging to Δ. This defines $\overline{Z}$ for every Z not on Σ. If Z is on Σ, let $\overline{Z} = Z$. The mapping $Z \to \overline{Z}$ is called an *involution of the first kind*.

If Σ has the form (10) and
$$Z = \begin{pmatrix} a & b \\ b & c \end{pmatrix},$$
then
$$r\left(Z, \begin{pmatrix} x & 0 \\ 0 & z \end{pmatrix}\right) \leqslant 1$$
implies
$$(a-x)(c-z) - b^2 = 0. \tag{11}$$
That is, Δ is the set of elements satisfying (11). Let
$$\overline{Z} = \begin{pmatrix} \bar{a} & \bar{b} \\ \bar{b} & \bar{c} \end{pmatrix},$$
then (11) implies
$$(\bar{a}-x)(\bar{c}-z) - \bar{b}^2 = 0.$$
Consequently $a = \bar{a}$, $c = \bar{c}$ and $b^2 = \bar{b}^2$. Then $\bar{b} = -b$.

Analytically, the mapping can therefore be expressed in the form
$$\overline{Z} = \begin{pmatrix} 1 & 0 \\ 0 & -1 \end{pmatrix} Z \begin{pmatrix} 1 & 0 \\ 0 & -1 \end{pmatrix}. \tag{12}$$

Theorem 3 *The mapping $Z \to \overline{Z}$ is an involution of the first kind if and only if*
$$\overline{Z} = AZA' + S, \quad A^2 = I, \quad ASA' + S = 0. \tag{13}$$

Using (12), let
$$Z = BWB' + T$$
and
$$\overline{Z} = B\overline{W}B' + T.$$

Then
$$\overline{W} = B^{-1}\begin{pmatrix} 1 & 0 \\ 0 & -1 \end{pmatrix} BWB'\begin{pmatrix} 1 & 0 \\ 0 & -1 \end{pmatrix} B'^{-1}$$
$$+ B^{-1}\left(\begin{pmatrix} 1 & 0 \\ 0 & -1 \end{pmatrix} T \begin{pmatrix} 1 & 0 \\ 0 & -1 \end{pmatrix} - T\right) B'^{-1}, \qquad (14)$$

which is evidently of the form (13).

Conversely, since $A^2 = I$, we can find B such that
$$BAB^{-1} = \begin{pmatrix} 1 & 0 \\ 0 & -1 \end{pmatrix}$$

(or $\pm I$, but these cases give only the identity mapping). Further
$$S = \frac{1}{2}(S - ASA') = B^{-1}\left(\begin{pmatrix} 1 & 0 \\ 0 & -1 \end{pmatrix} T \begin{pmatrix} 1 & 0 \\ 0 & -1 \end{pmatrix} - T\right) B'^{-1},$$

where
$$T = -\frac{1}{2} BSB'.$$

That is, (13) can be expressed as (14), which proves the theorem.

Theorem 3 shows that an involution of the first kind is uniquely determined by a reducible subspace and that every point of this reducible space is invariant under the involution.

Two different involutions of the first kind are said to be commutative if one carries the reducible subspace of the other onto itself.

Theorem 4 *Any pair of commutative involutions of the first kind can be carried simultaneously into* (12) *and*
$$\overline{Z} = \begin{pmatrix} 0 & 1 \\ 1 & 0 \end{pmatrix} Z \begin{pmatrix} 0 & 1 \\ 1 & 0 \end{pmatrix}. \qquad (15)$$

Proof Let one of the involutions be (12). The other carries the reducible subspace
$$\begin{pmatrix} x & 0 \\ 0 & z \end{pmatrix}$$
onto itself. Let
$$A = \begin{pmatrix} a_1 & a_2 \\ a_3 & a_4 \end{pmatrix}, \quad S = \begin{pmatrix} s_1 & s_2 \\ s_2 & s_3 \end{pmatrix}$$

in (13) and $Z = \begin{pmatrix} x & y \\ y & z \end{pmatrix}$. Then we have

$$\bar{y} = a_1 a_3 x + (a_2 a_3 + a_1 a_4) y + a_2 a_4 z + s_2 = 0$$

for $y = 0$ and arbitrary x and z. Thus

$$a_1 a_3 = a_2 a_4 = s_2 = 0.$$

Since A is nonsingular, we have either

$$a_1 = a_4 = 0$$

or

$$a_2 = a_3 = 0.$$

The second case cannot happen, as otherwise the second involution coincides with the first. Thus we have $a_1 = a_4 = 0$, and

$$\bar{Z} = \begin{pmatrix} 0 & b \\ 1/b & 0 \end{pmatrix} Z \begin{pmatrix} 0 & b \\ 1/b & 0 \end{pmatrix}' + \begin{pmatrix} -sb^2 & 0 \\ 0 & s \end{pmatrix}. \tag{16}$$

Using

$$\bar{Z} = \begin{pmatrix} 0 & b \\ 1 & 0 \end{pmatrix} \overline{W} \begin{pmatrix} 0 & b \\ 1 & 0 \end{pmatrix}' + \begin{pmatrix} -b^2 s & 0 \\ 0 & 0 \end{pmatrix},$$

$$Z = \begin{pmatrix} 0 & b \\ 1 & 0 \end{pmatrix} W \begin{pmatrix} 0 & b \\ 1 & 0 \end{pmatrix}' + \begin{pmatrix} -b^2 s & 0 \\ 0 & 0 \end{pmatrix},$$

we have (12) and (15) with W instead of Z.

The product of two commutative involutions of the first kind is called *an involution of the second kind*. Therefore an involution of the second kind can be carried into

$$\bar{Z} = \begin{pmatrix} 0 & 1 \\ -1 & 0 \end{pmatrix} Z \begin{pmatrix} 0 & -1 \\ 1 & 0 \end{pmatrix}. \tag{17}$$

The fixed elements of this transformation are of the form

$$t \begin{pmatrix} 1 & 0 \\ 0 & 1 \end{pmatrix}. \tag{18}$$

The manifold of the fixed elements of an involution of the second kind is called a *chain*.

We are now going to examine all the chains in a reducible subspace. Notice that (16) represents all the involutions of the first kind which are commutative with (12). The product of (12) and (16) is

$$\overline{Z} = \begin{pmatrix} 0 & b \\ -1/b & 0 \end{pmatrix} Z \begin{pmatrix} 0 & b \\ -1/b & 0 \end{pmatrix}' + \begin{pmatrix} -sb^2 & 0 \\ 0 & s \end{pmatrix}.$$

The chain in the reducible space $\begin{pmatrix} x & 0 \\ 0 & z \end{pmatrix}$ is therefore of the form

$$\begin{pmatrix} b^2(t-s) & 0 \\ 0 & t \end{pmatrix},$$

where t runs over all elements of Φ. We put in more parameters to make it similar to a straight line. The chain is also equivalent to

$$\begin{pmatrix} at+b & 0 \\ 0 & ct+d \end{pmatrix} \tag{19}$$

if $a = cx^2$ is soluble.

In the next section, we shall obtain harmonic separation by means of the construction of complete quadrateral. But we should notice that we have neither the fact that passing through any two points there is a "line" nor the fact that any two "lines" have a point of intersection.

The chains passing through a point $\begin{pmatrix} b & 0 \\ 0 & d \end{pmatrix}$ are of the form (19). The chain passing through two points

$$\begin{pmatrix} a & 0 \\ 0 & b \end{pmatrix} \text{ and } \begin{pmatrix} c & 0 \\ 0 & d \end{pmatrix}$$

is of the form

$$\lambda \begin{pmatrix} a & 0 \\ 0 & b \end{pmatrix} + (1-\lambda) \begin{pmatrix} c & 0 \\ 0 & d \end{pmatrix}, \tag{20}$$

if it exists. The condition for existence is that

$$(a-c)/(b-d) \tag{21}$$

is a square element in Φ.

4. Proof of the theorem for $n = 2$

Theorem 5 *Every mapping carrying symmetric matrices into symmetric matrices and keeping the rank of differences invariant is a combination of a homothetic transformation*
$$Z_1 = aZ,$$
a translation
$$Z_1 = Z + S, \qquad S' = S,$$
an automorphism of the field
$$Z_1 = Z^\sigma$$
and a congruence relation
$$Z_1 = AZA', \qquad \det(A) \neq 0.$$

This is practically the theorem stated in the introduction. In this section, we shall give a proof for $n = 2$.

Notice that the notion of reduced subspaces and chains depends only on the notion of arithmetic distance.

Let $\Gamma(Z)$ be a mapping of the type under consideration. It is evidently one to one. Without loss of generality, we may assume that it carries the reducible subspace
$$\begin{pmatrix} x & 0 \\ 0 & z \end{pmatrix}$$
onto itself and the chain $t \begin{pmatrix} 1 & 0 \\ 0 & 1 \end{pmatrix}$ onto itself, and moreover, that
$$\Gamma(0) = 0, \qquad \Gamma(I) = I.$$

Let
$$\Gamma(aI) = a^\sigma I.$$

We shall prove that σ is an automorphism of the field Φ.

1) $(a+b)^\sigma = a^\sigma + b^\sigma$.

Consider only the chains in the reducible subspace
$$\begin{pmatrix} x & 0 \\ 0 & z \end{pmatrix}.$$

Any chain, different from tI, passing through 0 can be expressed as

$$\begin{pmatrix} 1 & 0 \\ 0 & p^2 \end{pmatrix} t, \quad t \in \Phi. \tag{22}$$

Any chain which does not intersect tI can be expressed as

$$tI + \begin{pmatrix} q & 0 \\ 0 & 0 \end{pmatrix}, \quad t \in \Phi. \tag{23}$$

The intersection of (22) and (23) is given by

$$\frac{q}{1-p^2} \begin{pmatrix} 1 & 0 \\ 0 & p^2 \end{pmatrix}. \tag{24}$$

The chain passing through aI and (24) is of the form

$$atI + \frac{q}{1-p^2} \begin{pmatrix} 1 & 0 \\ 0 & p^2 \end{pmatrix} (1-t), \quad t \in \Phi, \tag{25}$$

provided that

$$\left(a - \frac{q}{1-p^2}\right) \bigg/ \left(a - \frac{qp^2}{1-p^2}\right) \tag{26}$$

is a square element in Φ.

From bI, we draw a chain which does not intersect (22); it has the form

$$\begin{pmatrix} 1 & 0 \\ 0 & p^2 \end{pmatrix} t + bI, \quad t \in \Phi. \tag{27}$$

The intersection of (27) and (23) is given by

$$\frac{q}{1-p^2} \begin{pmatrix} 1 & 0 \\ 0 & p^2 \end{pmatrix} + bI. \tag{28}$$

From (28), we draw a chain which does not intersect (25); it has the form

$$\left(aI - \begin{pmatrix} 1 & 0 \\ 0 & p^2 \end{pmatrix} \frac{q}{1-p^2}\right) t + \frac{q}{1-p^2} \begin{pmatrix} 1 & 0 \\ 0 & p^2 \end{pmatrix} + bI, \quad t \in \Phi. \tag{29}$$

The existence of (29) is assured by (26). This chain intersects tI at the point $(a+b)I$.

That is, for any p and q satisfying (26), we can construct $(a+b)I$ by means of the processes of "meet" and "join". This property is certainly carried over by the

operation Γ. The existence of p and q satisfying (26) is assured since (26) is linear in q. Therefore we have
$$\Gamma(aI + bI) = \Gamma(aI) + \Gamma(bI).$$

2) $(ab)^\sigma = a^\sigma b^\sigma$.

From 0 we draw a chain
$$\begin{pmatrix} 1 & 0 \\ 0 & p^2 \end{pmatrix} t, \quad t \in \Phi. \tag{30}$$

From I, we draw another chain
$$\begin{pmatrix} 1 & 0 \\ 0 & q^2 \end{pmatrix} t + I, \quad t \in \Phi. \tag{31}$$

Their intersection is given by
$$\frac{q^2 - 1}{q^2 - p^2} \begin{pmatrix} 1 & 0 \\ 0 & p^2 \end{pmatrix}. \tag{32}$$

Connecting (32) and bI, we have
$$t \frac{q^2 - 1}{q^2 - p^2} \begin{pmatrix} 1 & 0 \\ 0 & p^2 \end{pmatrix} + (1 - t)bI, \quad t \in \Phi, \tag{33}$$

provided that
$$\left(\frac{q^2 - 1}{q^2 - p^2} - b \right) \bigg/ \left(\frac{q^2 - 1}{q^2 - p^2} p^2 - b \right) \tag{34}$$

is a square element r^2.

From aI we draw a chain not intersecting (31),
$$\begin{pmatrix} 1 & 0 \\ 0 & q^2 \end{pmatrix} t + aI, \quad t \in \Phi, \tag{35}$$

which intersects (30) at
$$\frac{q^2 - 1}{q^2 - p^2} a \begin{pmatrix} 1 & 0 \\ 0 & p^2 \end{pmatrix}. \tag{36}$$

From (36), we draw a chain not intersecting (33),
$$t \left(\frac{q^2 - 1}{q^2 - p^2} \begin{pmatrix} 1 & 0 \\ 0 & p^2 \end{pmatrix} - bI \right) + \frac{q^2 - 1}{q^2 - p^2} a \begin{pmatrix} 1 & 0 \\ 0 & p^2 \end{pmatrix}. \tag{37}$$

This intersects the chain tI at abI. The existence of (37) is assured by (34).

Therefore we have
$$(ab)^\sigma = a^\sigma b^\sigma,$$
provided that we can justify that (34) is always soluble in p, q, r. That is, for any b, we need p, q, r in the field such that
$$b = \frac{q^2 - 1}{q^2 - p^2} \frac{1 - p^2 r^2}{1 - r^2}.$$

For any b, there exist two elements u and v of the field such that
$$b = uv, \quad -2uv + u + v \neq 0.$$

Thus, we have a solution
$$r = 0, \quad p = \frac{v - u}{u + v - 2uv}, \quad q = \frac{2 - u - v}{u + v - 2uv}.$$

We may assume without loss of generality that
$$\Gamma(aI) = aI \tag{38}$$
for all a belonging to Φ.

Let $\begin{pmatrix} x & 0 \\ 0 & z \end{pmatrix}$ be any point of the reducible space which is not on the chain tI, and
$$\Gamma \begin{pmatrix} x & 0 \\ 0 & z \end{pmatrix} = \begin{pmatrix} x^* & 0 \\ 0 & z^* \end{pmatrix}. \tag{39}$$

Since there are two and only two points on the chain tI, namely
$$\begin{pmatrix} x & 0 \\ 0 & x \end{pmatrix} \text{ and } \begin{pmatrix} z & 0 \\ 0 & z \end{pmatrix}$$
such that
$$r\left(\begin{pmatrix} x & 0 \\ 0 & x \end{pmatrix}, \begin{pmatrix} x & 0 \\ 0 & z \end{pmatrix}\right) = r\left(\begin{pmatrix} z & 0 \\ 0 & z \end{pmatrix}, \begin{pmatrix} x & 0 \\ 0 & z \end{pmatrix}\right) = 1,$$
we have then
$$\Gamma \begin{pmatrix} x & 0 \\ 0 & z \end{pmatrix} = \begin{pmatrix} x & 0 \\ 0 & z \end{pmatrix} \text{ or } \begin{pmatrix} z & 0 \\ 0 & x \end{pmatrix}. \tag{40}$$

We may assume that
$$\Gamma\begin{pmatrix} 1 & 0 \\ 0 & 0 \end{pmatrix} = \begin{pmatrix} 1 & 0 \\ 0 & 0 \end{pmatrix},$$
for otherwise, we can use
$$\Gamma_1(z) = \begin{pmatrix} 0 & 1 \\ 1 & 0 \end{pmatrix} \Gamma(z) \begin{pmatrix} 0 & 1 \\ 1 & 0 \end{pmatrix}$$
instead of Γ, and Γ_1 satisfies our requirement. Consequently
$$\Gamma\begin{pmatrix} 0 & 0 \\ 0 & 1 \end{pmatrix} = \begin{pmatrix} 0 & 0 \\ 0 & 1 \end{pmatrix}.$$
Since $\begin{pmatrix} x & 0 \\ 0 & 0 \end{pmatrix}$ lies on the subspace $\Pi\left(0, \begin{pmatrix} 1 & 0 \\ 0 & 0 \end{pmatrix}\right)$, we have
$$\Gamma\begin{pmatrix} x & 0 \\ 0 & 0 \end{pmatrix} = \begin{pmatrix} x & 0 \\ 0 & 0 \end{pmatrix}$$
for all $x \in \Phi$. Similarly, we have
$$\Gamma\begin{pmatrix} 0 & 0 \\ 0 & z \end{pmatrix} = \begin{pmatrix} 0 & 0 \\ 0 & z \end{pmatrix}.$$

Since
$$r\left(\begin{pmatrix} x & 0 \\ 0 & z \end{pmatrix}, \begin{pmatrix} x & 0 \\ 0 & 0 \end{pmatrix}\right) = 1,$$
but
$$r\left(\begin{pmatrix} z & 0 \\ 0 & x \end{pmatrix}, \begin{pmatrix} x & 0 \\ 0 & 0 \end{pmatrix}\right) \neq 1,$$
we have inmediately
$$\Gamma\begin{pmatrix} x & 0 \\ 0 & z \end{pmatrix} = \begin{pmatrix} x & 0 \\ 0 & z \end{pmatrix}. \tag{41}$$

Now we let
$$\Gamma\begin{pmatrix} a & b \\ b & c \end{pmatrix} = \begin{pmatrix} a^* & b^* \\ b^* & c^* \end{pmatrix}, \tag{42}$$
we consider these $\begin{pmatrix} x & 0 \\ 0 & z \end{pmatrix}$ satisfying
$$r\left(\begin{pmatrix} a & b \\ b & c \end{pmatrix}, \begin{pmatrix} x & 0 \\ 0 & z \end{pmatrix}\right) \leqslant 1,$$

that is
$$(a-x)(c-z) - b^2 = 0.$$

By (41) and (42), this equation implies
$$(a^* - x)(c^* - z) - b^{*2} = 0$$

and vice versa. Thus we have
$$a = a^*, \quad c = c^*, \quad b = \pm b^*.$$

Since
$$\begin{pmatrix} 1 & 0 \\ 0 & -1 \end{pmatrix} \begin{pmatrix} a & b \\ b & c \end{pmatrix} \begin{pmatrix} 1 & 0 \\ 0 & -1 \end{pmatrix} = \begin{pmatrix} a & -b \\ -b & c \end{pmatrix},$$

we may assume that
$$\Gamma \begin{pmatrix} x & 1 \\ 1 & z \end{pmatrix} = \begin{pmatrix} x & 1 \\ 1 & z \end{pmatrix}. \tag{43}$$

Suppose that there exists $b \neq 0$, such that
$$\Gamma \begin{pmatrix} a & b \\ b & c \end{pmatrix} = \begin{pmatrix} a & -b \\ -b & c \end{pmatrix}.$$

By (43), we should then have
$$(x-a)(z-c) - (1-b)^2 = 0,$$

which implies
$$(x-a)(z-c) - (1+b)^2 = 0.$$

This is impossible, since the characteristic of the field is different from 2. Therefore
$$\Gamma \begin{pmatrix} a & b \\ b & c \end{pmatrix} = \begin{pmatrix} a & b \\ b & c \end{pmatrix}$$

and the theorem is proved for $n = 2$.

Remark The theorem seems still to be true for the field with characteristic 2. For the prime field with characteristic 2, the proof is quite simple. The space contains only eight points
$$\begin{pmatrix} 0 & 0 \\ 0 & 0 \end{pmatrix}, \begin{pmatrix} 1 & 0 \\ 0 & 0 \end{pmatrix}, \begin{pmatrix} 0 & 0 \\ 0 & 1 \end{pmatrix}, \begin{pmatrix} 1 & 1 \\ 1 & 1 \end{pmatrix},$$
$$\begin{pmatrix} 1 & 0 \\ 0 & 1 \end{pmatrix}, \begin{pmatrix} 1 & 1 \\ 1 & 0 \end{pmatrix}, \begin{pmatrix} 0 & 1 \\ 1 & 1 \end{pmatrix}, \begin{pmatrix} 0 & 1 \\ 1 & 0 \end{pmatrix}.$$

We may assume that

$$\Gamma\begin{pmatrix} 0 & 0 \\ 0 & 0 \end{pmatrix} = \begin{pmatrix} 0 & 0 \\ 0 & 0 \end{pmatrix}, \quad \Gamma\begin{pmatrix} 1 & 0 \\ 0 & 0 \end{pmatrix} = \begin{pmatrix} 1 & 0 \\ 0 & 0 \end{pmatrix},$$

$$\Gamma\begin{pmatrix} 0 & 0 \\ 0 & 1 \end{pmatrix} = \begin{pmatrix} 0 & 0 \\ 0 & 1 \end{pmatrix}.$$

The only singular matrix which has not been taken into consideration is $\begin{pmatrix} 1 & 1 \\ 1 & 1 \end{pmatrix}$. Therefore

$$\Gamma\begin{pmatrix} 1 & 1 \\ 1 & 1 \end{pmatrix} = \begin{pmatrix} 1 & 1 \\ 1 & 1 \end{pmatrix}.$$

Since $\begin{pmatrix} 1 & 0 \\ 0 & 1 \end{pmatrix}$ is the only element satisfying

$$r\left(\begin{pmatrix} 1 & 0 \\ 0 & 1 \end{pmatrix}, \begin{pmatrix} 1 & 0 \\ 0 & 0 \end{pmatrix}\right) = r\left(\begin{pmatrix} 1 & 0 \\ 0 & 1 \end{pmatrix}, \begin{pmatrix} 0 & 0 \\ 0 & 1 \end{pmatrix}\right) = 1,$$

we have

$$\Gamma\begin{pmatrix} 1 & 0 \\ 0 & 1 \end{pmatrix} = \begin{pmatrix} 1 & 0 \\ 0 & 1 \end{pmatrix}.$$

Similarly

$$\Gamma\begin{pmatrix} 1 & 1 \\ 1 & 0 \end{pmatrix} = \begin{pmatrix} 1 & 1 \\ 1 & 0 \end{pmatrix}, \quad \Gamma\begin{pmatrix} 0 & 1 \\ 1 & 1 \end{pmatrix} = \begin{pmatrix} 0 & 1 \\ 1 & 1 \end{pmatrix}.$$

Finally, we have

$$\Gamma\begin{pmatrix} 0 & 1 \\ 1 & 0 \end{pmatrix} = \begin{pmatrix} 0 & 0 \\ 1 & 0 \end{pmatrix}.$$

5. Subspaces

In order to extend our notion of subspaces to higher dimensions, we introduce the concept of a dieder manifold.

Definition Let P and Q be two symmetric matrices of order n. The symmetric matrices X satisfying

$$r(P, X) + r(X, Q) = r(P, Q) \tag{44}$$

form a manifold which is called a *dieder manifold* spanned by the points P and Q. The arithmetic distance between P and Q is called the *extent* of the dieder manifold.

Since the pairs of points having a fixed arithmetic distance form a transitive set[①], the dieder manifolds of the same extent form a transitive set.

Theorem 6 *If $p < n$, the points on the dieder manifold spanned by*

$$0, \quad \begin{pmatrix} I^{(p)} & 0 \\ 0 & 0 \end{pmatrix}$$

are of the form

$$\begin{pmatrix} X^{(p)} & 0 \\ 0 & 0 \end{pmatrix}. \tag{45}$$

Proof Let X be a point on the dieder manifold and

$$X = \begin{pmatrix} X_{11}^{(p)} & X_{12} \\ X_{12}^1 & X_{22} \end{pmatrix}, \quad X_{12} = X_{12}^{(p,n-p)}, \quad X_{22} = X_{22}^{(n-p)}.$$

Since

$$p = r(P, X) + r(X, Q) \geqslant r(0, X_{11}) + r(X_{11}, I^{(p)}) \geqslant r(0, I^{(p)}) = p,$$

we have

$$r(0, X_{11}) + r(X_{11}, I^{(p)}) = p. \tag{46}$$

There is a nonsingular p-rowed matrix Γ such that $\Gamma X_{11} \Gamma^{-1}$ is of the normal form. In virtue of (46), we find that the normal form becomes

$$\Gamma X_{11} \Gamma^{-1} = \begin{pmatrix} I^{(p)} & 0 \\ 0 & 0 \end{pmatrix}, \quad 0 \leqslant q \leqslant p. \tag{47}$$

We shall now prove that $X_{12} = 0, X_{22} = 0$. Let

$$X_0 = \begin{pmatrix} X_{11}^{(p)} & v' \\ v & a \end{pmatrix}$$

be a $(p+1)$-rowed principal minor of X; it is sufficient to prove that $v = 0, a = 0$. From (44), we deduce in the same way used to establish (46), that

$$p = r(p, X) + r(X, Q) \geqslant r(0^{(p+1)}, X_0) + r\left(X_0, \begin{pmatrix} I^{(p)} & 0 \\ 0 & 0^{(1)} \end{pmatrix}\right) \geqslant p.$$

Consequently, we have

$$r(0^{(p+1)}, X_0) = r(0^{(p)}, X_{11}) = q \tag{48}$$

---

① I, Theorem 2.

and
$$r\left(X_0, \begin{pmatrix} I^{(p)} & 0 \\ 0 & 0 \end{pmatrix}\right) = r(X_{11}, I^{(p)}) = p - q. \tag{49}$$

From (48), it follows that

$$X_0 = \begin{pmatrix} \Gamma^{-1} & 0 \\ 0 & 0 \end{pmatrix} \begin{pmatrix} I^{(q)} & 0 & w'_1 \\ 0 & 0^{(p-q)} & w'_2 \\ u_1 & u_2 & 0 \end{pmatrix} \begin{pmatrix} \Gamma & 0 \\ 0 & 1 \end{pmatrix}$$

is of rank q, where

$$(u_1, u_2)\Gamma = v, \qquad (w_1, w_2)\Gamma'^{-1} = v.$$

Since

$$\begin{pmatrix} I^{(q)} & 0 & 0 \\ 0 & I^{(p-q)} & 0 \\ -u_1 & 0 & 1 \end{pmatrix} \begin{pmatrix} I^{(q)} & 0 & w'_1 \\ 0 & 0 & w'_2 \\ u_1 & u_2 & a \end{pmatrix} \begin{pmatrix} I^{(q)} & 0 & -w'_1 \\ 0 & I & 0 \\ 0 & 0 & 1 \end{pmatrix}$$

$$= \begin{pmatrix} I^{(q)} & 0 & 0 \\ 0 & 0 & w'_2 \\ 0 & u_2 & a - u_1 w'_1 \end{pmatrix}$$

is of rank q, we deduce immediately that $u_2 = 0$, $w_2 = 0$, and $a - u_1 w'_1 = 0$.

Using (49), we see also that $u_1 = 0$, $w_1 = 0$ and $a - u_2 w'_2 = 0$. Therefore $v = 0$ and $a = 0$. This establishes our theorem.

Definition Let $p < n$. A set of points is said to form a *normal subspace of rank p*, if every pair of points of the set has arithmetic distance $\leqslant p$, if it contains two points of arithmetic distance p, and if it contains all dieder manifolds spanned by any two points of the set.

Theorem 7 *Normal subspaces of the same rank form a transitive set; more precisely, every normal subspace of rank p can be transformed by (8) into the normal subspace*

$$\begin{pmatrix} X^{(p)} & 0 \\ 0 & 0 \end{pmatrix}, \tag{50}$$

where $X^{(p)}$ runs over all p-rowed symmetric matrices.

For the proof of this theorem see III, Theorem 2, p. 231(2).

6. Proof of Theorem 5 in general

We now suppose that $n \geqslant 3$. Let

$$\Gamma(Z) = Z_1$$

be the mapping under consideration. Suppose that

$$\Gamma(0) = 0.$$

The points of the form

$$\begin{pmatrix} X_1^{(n-1)} & 0 \\ 0 & 0 \end{pmatrix}$$

form a normal subspace of rank $n-1$. Since the arithmetic distance is invariant, the set of points

$$\Gamma \begin{pmatrix} X^{(n-1)} & 0 \\ 0 & 0 \end{pmatrix}$$

constitutes also a normal subspace of rank $n-1$. Since the set of all normal subspaces forms a transitive set (Theorem 7), we may suppose that

$$\Gamma \begin{pmatrix} W^{(n-1)} & 0 \\ 0 & 0 \end{pmatrix} = \begin{pmatrix} W_1'^{(n-1)} & 0 \\ 0 & 0 \end{pmatrix}.$$

Thus Γ induces a mapping on $(n-1)$-rowed symmetric matrices, and it keeps arithmetic distance invariant. By the hypothesis of induction, we have

$$W = a\alpha W_1^\sigma \alpha' + \mu, \qquad \alpha = \alpha^{(n-1)}, \qquad \mu' = \mu = \mu^{(n-1)}.$$

Then the mapping

$$Z = aAZ_1^\sigma A' + S,$$

where

$$A = \begin{pmatrix} \alpha & 0 \\ 0 & 1 \end{pmatrix}, \qquad S = \begin{pmatrix} \mu & 0 \\ 0 & 0 \end{pmatrix}$$

carries the mapping $\Gamma(Z)$ into a new one with

$$\Gamma \begin{pmatrix} W^{(n-1)} & 0 \\ 0 & 0 \end{pmatrix} = \begin{pmatrix} W^{(n-1)} & 0 \\ 0 & 1 \end{pmatrix}. \tag{51}$$

Since

$$\begin{pmatrix} 0^{(n-1)} & 0 \\ 0 & 1 \end{pmatrix}$$

is of rank 1, we may let

$$\Gamma \begin{pmatrix} 0^{(n-1)} & 0 \\ 0 & 1 \end{pmatrix} = \varepsilon(a_1, \cdots, a_n)'(a_1, \cdots, a_n), \qquad (52)$$

where ε is either 1 or $x^2 = \varepsilon$ is not soluble. Since the arithmetic distance between

$$\begin{pmatrix} I^{(n-1)} & 0 \\ 0 & 0 \end{pmatrix} \text{ and } \begin{pmatrix} 0^{(n-1)} & 0 \\ 0 & 0 \end{pmatrix}$$

is n, by (51), we have $a_n \neq 0$. Let

$$A = \begin{pmatrix} I^{(n-1)} & 0 \\ -a_n^{-1}v & a_n^{-1} \end{pmatrix}, \qquad v = (a_1, \cdots, a_{n-1}).$$

Then

$$Z_1 = A'ZA$$

carries (51) into itself and

$$A'(a_1, \cdots, a_n)'(a_1, \cdots, a_n)A = (0, 0, \cdots, 1)'(0, 0, \cdots, 1).$$

Thus we may assume further that

$$\Gamma \begin{pmatrix} 0^{(n-1)} & 0 \\ 0 & 1 \end{pmatrix} = \varepsilon \begin{pmatrix} 0^{(n-1)} & 0 \\ 0 & 1 \end{pmatrix}. \qquad (53)$$

The points

$$\begin{pmatrix} 1 & 0 & \cdots & 0 \\ 0 & 0 & \cdots & 0 \\ \vdots & \vdots & & \vdots \\ 0 & 0 & \cdots & 0 \end{pmatrix} \text{ and } \begin{pmatrix} 0 & 0 & \cdots & 0 \\ 0 & 0 & \cdots & 0 \\ \vdots & \vdots & & \vdots \\ 0 & 0 & \cdots & \varepsilon \end{pmatrix}, \qquad \varepsilon \neq 0$$

span a normal subspace

$$\begin{pmatrix} a_{11} & 0 & \cdots & a_{1n} \\ 0 & 0 & \cdots & 0 \\ \vdots & \vdots & & \vdots \\ 0 & 0 & \cdots & 0 \\ a_{1n} & 0 & \cdots & a_{nn} \end{pmatrix}$$

of rank 2. The mapping Γ induces an automorphic mapping on two-rowed symmetric matrices
$$\begin{pmatrix} a_{11} & a_{1n} \\ a_{1n} & a_{nn} \end{pmatrix}.$$
Let it be
$$\gamma \begin{pmatrix} a_{11} & a_{1n} \\ a_{1n} & a_{nn} \end{pmatrix} = \begin{pmatrix} a_{11}^* & a_{1n}^* \\ a_{1n}^* & a_{nn}^* \end{pmatrix},$$
and it satisfies
$$\gamma(0) = 0, \quad \gamma \begin{pmatrix} x & 0 \\ 0 & 0 \end{pmatrix} = \begin{pmatrix} x & 0 \\ 0 & 0 \end{pmatrix}, \quad \gamma \begin{pmatrix} 0 & 0 \\ 0 & 1 \end{pmatrix} = \begin{pmatrix} 0 & 0 \\ 0 & \varepsilon \end{pmatrix}. \quad (54)$$
Since the theorem is true for $n = 2$, we have, by the first equation of (54),
$$\gamma \begin{pmatrix} a_{11} & a_{1n} \\ a_{1n} & a_{nn} \end{pmatrix} = aA \begin{pmatrix} a_{11} & a_{1n} \\ a_{1n} & a_{nn} \end{pmatrix}^\sigma A', \quad A = A^{(2)},$$
where a is either 1 or a nonsquare element of Φ. From the second and third equations of (54), we deduce
$$A = \begin{pmatrix} b & 0 \\ 0 & c \end{pmatrix}$$
and
$$ab^2 x^\sigma = x, \quad ac^2 = \varepsilon^\sigma$$
for all x. If in particular $x = 1$, we see that a and ε must be square elements, whence $a = \varepsilon = 1$, and $b^2 = c^2 = 1$. Consequently $x^\sigma = x$, that is, σ is the identity automorphism. Substantially, we have only two cases: (1) $b = c = 1$ and (2) $b = -c = 1$.

Thus
$$\Gamma \begin{pmatrix} a_{11} & 0 & \cdots & 0 & a_{1n} \\ 0 & 0 & \cdots & 0 & 0 \\ \vdots & \vdots & & \vdots & \vdots \\ a_{1n} & 0 & \cdots & 0 & a_{nn} \end{pmatrix} = \begin{pmatrix} a_{11} & 0 & \cdots & 0 & \pm a_{1n} \\ 0 & 0 & \cdots & 0 & 0 \\ \vdots & \vdots & & \vdots & \vdots \\ \pm a_{1n} & 0 & \cdots & 0 & a_{nn} \end{pmatrix}. \quad (55)$$

The mapping Γ leaves
$$\begin{pmatrix} 0^{(1)} & 0 & 0 \\ 0 & I^{(n-2)} & 0 \\ 0 & 0 & 1^{(1)} \end{pmatrix}, \quad \begin{pmatrix} 0^{(1)} & 0 & 0 \\ 0 & 0^{(n-2)} & 0 \\ 0 & 0 & 1^{(1)} \end{pmatrix} \quad (56)$$

invariant and
$$\Gamma\begin{pmatrix} 0 & 0 \\ 0 & W^{(n-1)} \end{pmatrix} = \begin{pmatrix} 0 & 0 \\ 0 & W^{(n-1)} \end{pmatrix},$$
since two points of (56) span the normal subspace in the bracket. We have by the hypothesis of induction that
$$W = aAW^\tau A' + S, \qquad A = A^{(n-1)}, \qquad S = S^{(n-1)} = S',$$
where a is either 1 or a non-square element. By (51) and (55), it leaves
$$W = \begin{pmatrix} X^{(n-2)} & 0 \\ 0 & 0 \end{pmatrix}, \qquad \begin{pmatrix} 0^{(n-2)} & 0 \\ 0 & x \end{pmatrix}$$
pointwise invariant, whence $S = 0$, $a = 1, \tau = 1$ and
$$A^{(n-1)} = I \quad \text{or} \quad A^{(n-1)} = \begin{pmatrix} I^{(n-2)} & 0 \\ 0 & -1 \end{pmatrix}.$$

For the former case we have
$$\Gamma\begin{pmatrix} 0 & 0 \\ 0 & W^{(n-1)} \end{pmatrix} = \begin{pmatrix} 0 & 0 \\ 0 & W^{(n-1)} \end{pmatrix}. \tag{57}$$

For the latter case
$$Z_1 = \begin{pmatrix} I^{(n-1)} & 0 \\ 0 & -1 \end{pmatrix} Z \begin{pmatrix} I^{(n-1)} & 0 \\ 0 & -1 \end{pmatrix}$$
carries Γ into a new transformation which satisfies (51), (55) and (57).

As in III, p. 238, we deduce
$$\Gamma((z_{ij})) = (z_{ij}^1), \qquad z_{ii} = z_{ii}^1, \qquad z_{1n} = \pm z_{1n}^1$$
and
$$z_{ij} = z_{ij}^1 \qquad \text{for } (i,j) \neq (1,n).$$
Suppose that there exists a $Z = (z_{ij})$ such that
$$\Gamma\begin{pmatrix} z_{11} & \cdots & z_{1n} \\ \vdots & & \vdots \\ z_{1n} & \cdots & z_{nn} \end{pmatrix} = \begin{pmatrix} z_{11} & \cdots & z_{1,n-1} & -z_{1n} \\ \vdots & & \vdots & \vdots \\ -z_{1n} & \cdots & z_{n-1,n} & -z_{nn} \end{pmatrix}, \quad z_{1n} \neq 0.$$

Since

$$\Gamma\begin{pmatrix} z_{11} - z_{1n} & \cdots & z_{1,n-1} - z_{1n} & 0 \\ \vdots & & \vdots & \vdots \\ 0 & \cdots & z_{n-1,n} - z_{1n} & z_{nn} - z_{1n} \end{pmatrix}$$

$$= \begin{pmatrix} z_{11} - z_{1n} & \cdots & z_{1,n-1} - z_{1n} & 0 \\ \vdots & & \vdots & \vdots \\ 0 & \cdots & z_{n-1,n} - z_{1n} & z_{nn} - z_{1n} \end{pmatrix}$$

and since arithmetic distance is invariant, we see that the matrices

$$\begin{pmatrix} z_{1n} & \cdots & z_{1n} & z_{1n} \\ \vdots & & \vdots & \vdots \\ z_{1n} & \cdots & z_{1n} & z_{1n} \end{pmatrix} \text{ and } \begin{pmatrix} z_{1n} & \cdots & z_{1n} & -z_{1n} \\ z_{1n} & \cdots & z_{1n} & z_{1n} \\ \vdots & & \vdots & \vdots \\ -z_{1n} & \cdots & z_{1n} & z_{1n} \end{pmatrix}$$

have the same rank. This is a contradiction, since the former is of rank 1 and the latter is of rank 3. Thus $z_{1n}^1 = z_{1n}$. That is,

$$\Gamma(Z) = Z.$$

The theorem follows.

7. Proof of the fundamental theorem of the projective geometry of symmetric matrices

A point (X, Y), in homogeneous coordinate, is called finite if Y is nonsingular, otherwise, it is called infinite. The patricular infinite point $(I, 0)$ is denoted by ∞. Evidently a necessary and sufficient condition for a point P to be finite is that

$$r(P, \infty) = n.$$

Since the projective space of symmetric matrices is transitive, we may assume that the automorphic mapping of the space carries ∞ into itself. If it keeps arithmetic distance invariant, the mapping carries finite points into finite points and keeps the arithmetic distance of finite points invariant. By Theorem 5, we have

Theorem 8 Let $n > 1$. An automorphic mapping of the projective space keeping arithmetic distance invariant is of the form

$$(X_1, Y_1) = Q(X^\sigma, Y^\sigma)\mathfrak{T},$$

where

$$\mathfrak{T}\mathfrak{F}\mathfrak{T}' = a\mathfrak{F}.$$

8. Remarks

1°. The condition stated in the theorem can be somewhat weakened since the invariance of the arithmetic distance 1 implies the invariance of arithmetic distances > 1. In fact we have

Theorem 9 Two points P and Q are of arithmetic distance p if and only if there exists $p-1$ points $X_1, \cdots, X_{p-1}$ such that

$$r(P, X_1) = r(X_1, X_2) = \cdots = r(X_{p-2}, X_{p-1}) = r(X_{p-1}, Q) = 1, \qquad (58)$$

p being the least integer with this property.

Proof 1) If P and Q have the arithmetic distance p, we may assume without loss of generality that they are

$$0, \quad \begin{pmatrix} I^{(p)} & 0 \\ 0 & 0 \end{pmatrix}$$

respectively. Then,

$$X_q = \begin{pmatrix} I^{(q)} & 0 \\ 0 & 0 \end{pmatrix}, \quad q = 1, 2, \cdots, p-1$$

are the points required.

2) By the triangle inequality for arithmetic distances, we have

$$r(P, Q) \leqslant r(R, X_1) + r(X_1, X_2) + \cdots + r(X_{p-1}, Q) \leqslant p.$$

If $r(P, Q) < p$, then we should have fewer X's such that (58) holds. Therefore $r(P, Q) = p$.

From this we deduce immediately the following sharper result:

Theorem 10 Let Σ be the set of all symmetric matrices of order $n \geqslant 2$ over a field Φ with characteristic $\neq 2$. Any one to one mapping of Σ onto itself carrying a pair of symmetric matrices with arithmetic distance one into a pair with the same property is of the form

$$Z_1 = aAZ^\sigma A' + S, \quad S = S',$$

where $a \in \Phi$ and σ is an automorphism of Φ.

2°. The previous theorem seems to remain true for a field with characteristic 2.

3°. The method for the construction of involutions has not been used to its full capacity. In fact, it is very likely that we can construct the whole symplectic group by means of this procedure.

On the multiplicative group of a field*

Theroem 1 *The multiplicative group of a field is either abelian or not metabelian.*

Let K be the field under consideration and let Z be its center. If K is identical to Z, then the multiplicative group K_0 is abelian. Now we are going to prove that if Z is a proper subfield of K, then the multiplicative group K_0 is not metabelian. By the words "not metabelian", we mean that the derived series of K_0,

$$K_0 \supset K_1 \supset K_2 \supset \cdots \supset K_r \supset \cdots$$

never ends at identity, where K_r is the commutator subgroup of K_{r-1}. We use

$$(a,b) = a^{-1}b^{-1}ab$$

to denote the commutator of a and b, then K_r is generated by elements of the form (a,b), where a and b run over all the elements of K_{r-1}.

To prove the theorem we require several lemmas:

Lemma 1 *Let $\Delta = C(x,y)$ be a field obtained from its center C by the adjunction of two elements x and y satisfying*

$$xy = ryx,$$

where r ($\neq 1$) belongs to the center. A necessary and sufficient condition for x being algebraic with respect to C is that r is a root of unity.

Proof 1) Suppose that r is a q-th root of unity ($q > 1$). Then x^q commutes with y, therefore it commutes with every element of Δ. It belongs to C. Then x is algebraic.

2) Suppose x is algebraic, say the equation

$$f(x) = a_k x^k + \cdots + a_0 = 0 \tag{1}$$

has the least degree k. Transforming (1) by y, we have

$$y^{-1}f(x)y = a_k r^k x^k + \cdots + a_1 rx + a_0 = 0. \tag{2}$$

---

* Received June 7, 1950.
Reprinted from *Science Record*, 1950, **3**(1): 1–6.

Substracting (1) from (2) and dividing by x, we have a polynomial of degree less than k unless $a_h(r^h - 1) = 0$ for $1 \leqslant h \leqslant k$. Since $a_k \neq 0$, we have $r^k = 1$. The lemma is therefore proved.

Lemma 2 Let Σ be a field processing two elements x and y satisfying $xy = ryx$, where r belongs to the center of Σ. We define

$$x_1 = (y, 1-x) = y^{-1}(1-x)^{-1}y(1-x) = (1-rx)^{-1}(1-x)$$

and $x_l = (y, x_{l-1})$. Then

$$x_l = x_l(x) = \prod_{t=0}^{l}(1 - r^t x)^{(-1)^t \binom{l}{t}}.$$

Proof The lemma is evidently true for $l = 1$. By the hypothesis of induction, we have

$$x_l = y^{-1} x_{l-1}^{-1}(x) y x_{l-1}(x) = (x_{l-1}(rx))^{-1} x_{l-1}(x)$$

$$= \prod_{t=1}^{l}(1 - r^t x)^{(-1)^t \binom{l-1}{t-1}} \prod_{t=0}^{l-1}(1 - r^t x)^{(-1)^t \binom{l-1}{t})}$$

$$= \prod_{t=0}^{l}(1 - r^t x)^{(-1)^t (\binom{l-1}{t-1} + \binom{l-1}{t}))}.$$

Thus the lemma is proved.

Proof of the theorem In a previous paper the author[1] proved that K_1 is not contained in the center Z. We suppose that K_{r-1} is not contained in the center Z, but K_r is contained in the center Z.

Let

$$a_1, a_2, \cdots$$

be a representative system of K_{r-1} by K_r. By a theorem of the author[2], K_{r-1} generates the field K, thus we have

$$K = Z(a_1, a_2, \cdots),$$

where a_i belongs to K_{r-1} and (a_i, a_j) belongs to K_r, then to Z.

Since K is not commutative, we have $x = a_i$ and $y = a_j$ such that

$$(x, y) = r,$$

which belongs to Z but $r \neq 1$.

We define x_l as in Lemma 2. Then for $l \geqslant r$, x_l belongs to K_r, since y and x_s belongs to K_{r-1} and K_s respectively.

1) r is not a root of unity. By Lemma 1, x is transcendental with respect to Z. Then, for $l \geqslant r$

$$x_l = \prod_{t=0}^{l}(1-r^t x)^{(-1)^t \binom{l}{t}}$$

cannot belong to Z. For otherwise, from

$$x_l \prod_{0 \leqslant 2t+1 \leqslant l}(1-r^{2t+1}x)^{\binom{l}{2t+1}} = \prod_{0 \leqslant 2t \leqslant l}(1-r^{2t}x)^{\binom{l}{2t}},$$

we deduce that $x_l = 1$ and

$$\sum_{0 \leqslant 2t+1 \leqslant l} \binom{l}{2t+1} r^{2t+1} = \sum_{0 \leqslant 2t \leqslant l} \binom{l}{2t} r^{2t} \tag{3}$$

by comparing the constant term and the coefficient of x. (3) can be written as

$$(1-r)^l = \sum_{t=0}^{l}(-1)^t \binom{l}{t} r^t = 0,$$

which is impossible.

2) r is a q-th primitive root of unity. Let us consider the field $\Delta = Z(x,y)$. Let C be the center of Δ, then, evidently we have $\Delta = C(x,y)$. If C contains only a finite number of elements, then Δ is a finite field, and consequently it is commutative. This is a contradiction. Now we assume that C contains infinitely many elements.

Since x^q is commutative with y, it belongs to C, say

$$x^q = \alpha.$$

This is a polynomial of the lowest degree satisfied by x, for otherwise from

$$f(x) - y^{-1}f(x)y = f(x) - f(rx) = 0,$$

we deduce that the order of r is less than q.

As a runs through C,

$$x_l(as) = \prod_{t=0}^{l}(1-r^t ax)^{(-1)^t \binom{l}{t}} = \prod_{t=0}^{q-1}(1-r^t ax)^{\lambda_t},$$

where

$$\lambda_t = \sum_{0 \leqslant t+qs \leqslant l}(-1)^{t+qs} \binom{l}{t+qs}.$$

For $l \geqslant r$, $x_l(ax)$ belongs to C for all a. Notice that

$$\prod_{t=0}^{q-1}(1 - r^t ax) = 1 - (ax)^q = 1 - \alpha a^q$$

belongs to the center. Let

$$-N = \min_{0 \leqslant t \leqslant q-1} \lambda_t.$$

Since $\sum_{t=0}^{q-1} \lambda_t = 0$, N is a positive integer. Write

$$x_l(ax)(1 - \alpha a^q)^N = \prod_{t=0}^{q-1}(1 - r^t ax)^{\lambda_t + N}$$

$$= A_0(a) + A_1(a)x + \cdots + A_{q-1}(a)x^{q-1}, \tag{4}$$

where the coefficient of a in $A_1(a)$ is equal to

$$\sum_{t=0}^{q-1}(\lambda_t + N)r^t = \sum_{t=0}^{q-1} \lambda_t r^t = \sum_{0 \leqslant m \leqslant l}(-r)^m \binom{l}{m} = (1-r)^l \neq 0. \tag{5}$$

Since the left hand side of (4) belongs to C and does not satisfy any polynomial of degree less than q, we deduce

$$A_1(a) = 0,$$

for all a belonging to C. Since C contains infinitely many elements, the coefficient of a in $A_1(a)$ must be zero. This contradicts (5). The theorem is completely proved.

As corollaries, we have the following theorems:

Theorem 2 *Let K_r be the r-th derived group of the multiplicative group of a not commutative field. Then the field is generated by K_r.*

Since

$$a^{-1}b^{-1}ab = a^{-2}(ab^{-1})^2 b^2,$$

we have

Theorem 3 *Every field which is not commutative has no proper subfield containing all the squares of the r-th commutator.*

This theorem was proved previously by the author[3] for $r = 1$.

Remark 1 Theorem 2 can be easily extended to all sorts of commutator subgroups.

In fact, let us define by induction a form of r steps by (a,b) where the greater of the steps of a and b is equal to $r-1$. We fix a form of r steps, and let each alphabet involved run over the elements of the group K_0, then we obtain a system of elements L. The group G generated by all the elements of L is called a commutator subgroup of step r. Then the field K, not commutative, is generated by G. For the set L contains all r-th commutators, this can be proved easily by induction.

Remark 2 The corresponding theorem about the multiplicative group of invertible elements of a simple ring with descending chain condition is known. In fact, we can prove that $K_1 = K_2$ except the 2-rowed total matrix algebraic over field of two or three elements.

References

[1] Hua L K. On the automorphisms of classical group. *Monographs of Amer. Math. Soc.*, 1950.

[2] ———. Some properties of a field. *Proc. of Nat. Acad. of Sciences*, 1949, **35**: 533-537, Theorem 1.

[3] ———. *ibid*, Theorem 6.

环之准同构及对射影几何的一应用[*]

于一九四九年著者[1]曾解决域之广义同构 (semi-automorphism of a field) 问题. 关于此问题之历史, 可参阅 Reidemeister[2], Wachs[3], Ancochea[4-6], Kaplansky[7]. 本文将简化原证, 并将其推展至无零因子之环 (ring without zero divisor), 然后再证明所谓 "射影几何之基本定理". 前仅 Ancochea[5] 证得一特例, 该例中限定四元域 (quaternion); 对任何域, 既未证明该条件为充分, 亦未能证其为必要. 本文将证明此为一必要且充分之条件.

定义 命 R 及 R' 为二环, 且设 R' 无零因子, 其间有一对应之关系: 对 R 之一元素 a, R' 中有一元素 a^σ 对应之. 若此对应适合下列之性质, 则谓之准同构 (homomorphism),

$$(a+b)^\sigma = a^\sigma + b^\sigma, \tag{1}$$

$$(ab)^\sigma = a^\sigma b^\sigma. \tag{2}$$

如换 (2) 为

$$(ab)^\sigma = b^\sigma a^\sigma, \tag{3}$$

则该对应谓之反准同构 (anti-homomorphism); 若换 (2) 为

$$(a^2)^\sigma = (a^\sigma)^2 \tag{4}$$

及

$$(aba)^\sigma = a^\sigma b^\sigma a^\sigma, \tag{5}$$

则谓之广义准同构 (semi-homomorphism). 显然准同构及反准同构皆为广义准同构. 但其逆如何? 换言之, 是否有异于准同构及反准同构之广义准同构存在? 此为一前所未决之问题, 今解决之如下:

定理一 广义准同构必为准同构或反准同构.

证明 由 (5) 及 (1) 可得

$$\begin{aligned}(abc+cba)^\sigma &= [(a+c)b(a+c) - aba - cbc]^\sigma \\ &= (a^\sigma + c^\sigma)b^\sigma(a^\sigma + c^\sigma) - a^\sigma b^\sigma a^\sigma - c^\sigma b^\sigma c^\sigma \\ &= a^\sigma b^\sigma c^\sigma + c^\sigma b^\sigma a^\sigma. \end{aligned} \tag{6}$$

---

[*] 1950 年 4 月收到. 发表于《中国科学》, 1950, **1**(1): 1–6.

由 (4) 及 (6) 可得

$$[(ab)^\sigma - a^\sigma b^\sigma][(ab)^\sigma - b^\sigma a^\sigma]$$
$$=[(ab)^\sigma]^2 + a^\sigma(b^\sigma)^2 a^\sigma - [a^\sigma b^\sigma(ab)^\sigma + (ab)^\sigma b^\sigma a^\sigma]$$
$$=[(ab)^2 + ab^2 a - ab(ab) - (ab)ba]^\sigma = 0.$$

因 R' 无零因子, 故应得

$$(ab)^\sigma = \begin{cases} a^\sigma b^\sigma \text{ 或} \\ b^\sigma a^\sigma. \end{cases} \tag{7}$$

若有一对元素 a 及 b 使

$$(ab)^\sigma = a^\sigma b^\sigma \neq b^\sigma a^\sigma, \tag{8}$$

吾往证, 对任一元素 c 常有

$$(cb)^\sigma = c^\sigma b^\sigma. \tag{9}$$

由此式不真, 则由 (7) 可知

$$(cb)^\sigma = b^\sigma c^\sigma \neq c^\sigma b^\sigma. \tag{10}$$

再由 (7) 及 (8), (10) 可知

$$a^\sigma b^\sigma + b^\sigma c^\sigma = (ab)^\sigma + (cb)^\sigma = [(a+c)b]^\sigma = \begin{cases} (a^\sigma + c^\sigma)b^\sigma \text{ 或} \\ b^\sigma(a^\sigma + c^\sigma). \end{cases}$$

此式或与 (8) 矛盾或与 (10) 矛盾, 是以 (10) 不能成立, 故得 (9) 式. 同法可证, 对任一元素 d 吾人常有

$$(ad)^\sigma = a^\sigma d^\sigma. \tag{11}$$

若能证明对任二元素 c 及 d 常有

$$(cd)^\sigma = c^\sigma d^\sigma,$$

则定理已明. 今假定有一对 c, d 使

$$(cd)^\sigma = d^\sigma c^\sigma \neq c^\sigma d^\sigma, \tag{12}$$

由上述之法泡制, 由 (12) 可得

$$(ad)^\sigma = d^\sigma a^\sigma, \qquad (cb)^\sigma = b^\sigma c^\sigma. \tag{13}$$

引用 (7), (8), (9), (11), (12) 及 (13),

$$a^\sigma b^\sigma + a^\sigma d^\sigma + c^\sigma b^\sigma + d^\sigma c^\sigma = [(a+c)(b+d)]^\sigma = \begin{cases} (a^\sigma + c^\sigma)(b^\sigma + d^\sigma) \text{ 或} \\ (b^\sigma + d^\sigma)(a^\sigma + c^\sigma). \end{cases}$$

此二式皆不可能, 定理于是证明.

若 R 及 R' 为域, 则准同构必为同构 (isomorphism), 因之得

定理二 凡域之广义同构必为同构或反同构, 凡域之广义自同构, 必为一自同构或反自同构.

附记一 假定 R 及 R' 各有一单位元素 1 及 $1'$, 若 $1^\sigma = 1'$, 则可由 (5) 以得 (4).

附记二 若以

$$(ab + ba)^\sigma = a^\sigma b^\sigma + b^\sigma a^\sigma \tag{14}$$

代 (2), 则该对应谓之 Jordan 准同构. 假定环 R' 有如次之性质: 由 $2a = 0$ 可得 $a = 0$, 如是则 Jordan 准同构与广义准同构同. 其证明如下: 由 (4) 及 (1) 立得

$$(ab + ba)^\sigma = [(a+b)^2 - a^2 - b^2]^\sigma = (a^\sigma + b^\sigma)^2 - (a^\sigma)^2 - (b^\sigma)^2$$
$$= a^\sigma b^\sigma + b^\sigma a^\sigma.$$

此即 (14). 反之, 于 (1) 及 (14) 中命 $b = a$ 各得

$$(2a)^\sigma = 2a^\sigma, \qquad (2a^2)^\sigma = 2(a^\sigma)^2,$$

故得 (4) 式. 再由 (14), (1) 及 (4) 可知

$$2(aba)^\sigma = (2aba)^\sigma = [(ab+ba)a + a(ab+ba) - (a^2b + ba^2)]^\sigma$$
$$= [(a^\sigma b^\sigma + b^\sigma a^\sigma)a^\sigma + a^\sigma(a^\sigma b^\sigma + b^\sigma a^\sigma) - (a^{\sigma 2} b^\sigma + b^\sigma a^{\sigma 2})]$$
$$= 2a^\sigma b^\sigma a^\sigma.$$

此即 (5) 式.

兹应用定理二以解决非交换域之一度射影几何之基本定理.

设 k 为一域 (并不限定 $ab = ba$), 其特征非二. 以 k 中之任一非全零之元素对 (x_1, x_2) 表一点. 若 k 中有一元素 h 使 $y_1 = hx_1, y_2 = hx_2$, 则此二元素对谓之表同一点. 特如 $x_2 \neq 0$, 则 (x_1, x_2) 所表之点 P 亦可表为 $(x_2^{-1}x_1, 1)$. 此 $x_2^{-1}x_1$ 名为该点 P 之非齐次坐标. $x_2 = 0$ 之点谓之无穷点. 显然凡非无穷点必对应于 k 中之一元素. 反之, 凡 k 中之一元素, 必有一非无穷点以之为其非齐次坐标, 所有的点成一个一度射影空间 π.

若数阵
$$\begin{pmatrix} a & b \\ c & d \end{pmatrix}$$
有可逆性，设其逆由次式决定
$$\begin{pmatrix} a' & b' \\ c' & d' \end{pmatrix} \begin{pmatrix} a & b \\ c & d \end{pmatrix} = I, \tag{15}$$
则变形
$$z' = (az+b)(cz+d)^{-1} \tag{16}$$
变 π 为其自己. 此种变换成一种一一对应. 由 (15) 可得恒等式
$$(-zc'+a')(az+b) = (zd'-b')(cz+d),$$
故 (16) 亦可写为
$$z' = (-zc'+a')^{-1}(zd'-b'). \tag{17}$$
更普遍些：若 σ 为 k 域之自同构或反自同构，则可导入变形
$$z' = (az^{\sigma}+b)(cz^{\sigma}+d)^{-1}. \tag{18}$$
此名为广义射影变形. 显然, 此亦建立 π 之一自己的一一对应.

如有四点 z_1, z_2, z_3, z_4 适合如下之关系，则名为调和点列,
$$(z_2-z_4)^{-1}(z_2-z_3)(z_1-z_3)^{-1}(z_1-z_4) = -1. \tag{19}$$
关系 (19) 名为调和关系.

定理三 凡广义射影变形变调和点列为调和点列. 反之，凡使调和点列仍变为调和点列之变形必为一广义射影变形.

命 z_i' 及 z_j' 各为 z_i 及 z_j 之对应点. 由 (16) 及 (17) 及关系 (15) 立得
$$\begin{aligned} z_i' - z_j' &= (-z_ic'+a')^{-1}(z_id'-b') - (az_j+b)(cz_j+d)^{-1} \\ &= (-z_ic'+a')^{-1}[(z_id'-b')(cz_j+d) - (-z_ic'+a')(az_j+b)](cz_j+d)^{-1} \\ &= (-z_ic'+a')^{-1}(z_i-z_j)(cz_j+d)^{-1}, \end{aligned}$$
代入 (19), 可见 z_1', z_2', z_3', z_4', 仍成一调和点列.

更由恒等式
$$a^{-1}(a-b)b^{-1} = b^{-1}(a-b)a^{-1}$$

可得
$$(z_2 - z_4)^{-1}(z_2 - z_3)(z_3 - z_4)^{-1} = (z_3 - z_4)^{-1}(z_2 - z_3)(z_2 - z_4)^{-1},$$

及
$$(z_1 - z_4)^{-1}(z_1 - z_3)(z_3 - z_4)^{-1} = (z_3 - z_4)^{-1}(z_1 - z_3)(z_1 - z_4)^{-1}.$$

故推得
$$(z_2 - z_4)^{-1}(z_2 - z_3)(z_1 - z_3)^{-1}(z_1 - z_4)$$
$$= (z_3 - z_4)^{-1}[(z_2 - z_3)(z_2 - z_4)^{-1}(z_1 - z_4)(z_1 - z_3)^{-1}](z_3 - z_4).$$

由 (19) 可知
$$(z_2 - z_3)(z_2 - z_4)^{-1}(z_1 - z_4)(z_1 - z_3)^{-1} = -1,$$

即得
$$(z_1 - z_4)(z_1 - z_3)^{-1}(z_2 - z_3)(z_2 - z_4)^{-1} = -1. \tag{20}$$

显然 (19) 经 k 之自同构不变. 由 (20) 可知调和关系, 经反自同构亦不变.

定理三之逆部分乃次定理之引申:

定理四 凡一域 k 之自己一一对应, 且使调和关系不变, 此对应必为如下之形态:
$$z' = q z^\sigma r + s, \tag{21}$$

此处 q 及 r 为 k 中非零之元素, σ 为 k 之自同构或反自同构.

因 (21) 可变任何二元素为 0 及 1, 故仅需讨论对应之使 0 及 1 不变者. 命 $z \to z^\sigma$ 为此种对应, 即 $1^\sigma = 1, 0^\sigma = 0$. 由 (19) 式解出 z_4, 则

$$z_4 = z_1 + (z_1 - z_2)(z_1 + z_2 - 2z_3)^{-1}(z_1 - z_3).$$

若 $z_3 = \frac{1}{2}(z_1 + z_2)$, 则 z_4 不能存在, 故可知

$$z_3^\sigma = \left[\frac{1}{2}(z_1 + z_2)\right]^\sigma = \frac{1}{2}(z_1^\sigma + z_2^\sigma). \tag{22}$$

命 $z_4 = 0$, 则

$$z_3^\sigma = (2z_1(z_1 + z_2)^{-1} z_2)^\sigma = 2z_1^\sigma (z_1^\sigma + z_2^\sigma)^{-1} z_2^\sigma. \tag{23}$$

由 (22) 中命 $z_1 = 2x$ 及 $z_2 = 2y$, 则得

$$2(x + y)^\sigma = (2x)^\sigma + (2y)^\sigma. \tag{24}$$

命 $y=0$, 则 $2x^\sigma = (2x)^\sigma$. 由 (24) 可得

$$(x+y)^\sigma = x^\sigma + y^\sigma. \tag{25}$$

于 (23) 中命 $z_1 = a$, $z_2 = 1-a$, 则

$$[2a(1-a)]^\sigma = 2a^\sigma(1-a^\sigma).$$

由 (25) 可知

$$(a^2)^\sigma = (a^\sigma)^2.$$

由此式立得

$$(xy+yx)^\sigma = [(x+y)^2 - x^2 - y^2]^\sigma = x^\sigma y^\sigma + y^\sigma x^\sigma.$$

由附记二及定理二, 故得定理.

参 考 文 献

[1] Hua L K. On semi-antomorphism of a field. *Proc. of Nat. Acad.*, 1949, **35**: 387−389.

[2] Reidemeister K. *Grundlagen der Geometrie*, Chapt. 7. Berlin, 1930.

[3] Wachs S. Essai sur la géométrie projective quaternienne. *Ac. r. de Belgique, Cl. d. Sc. Mém.*, 1936, **8**: 1−134.

[4] Ancochea E. Sobre el teorema fundamental de la geometria projectiva. *Rev. Mat. Hisp.−Amer.*, 1941, **1**(4).

[5] ——— Théoréme de Staudt en géométrie quaternienne. *Jour. für Math.*, 1942, **184**: 193−198.

[6] ——— *Annals Math.*, 1947, **48**: 147−153.

[7] Kaplansky I. *Duke Math. Jour.*, 1947, **14**: 521−525.

A theorem on matrices over a sfield and its applications*

1. Introduction

Let Φ be a sfield (or a division ring). Let m and n be two positive integers such that $n \leqslant m$. We use $M = M^{(n,m)}$ to denote an $n \times m$ matrix over Φ and $M^{(n)} = M^{(n,n)}$ and M' to denote the transposed matrix of M.

Two $n \times m$ matrices Z and W are said to be *coherent*, if the rank of their difference $Z - W$ is one.

The aim of part I of the paper is to establish the following theorem:

Theorem 1 *Suppose $1 < n \leqslant m$. Any one-to-one mapping which carries $n \times m$ matrices into $n \times m$ matrices and leaves the coherence invariant is of the form*

$$Z_1 = PZ^\sigma Q + R, \tag{1}$$

where $P (= P^{(n)})$ and $Q (= Q^{(m)})$ are non-singular, R is an $n \times m$ matrix and σ is an automorphism of the sfield Φ. When $n = m$, in addition to (1), we have also

$$Z_1 = PZ^{t\sigma}Q + R, \tag{2}$$

where σ is an anti-automorphism of the sfield.

This asserts that the equivalence relation used in the theory of matrices has a unique position.

In a previous paper[1] the author established a corressponding theorem for symmetric matrices by the construction of involutions. The method to be used in the present investination is entirely different. The main idea is based upon the method of "meet and join". It suggests a possibility for the formulation of this geometry in terms of lattice theory.

In case $m = n = 1$, the problem is closely related to the fundamental theorem of projective geometry over a sfield, which has been discussed, though very incompletely, by Ancochea[5]. In a short note the author[6] proved the fundamental theorem without restriction. There seems to have no essential difficulty to build up a theory of projective geometry over a simple ring with chain condition, but now instead of the

---

* *Academia Sinica.* Reprinted from *Chinese Mathematical Society*, 1951, **1**(2): 109–163.

harmonic relation, the weaker concept of coherence plays an important role. Later we shall treat the geometry of rectangular matrices which is practically more general than this.

In part II and part III, we shall give several comparatively more direct applications.

It is well-known that every automorphism of a simple algebra is an inner one (see, e.g. Albert [2], p.51). It was extended by Ancochea[3] and Kaplansky[4] to the so-called semi-automorphisms. As an easy consequence of our present theorem with $m = n\,(> 1)$, we solve the problem about semi-automorphisms of a simple ring with descending chain condition for the left ideals.

It can be readily seen that a semi-automorphism is just a Jordan automorphism, provided that the charactaristic of the sfield is different from 2. Incidentally we solve the problem of Jordan isomorphism for simple ring with chain condition. How is the analogous problem for Lie isomorphism? The answer is not so perfect as the corresponding result for Jordan isomorphism. The author can only establish the result for sfield with characteristic different from 2 and 3. Nevertheless, in case the sfield is commutative our result is better that the known ones (Jacobson[7]).

Another application is to the Grassmann geometry, namely:

Let us consider the space formed by $(n-1)$-dimensional linear manifolds of the $(m+n-1)$-dimensional projective space. Two different $(n-1)$-dimensional manifolds are said to be *coherent*, if both of them are contained in an n-dimensional linear manifold. By means of Theorem 1, we can obtain all the one-to-one transformations keeping coherence invariant.

More precisely, let
$$(Z_{i1}, \cdots, Z_{i,m+n}), \quad 1 \leqslant i \leqslant n$$
be n points in the $(m+n-1)$-dimensional space. Let the matrix
$$Z = (Z_{ij})_{1 \leqslant i \leqslant n, 1 \leqslant j \leqslant m+n}$$
denote the $(n-1)$-dimensional manifold spanned by these n points, where Z is assumed of rank n.

Two $n \times (m+n)$ matrices Z and Z_1, both of rank n, represent the same manifold if and only if there is a matrix $Q = Q^{(n)}$, naturally nonsingular, such that
$$Z_1 = QZ. \tag{3}$$

We shall study the space formed by all the $(n-1)$-dimensional linear manifolds. The space admits evidently the mapping

$$Z_1 = QZ^\sigma T, \qquad (4)$$

where $Q = Q^{(n)}$ and $T = T^{(m+n)}$ are both nonsingular.

Analytically, two $(n-1)$-dimensional linear manifolds W and Z are coherent, if and only if the rank of the $2n \times (m+n)$-matrix

$$\begin{pmatrix} Z \\ W \end{pmatrix} \qquad (5)$$

is equal to $n+1$.

It will be established as a consequence of Theorem 1 that any one-to-one transformation which carries the space formed by all $(n-1)$-dimensional manifolds onto itself and leaves the coherence invariant is of the form (4). However, in case $m = n$, besides (4), it admits another type of transformations which will be described in the text.

Part I. Algebraic Part

2. Affine geometry

In this section we shall assume that $n = 1$.

Theorem 1 fails for $n = 1$, because the coherence is now merely a condition which asserts that the correspondence is one-to-one. For the sake of completeness, we give here the following corresponding theorem with an additional condition. Now we suppose $m > 1$.

A *line* (or left line) is defined by those points given by

$$t(a_1, \cdots, a_m) + (b_1, \cdots, b_m), \qquad (6)$$

where the a's and the b's are fixed and t runs over all the elements of the sfield Φ. Two lines are called intersecting if they have a point in common. This is called the *coherence relation betwen lines*.

Evidently, passing through two points $(x_1, \cdots, x_m)$ and $(y_1, \cdots, y_m)$, we have a unique line

$$(1-t)(x_1, \cdots, x_m) + t(y_1, \cdots, y_m). \qquad (7)$$

Theorem 2 *Let $m > 1$. Every one-to-one mapping of the m-dimensional space (or left space) on to itself carrying lines into lines is of the form*

$$(z_1, \cdots, z_m) = (w_1, \cdots, w_m)^\sigma Q + (r_1, \cdots, r_m), \qquad (8)$$

where $Q = Q^{(m)}$ is a non-singular matrix and σ is an automorphism of the sfield.

The converse of the theorem is also true, since it is evident that (8) carries lines into lines. The theorem should have been known to geometers, since its proof depends on the classical method of meet and join. However I could not find an exact reference for it. Nevertheless, the theorem is known at least for the real field (Veblen and Whitehead[8]). The proof is comparatively simpler there, since the identity is the only automorphism. For completeness a proof of Theorem 2 is given in the next section.

3. Proof of Theorem 2

1) $m = 2$. Suppose that a pair of point (a_1, a_2) and (b_1, b_2) is carried into (c_1, c_2) and (d_1, d_2) respectively. Then the line joining the previous pair.
$$(1-x)(a_1, a_2) + x(b_1, b_2)$$
is carried into the line joining the second pair
$$(1-x^\sigma)(c_1, c_2) + x^\sigma(d_1, d_2), \tag{9}$$
we shall now prove that $x \to x^\sigma$ is an automorphism of the sfield. Since there is a mapping of the form (8) carrying any pair of points, into $(0,0)$ and $(1,0)$ respectively, it is enough to prove our assertion for $(a_1, a_2) = (c_1, c_2) = (0, 0)$ and $(b_1, b_2) = (d_1, d_2) = (1, 0)$. That is, if the line $(t, 0)$ is carried into $(t^\sigma, 0)$ and $1^\sigma = 1$, $0^\sigma = 0$, then
$$(s+t)^\sigma = s^\sigma + t^\sigma, \quad (st)^\sigma = s^\sigma t^\sigma.$$
These follow from the classical method of "meet and join" as illustrated respectively in the following figures:

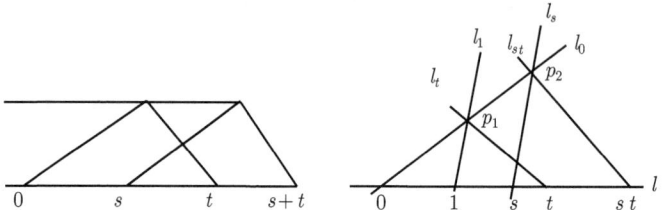

Since the sfield is not necessarily commutative, it might cause some difficulties for the second case, let us work it out in detail.

Let l be the line $(x, 0)$ which is determined by the points $(0, 0)$ and $(1, 0)$. We use s simply to denote the point $(s, 0)$. From 0, we draw a line $l_0 : ((xl_0, x), x \in \Phi)$ which is different from l, and from 1, we draw a line $l_1 : ((xl_1 + 1, x), x \in \Phi)$ which is different from l and is not parallel① to l_0 (i.e., $l_0 \neq l_1$). The intersection of the

---

① By parallel, we mean that the two lines do not intersect.

lines l_0 and l_1 is denoted by p_1 $((l_0-l_1)^{-1}l_0, (l_0-l_1)^{-1})$. Parallel to l_1 and passing through s, we draw a line $l_s : (xl_1+s, x)$. The intersection of l_s and l_0 is denoted by p_2 $(s(l_0-l_1)^{-1}l_0, s(l_0-l_1)^{-1})$. The line l_t joining p_1 and t is denoted by

$$((1-y)((l_0-l_1)^{-1}l_0, (l_0-l_1)^{-1}) + y(l, 0)).$$

Parallel to l_t and passing through p_2 we draw a line l_{st}:

$$((s(l_0-l_1)^{-1}l_0, s(l_0-l_1)^{-1}) + x(t - (l_0-l_1)^{-1}l_0, -(l_0-l_1)^{-1})).$$

The intersection of l_{st} and the basic line l is given by $x = st$, that is, it is the point $(st, 0)$. This completes the proof of the statement.

Let $A(x_1, x_2)$ be the mapping under consideration. Since any three points not on a line can be carried into $(0,0)$, $(1,0)$ and $(0,1)$, we may assume that

$$A(0,0) = (0,0), \quad A(1,0) = (1,0), \quad A(0,1) = (0,1).$$

After an automorphism, we may assume that, by the result proved before,

$$A(x, 0) = A(x, 0)$$

holds for all x belonging to Φ and,

$$A(0, y) = (0, y^\sigma)$$

for all y belonging to Φ. Then the point (x, y), being the intersection of a line passing through $(x, 0)$ and parallel to y-axis with a line passing through $(0, y)$ and parallel to x-axis, is mapped into (x, y^σ), i.e.,

$$A(x, y) = (x, y^\sigma). \tag{10}$$

By (9) with $(a_1, a_2) = (c_1, c_2) = 0$, $(b_1, b_2) = (x, y)$ and $(d_1, d_2) = (x, y^\sigma)$, we have

$$A(t(x, y)) = (t^\tau(x, y^\sigma)). \tag{11}$$

On the other hand, substituting tx and ty for x and y in (10) we obtain

$$A(tx, ty) = (tx, (ty)^\sigma). \tag{10_1}$$

Comparing (11) and (10_1), we have

$$\tau = \sigma = 1.$$

Therefore

$$A(x, y) = (x, y)$$

for all x and y in Φ. The theorem is thus proved.

2) Let $m > 2$. The theorem can be proved by induction, and we shall leave to the reader to generalize it.

Remark Without any essential difficulty, we can extend our theorem to the following form: let Φ_m ($m > 1$) be the m-dimensional affine (left or right) space over the sfield Φ. If there is a one-to-one mapping carrying Φ_m onto $\Phi'_{m'}$ and carrying lines into lines, then we have $m = m'$ and that either Φ and Φ' are isomorphic or they are anti-isomorphic. In the first case, Φ_m and Φ'_m are both left or right spaces; if we identify Φ' with Φ and suppose that Φ_m and Φ'_m are both left spaces, we obtain the transformation (2). For the second case, one of Φ_m and Φ'_m is the left and the other the right space; suppose Φ_m is the left space and Φ'_m is the right space the mapping can be written as

$$(w_1, \cdots, w_m) \to (Q(w_1, \cdots, w_m)^\tau)' + (r_1, \cdots, r_m)',$$

where τ denotes the anti-isomorphism.

4. Maximal set of rank 1

Definition A set of $n \times m$ matrices over Φ is called *maximal* (of rank 1), if any pair among them is coherent and if there is no other $n \times m$ matrix, outside of the set, which is coherent to each element of the set.

Theorem 3 *Every maximal set is equivalent under the group* (1) *to either*

$$\begin{pmatrix} x_1 & x_2 & \cdots & x_m \\ 0 & 0 & \cdots & 0 \\ \vdots & \vdots & & \vdots \\ 0 & 0 & \cdots & 0 \end{pmatrix} \qquad (12)$$

or

$$\begin{pmatrix} y_1 & 0 & \cdots & 0 \\ y_2 & 0 & \cdots & 0 \\ \vdots & \vdots & & \vdots \\ y_n & 0 & \cdots & 0 \end{pmatrix}. \qquad (13)$$

Proof We may suppose that the set contains two points 0 and

$$Z_0 = \begin{pmatrix} 1 & 0 & \cdots & 0 \\ 0 & 0 & \cdots & 0 \\ \vdots & \vdots & & \vdots \\ 0 & 0 & \cdots & 0 \end{pmatrix}.$$

Let Z be another point of the set. Since Z and 0 are coherent, the rank of Z is 1, we have therefore

$$Z = \begin{pmatrix} a_1b_1 & \cdots & a_1b_m \\ \vdots & & \vdots \\ a_nb_1 & \cdots & a_nb_m \end{pmatrix}.$$

Since $Z - Z_0$ is of rank 1, we have

$$a_ib_j = 0, \quad \text{for} \quad 2 \leqslant i \leqslant n, \quad 2 \leqslant j \leqslant m.$$

If $a_2, \cdots, a_n$ are not all zero, we have $b_2 = \cdots = b_m = 0$, that is

$$Z = \begin{pmatrix} a_1b_1 & 0 & \cdots & 0 \\ a_2b_1 & 0 & \cdots & 0 \\ \vdots & \vdots & & \vdots \\ a_nb_1 & 0 & \cdots & 0 \end{pmatrix}. \tag{14}$$

Otherwise, we have

$$Z = \begin{pmatrix} a_1b_1 & \cdots & a_1b_m \\ 0 & \cdots & 0 \\ \vdots & & \vdots \\ 0 & \cdots & 0 \end{pmatrix}. \tag{15}$$

The set of elements of the form

$$\begin{pmatrix} z & 0 & \cdots & 0 \\ 0 & 0 & \cdots & 0 \\ \vdots & \vdots & & \vdots \\ 0 & 0 & \cdots & 0 \end{pmatrix} \tag{16}$$

is evidently not a maximal set. If the set contains a point besides that of (16), say a point (14) with $a_2, \cdots, a_m$ not all zero, then none of the other elements of the set can be of the form (15). This asserts that the maximal set containing 0 and Z_0 is either (12) or (13), since evidently every pair of elements of (12) or (13) is coherent.

The above proof enables us also to establish the following theorem:

Theorem 4 *Given any two coherent matrices, there are two and only two maximal sets containing both of them.*

5. Intersections of maximal sets

Now we consider the intersection of two maximal sets. Supposing that their intersection is not empty, we may take 0 to be one of their common point. A maximal set containing 0 is either of the form

$$P \begin{pmatrix} y_1 & y_2 & \cdots & y_m \\ 0 & 0 & \cdots & 0 \\ \vdots & \vdots & & \vdots \\ 0 & 0 & \cdots & 0 \end{pmatrix} Q, \quad P = P^{(n)}, \quad Q = Q^{(m)},$$

or

$$P \begin{pmatrix} x_1 & 0 & \cdots & 0 \\ x_2 & 0 & \cdots & 0 \\ \vdots & \vdots & & \vdots \\ x_n & 0 & \cdots & 0 \end{pmatrix} Q.$$

In fact, let Z be a point of rank 1, then we have non-singular P and Q such that

$$P^{-1} Z Q^{-1} = \begin{pmatrix} 1 & 0 & \cdots & 0 \\ 0 & 0 & \cdots & 0 \\ \vdots & \vdots & & \vdots \\ 0 & 0 & \cdots & 0 \end{pmatrix}.$$

This establishes the assertion by the result of §4. Therefore we may assume that the two maximal sets are either

(i) $\begin{pmatrix} x_1 & \cdots & x_m \\ 0 & \cdots & 0 \\ \vdots & & \vdots \\ 0 & \cdots & 0 \end{pmatrix}, \quad P \begin{pmatrix} y_1 & \cdots & y_m \\ 0 & \cdots & 0 \\ \vdots & & \vdots \\ 0 & \cdots & 0 \end{pmatrix} Q,$

or

(ii) $\begin{pmatrix} x_1 & 0 & \cdots & 0 \\ x_2 & 0 & \cdots & 0 \\ \vdots & \vdots & & \vdots \\ x_n & 0 & \cdots & 0 \end{pmatrix}, \quad P \begin{pmatrix} y_1 & \cdots & y_m \\ 0 & \cdots & 0 \\ \vdots & & \vdots \\ 0 & \cdots & 0 \end{pmatrix} Q,$

or

(iii) $\begin{pmatrix} x_1 & 0 & \cdots & 0 \\ x_2 & 0 & \cdots & 0 \\ \vdots & \vdots & & \vdots \\ x_n & 0 & \cdots & 0 \end{pmatrix}, P \begin{pmatrix} y_1 & 0 & \cdots & 0 \\ y_2 & 0 & \cdots & 0 \\ \vdots & \vdots & & \vdots \\ y_n & 0 & \cdots & 0 \end{pmatrix} Q.$

The first and the third cases can be treated similarly.

Now we consider the case (i). Without loss of generality we may assume that $Q = I$. Let
$$P = (p_{rs})_{1 \leqslant r, s \leqslant n}.$$
The intersection is given by those y satisfying
$$(p_{21}, \cdots, p_{n1})'(y_1, \cdots, y_m) = 0.$$
If $(p_{21}, \cdots, p_{n1})$ is a zero vector, both maximal sets are identical. If $(p_{21}, \cdots, p_{n1})$ is not a zero vector, then $(y_1, \cdots, y_m)$ is a zero vector. That is, the maximal sets have only one point in common.

For the second case, the transformation $Z_1 = P^{-1}Z$ carries both maximal sets of (ii) into those of (ii) with $P = I$ and $Q = I$. Therefore the intersection of two maximal sets is given by

$$\begin{pmatrix} x & 0 & 0 & \cdots & 0 \\ 0 & 0 & 0 & \cdots & 0 \\ \vdots & \vdots & \vdots & & \vdots \\ 0 & 0 & 0 & \cdots & 0 \end{pmatrix}. \tag{17}$$

Therefore the intersection of two distinct maximal sets is either empty, a single point or a set equivalent to (17) under tho group (1).

Definition The intersection of two distinct maximal sets, which contains more than one point in common, is called a *line*.

Theorem 5 *Let X and Y be a pair of coherent matrices. There is one and only one line passing through them.*

Proof Without loss of generality, we may assume that
$$X = 0, \quad Y = \begin{pmatrix} 1 & 0 & \cdots & 0 \\ 0 & 0 & \cdots & 0 \\ \vdots & \vdots & & \vdots \\ 0 & 0 & \cdots & 0 \end{pmatrix}.$$

By Theorem 4, there aro two (and only two) maximal sets containing both; they are (12) and (13). The intersection is given by

$$\begin{pmatrix} t & 0 & \cdots & 0 \\ 0 & 0 & \cdots & 0 \\ \vdots & \vdots & & \vdots \end{pmatrix}.$$

Therfore we have the theorem.

The general equation of a line can be put into the form

$$p'tq + W,$$

where p and q are two vectors of dimension n and m respectively. Furthermore, the general equation of a line in the maximal set (12) is of the form

$$t \begin{pmatrix} a_1 & a_2 & \cdots & a_m \\ & 0 & & \end{pmatrix} + \begin{pmatrix} b_1 & b_2 & \cdots & b_m \\ & 0 & & \end{pmatrix} \qquad (18)$$

and in the maximal set (13) is of the form

$$\begin{pmatrix} a_1 & & \\ a_2 & & \\ \vdots & 0 & \\ a_n & & \end{pmatrix} t + \begin{pmatrix} b_1 & & \\ b_2 & & \\ \vdots & 0 & \\ b_n & & \end{pmatrix}. \qquad (19)$$

6. Linear subspaces

Let us now consider the geometry in a maximal set. We call, sometimes, a line a linear subspace of dimension 1. Let l be a line of the maximal set. Let p be a point of the maximal set outside of l. Draw all lines in the maximal set passing through p and intersecting l. The set of points on all these lines is defined to be a linear subspace of dimension 2. Let q be a linear subspace of dimension 2, and let p be a point of the maximal space but not in q. We construct the lines passing through p and intersecting q. The set of all the points on all these lines is called a linear subspace of dimension 3, etc. Evidently we have

Theorem 6 *Every linear subspace of dimension d can be constructed from any line of it by the previous process and the number of steps is exactly $d-1$. d is invariant under any mapping leaving the coherence invariant.*

Therefore, we have

Theorem 7 *The maximal set (12) is an affine space of dimension m and (13) is of dimension n.*

Evidently, for $m = n$, (12) is equivalent to (13) under a transformation of (2) which carries right lines into left lines and it can occur only when the field has an anti-automorphism. Moreover, if $m \neq n$, any mapping leaving the coherence invariant cannot carry (12) into (13).

7. Arithmetic distance

Definition Two matrices X_0 and X_r are said to be of arithmetic distance r, if there exist $r - 1$ matrices
$$X_1, \cdots, X_{r-1}$$
such that X_i and $X_{i-1}(1 \leqslant i \leqslant r)$ are coherent, and r is the least integer having this property.

Theorem 8 Two matrices X and Y are of arithmetic distance r, if and only if $X - Y$ is of rank r.

Proof If $X - Y$ is of rank r, we may assume that $X = 0$ and
$$Y = \begin{pmatrix} I^{(r)} & 0 \\ 0 & 0 \end{pmatrix},$$
where $I^{(r)}$ denotes the r-rowed identity. Then
$$X_\varrho = \begin{pmatrix} I^\varrho & 0 \\ 0 & 0 \end{pmatrix}, \quad 1 \leqslant \varrho \leqslant r - 1$$
are intermediate points, therefore the arithmetic distance between X and Y is $\leqslant r$.

Further, from the well-known property that
$$\mathrm{rank}(P + Q) \leqslant \mathrm{rank} P + \mathrm{rank} Q,$$
we deduce that, if X and Y are of arithmetic distance r, the rank of $X - Y$ is $\leqslant r$. The theorem is thus proved.

Theorem 9 Any r maximal sets of rank 1, which have a unique point in common and such that no two of them have any other common point can be carried by (1) simultaneously into

$$\begin{pmatrix} x_{11} & \cdots & x_{1m} \\ 0 & \cdots & 0 \\ \vdots & & \vdots \\ 0 & \cdots & 0 \end{pmatrix}, \begin{pmatrix} 0 & \cdots & 0 \\ x_{21} & \cdots & x_{2m} \\ 0 & \cdots & 0 \\ \vdots & & \vdots \end{pmatrix}, \cdots, \begin{pmatrix} 0 & \cdots & 0 \\ \vdots & & \vdots \\ x_{r1} & \cdots & x_{rm} \\ 0 & \cdots & 0 \end{pmatrix} \quad (20)$$

or into

$$\begin{pmatrix} y_{11} & 0 & \cdots & 0 \\ y_{21} & 0 & \cdots & 0 \\ \vdots & \vdots & & \vdots \\ y_{n1} & 0 & \cdots & 0 \end{pmatrix}, \begin{pmatrix} 0 & y_{12} & 0 & \cdots & 0 \\ 0 & y_{22} & 0 & \cdots & 0 \\ \vdots & \vdots & \vdots & & \vdots \\ 0 & y_{n2} & 0 & \cdots & 0 \end{pmatrix}, \cdots,$$

$$\begin{pmatrix} 0 & \cdots & 0 & y_{1r} & 0 & \cdots & 0 \\ 0 & \cdots & 0 & y_{2r} & 0 & \cdots & 0 \\ \vdots & & \vdots & \vdots & \vdots & & \vdots \\ 0 & \cdots & 0 & y_{nr} & 0 & \cdots & 0 \end{pmatrix}. \tag{21}$$

Proof In fact, let the common point be 0 and one of the maximal sets of rank 1 be

$$\begin{pmatrix} x_{11} & \cdots & x_{1m} \\ 0 & \cdots & 0 \\ \vdots & & \vdots \\ 0 & \cdots & 0 \end{pmatrix}$$

(owing to similarity, we omit the other case).

Let

$$Z_1 = \begin{pmatrix} a_1 b_1 & \cdots & a_1 b_m \\ \vdots & & \vdots \\ a_n b_1 & \cdots & a_n b_m \end{pmatrix}$$

be a point of the second maximal set of rank 1. Then, we have $P^{(n-1)}$ and $Q^{(m)}$ such that

$$\begin{pmatrix} 1 & 0 \\ 0 & P^{(n-1)} \end{pmatrix} Z_1 Q^{(m)} = \begin{pmatrix} c_1 & c_2 & \cdots & c_m \\ 1 & 0 & \cdots & 0 \\ 0 & 0 & \cdots & 0 \\ \vdots & \vdots & & \vdots \\ 0 & 0 & \cdots & 0 \end{pmatrix}.$$

Since it is of rank 1, we actually have $c_2 = \cdots = c_m = 0$ (notice that such a transformation leaves the first maximal set invariant). The transformation

$$Z_1 = \begin{pmatrix} 1 & -c_1 & 0 & \cdots & 0 \\ 0 & 1 & 0 & \cdots & 0 \\ 0 & 0 & 0 & \cdots & 0 \\ \vdots & \vdots & \vdots & & \vdots \\ 0 & 0 & 0 & \cdots & 0 \end{pmatrix} Z$$

leaves the first maximal set fixed and the previous point into

$$\begin{pmatrix} 0 & 0 & 0 & \cdots & 0 \\ 1 & 0 & 0 & \cdots & 0 \\ 0 & 0 & 0 & \cdots & 0 \\ \vdots & \vdots & \vdots & & \vdots \\ 0 & 0 & 0 & \cdots & 0 \end{pmatrix},$$

which spans with 0 a maximal set of rank 1:

$$\begin{pmatrix} 0 & 0 & \cdots & 0 \\ x_{21} & x_{22} & \cdots & x_{2m} \\ 0 & 0 & \cdots & 0 \\ \vdots & \vdots & & \vdots \end{pmatrix} \quad (22)$$

or

$$\begin{pmatrix} y_{11} & 0 & \cdots & 0 \\ y_{21} & 0 & \cdots & 0 \\ \vdots & \vdots & & \vdots \\ y_{n1} & 0 & \cdots & 0 \end{pmatrix}. \quad (23)$$

(23) cannot happen, since the intersection of the first and the second maximal sets contains more than one point.

Proceeding in this way, we have Theorem 9.

8. Proof of Theorem 1

We write

$$X = \begin{pmatrix} x_1 \\ \vdots \\ x_n \end{pmatrix} = (y_1, \cdots, y_m),$$

where x_i is a vector $(x_{i1}, \cdots, x_{im})$ and y_j is a column

$$y_j = \begin{pmatrix} y_{1j} \\ \vdots \\ y_{nj} \end{pmatrix},$$

we use also X_i and Y_j to denote respectively the matrices

$$\begin{pmatrix} 0 \\ \vdots \\ x_i \\ 0 \\ \vdots \end{pmatrix}, \quad (0,\cdots,0,y_j,0,\cdots,0).$$

Let $A(X)$ be a mapping considered in Theorem 1. By Theorem 9, we may assume that
$$A(X_i) = X_i^*, \quad A(0) = 0.$$
By the result of §6, A induces a mapping of x_i onto x_i^* carrying lines into lines. Therefore, by Theorem 2, we have
$$x_i^* = x^{\sigma_i} Q_i, \tag{24}$$
where $Q_i = Q_i^{(m)}$ is non-singular and σ_i is an automorphism of Φ. Without loss of generality we can assume
$$\sigma_1 = 1, \quad Q_1 = I^{(m)}. \tag{25}$$
The maximal set of rank 1, other than X_1 containing both 0 and $(0,0,\cdots,0,x_{1j},0,\cdots,0)$ is Y_j. Thus
$$A(Y_j) = Y_j^*.$$
By Theroem 2, we have
$$y_j^* = P_j y^{\tau_j}, \tag{26}$$
where $P_j = P_j^{(n)}$ is non-singular and τ_j is an automorphism of Φ.

Let $x_{ij} E_{ij}$ be the intersection of X_i and Y_j, then we have
$$A(x_{ij} E_{ij}) = x_{ij}^* E_{ij}.$$
Consequently, Q_i and P_j are diagonal matrix. Write
$$Q_i = [q_{i1},\cdots,q_{im}], \quad P_j = [p_{1j},\cdots,p_{nj}].$$
Comparing (24) and (26), we have
$$x_{ij}^{\sigma_i} q_{ij} = p_{ij} x_{ij}^{\tau_j}, \tag{27}$$
which holds for all x_{ij} of Φ. In particular, for $x_{ij} = 1$, we have $q_{ij} = p_{ij}$. Therefore, for $i = 1$, $p_{ij} = 1$ and $\tau_j = 1$ for all $j = 1,\cdots,n$. And then are inner automorphisms.

If we use $P_1^{-1}A$ instead of A we may assume without loss of generality, that A satisfies (24), (25), (26) and
$$P_1 = I^{(n)}.$$
Then from (27) for $j = 1$ we deduce that $q_{i1} = 1$ and $\sigma_i = 1$ for $i = 1, \cdots, m$. Further, the linear dependence of x_1 and x_i implies that of $x_1 = x_1 Q_1$ and $x_i Q_i$. Thus $Q_i = p_i I$. Since $Q_i = (1, q_{i2}, \cdots, q_{in})$, $Q_i = I$ for all i. Similarly, $P_j = 1$ for all j. Thus we have
$$A(X_i) = X_i, \quad A(Y_j) = Y_j. \tag{28}$$

Now we consider
$$A \begin{pmatrix} x_1 \\ \vdots \\ x_n \end{pmatrix} = \begin{pmatrix} x_1^* \\ \vdots \\ x_n^* \end{pmatrix}.$$

Suppose that the rank of $\begin{pmatrix} x_1 \\ \vdots \\ x_n \end{pmatrix}$ is n, so is $\begin{pmatrix} x_1^* \\ \vdots \\ x_n^* \end{pmatrix}$, by Theorem 7. Since

$$A \begin{pmatrix} x_1 - \lambda_2 x_2 - \cdots - \lambda_n x_n \\ 0 \\ \vdots \\ 0 \end{pmatrix} = \begin{pmatrix} x_1 - \lambda_2 x_2 - \cdots - \lambda_n x_n \\ 0 \\ \vdots \\ 0 \end{pmatrix},$$

the rank of
$$\begin{pmatrix} x_1^* - x_1 - \lambda_2 x_2 - \cdots - \lambda_n x_n \\ x_2^* \\ \vdots \\ x_n^* \end{pmatrix}$$

is less than n for any λ. Therefore
$$x_1 = x_1^* + \sum_{j=2}^n \mu_j x_j^*,$$
$$x_h = \sum_{j=2}^n \mu_{hj} x_j^*, \quad 2 \leqslant h \leqslant n.$$

Applying the same method to the i-th row instead of the first row, we have
$$x_i = x_i^* + \sum_{\substack{j=1 \\ j \neq i}}^n v_{ij}^{(i)} x_j^*,$$

$$x_k = \sum_{\substack{j=1 \\ j \neq i}}^{n} v_{kj}^{(i)} x_j^*, \quad k \neq i.$$

Eliminating x_1, we have

$$x_1^* = \sum_{\substack{j=1 \\ j \neq i}}^{n} v_{ij}^{(i)} x_j^* - \sum_{j=2}^{n} \mu_j x_j^*.$$

Because of the independence of $x_1^*, \cdots, x_n^*$, we have

$$\mu_j = 0, \quad 2 \leqslant j \leqslant n.$$

Thus

$$x_1 = x_1^*.$$

Similarly, we have $x_i = x_i^*$, that is, if X is of rank n, we have

$$AX = X. \tag{29}$$

As to the singular case, the theorem can be proved as follows: let

$$A \begin{pmatrix} x_1 \\ \vdots \\ x_n \end{pmatrix} = \begin{pmatrix} x_1^* \\ \vdots \\ x_n^* \end{pmatrix}.$$

If $x_{11} \neq 0$, we have, by (27), that

$$A \begin{pmatrix} x_{11} & x_{12} & x_{13} & \cdots & x_{1n} & x_{1n+1} & \cdots & x_{1m} \\ 0 & \lambda_2 & 0 & \cdots & 0 & 0 & \cdots & 0 \\ 0 & 0 & \lambda_3 & \cdots & 0 & 0 & \cdots & 0 \\ \vdots & \vdots & \vdots & & \vdots & \vdots & & \vdots \\ 0 & 0 & 0 & \cdots & \lambda_n & 0 & \cdots & 0 \end{pmatrix} = \begin{pmatrix} x_{11} & x_{12} & x_{13} & \cdots \\ 0 & \lambda_2 & 0 & \cdots \\ 0 & 0 & \lambda_3 & \cdots \\ \vdots & \vdots & \vdots & \\ 0 & 0 & \cdots & \cdots \end{pmatrix}.$$

Then

$$\begin{pmatrix} x_{11}^* - x_{11} & x_{12}^* - x_{12} & x_{13}^* - x_{13} & \cdots \\ x_{21}^* & x_{22}^* - \lambda_2 & x_{23}^* & \cdots \\ x_{31}^* & x_{32}^* & x_{33}^* - \lambda_3 & \cdots \\ \vdots & \vdots & \vdots & \end{pmatrix}$$

is of rank $< n$ for any $\lambda_2 \lambda_3 \cdots \neq 0$. Therefore the only possible case is that

$$x_{11}^* - x_{11} = 0.$$

If $x_{11}^* \neq 0$, we proceed in the reverse way. Thus we have $x_{11}^* = x_{11}$. Similarly, we have

$$x_{ij}^* = x_{ij}.$$

The theorem is thus proved completely.

Remark Without any essential difficulty, we can extend our theorem to the following form: let $\Phi_{n,m}$ ($l < n \leqslant m$) be the totality of all the $n \times m$ matrices over the sfield Φ. If there is a one to one mapping carrying $\Phi_{n,m}$ onto $\Phi'_{n',m'}$ and carrying the coherence relation of the one into the other, then we have either $n = n', m = m'$ or $n = m', m = n'$: For the first case, the sfields Φ and Φ' are isomorphic; if we identify them, we obtain the transformation (1). For the second case, the sfield Φ and Φ' are anti-isomorphisc, the mapping can be written as

$$(PZ^\tau Q + R)',$$

where τ denotes the anti-isomorphism.

Part II. Geometry of Rectangular Matrices

9. Projective geometry of rectangular matrices

We introduce the homogeneous coordinates of a rectangular matrix $Z\,(= Z^{(n,m)})$. We write

$$Z = (X, Y),$$

where $Y = Y^{(n)}$, $X = X^{(n,m)}$. The pair of matrices

$$(X, Y)$$

is called a homogeneous coordinate of the matrix Z. A necessary and sufficient condition that two pairs of matrices

$$(X, Y), \quad (X_1, Y_1)$$

represent the same matrix Z is that we have a non-singular $Q\,(= Q^{(n)})$ such that

$$Q(X, Y) = (X_1, Y_1). \tag{30}$$

Now we extend this notion to the pairs with singular Y.

We start with an $n \times (n + m)$-matrix W of rank n. Two matrices W and W_1 are said to be equivalent, if there is a matrix $Q\,(= Q^{(n)})$ (automatically non-singular) such that

$$W^* = QW. \tag{31}$$

This equivalence relation classifies $n \times (n + m)$ matrices into classes. We identify the matrices belonging to a class as a point. The totality of these points form the projective space of rectangular matrices. In case $n = 1$, this reduces to the ordinary projective geometry.

The space admits the transformation
$$W^* = QWP, \tag{32}$$
where $P = P^{(m+n)}$, $Q = Q^{(n)}$ are non-singular. The space is evidently transitive. Two different points W_1 and W_2 are said to be coherent, if the rank of
$$\begin{pmatrix} W_1 \\ W_2 \end{pmatrix} \tag{33}$$
is $n+1$. This is the smallest possible value for distinct W_1 and W_2. In fact, if (33) is of rank n, since W_1 is of rank n, each row of W_2 depends on those of W_1, and consequently.
$$W_2 = QW_1.$$
That is, W_1 and W_2 represent the same point.

Evidently, (32) leaves the coherence invariant, since
$$\begin{pmatrix} W_1^* \\ W_2^* \end{pmatrix} = \begin{pmatrix} Q_1 & O \\ O & Q_2 \end{pmatrix} \begin{pmatrix} W_1 \\ W_2 \end{pmatrix} P.$$

Let us break W up into two parts
$$W = (X, Y), \quad X = X^{(n,m)}, \quad Y = Y^{(n)}.$$
Those with non-singular Y are called finite points and those with singular Y are called points at infinity. If $W^* = (X^*, Y^*)$ and $W = (X, Y)$ are both finite points, we have the birational transformation
$$Z^* = Y^{*-1}X^* = (ZB + D)^{-1}(ZA + C), \tag{34}$$
where
$$P = \begin{pmatrix} A & B \\ C & D \end{pmatrix}, \quad A = A^{(m)}, \quad B = B^{(m,n)}, \quad C = C^{(n,m)}, \quad D = D^{(n)}.$$

Two finite points Z and Z_1 are coherent, if and only if the rank of $Z - Z_1$ is one. In fact, we have now
$$\begin{pmatrix} X & Y \\ X_1 & Y_1 \end{pmatrix} = \begin{pmatrix} Y & O \\ O & Y_1 \end{pmatrix} \begin{pmatrix} Z & I \\ Z_1 & I \end{pmatrix} = \begin{pmatrix} Y & O \\ O & Y_1 \end{pmatrix} \begin{pmatrix} I & I \\ O & I \end{pmatrix} \begin{pmatrix} Z - Z_1 & O \\ Z_1 & I \end{pmatrix}.$$

Theorem 10 *Any coherent pair of points can be carried simultaneously into*
$$(O, I^{(n)})$$
and
$$(N, I^{(n)}),$$
where N is a matrix with 1 at $(1,1)$-position and zero elsewhere.

Proof Since the space is transitive, we may assume that one of the points is
$$(O, I^{(n)}).$$
Break the coordinates of the other point into
$$(X^{(n,m)}, Y^{(n)}). \tag{35}$$
Then X is of rank 1. We have matrices $Q\,(= Q^{(n)})$ and $P\,(= P^{(m)})$ such that
$$QXP = N.$$
Then, we have
$$Q(X, Y)\begin{pmatrix} P & O \\ O & I \end{pmatrix} = (N, QY)$$
and
$$(O, I)\begin{pmatrix} P & O \\ O & I \end{pmatrix} = (O, I).$$
Therefore we may assume that $X = N$ in (35).

If Y is non-singular, then
$$(N, Y)\begin{pmatrix} I & O \\ O & Y^{-1} \end{pmatrix} = (N, I)$$
and
$$Y(O, I)\begin{pmatrix} I & O \\ O & Y^{-1} \end{pmatrix} = (O, I).$$
The theorem is proved.

If Y is singular, since (N, Y) is of rank n, we can write Y as
$$\begin{pmatrix} \lambda_2 y_2 + \cdots + \lambda_n y_n \\ y_2 \\ \vdots \\ y_n \end{pmatrix},$$

where $y_2, \cdots, y_n$ are independent vectors. There exists a vector z such that

$$\begin{pmatrix} z + \lambda_2 y_2 + \cdots + \lambda_n y_n \\ y_2 \\ \vdots \\ y_n \end{pmatrix}$$

if of rank n. Let

$$C = \begin{pmatrix} z \\ 0 \\ \vdots \\ 0 \end{pmatrix}.$$

Then

$$(O, I) \begin{pmatrix} I & C \\ O & I \end{pmatrix} = (O, I)$$

and

$$(N, Y) \begin{pmatrix} I & C \\ O & I \end{pmatrix} = (N, NC + Y),$$

where $NC + Y$ is non-singular. The theorem reduces to the cases considered previously.

10. Projective geometry of square matrices

In case $m = n$, there is an extra type of transformations which corresponds to the non-homogeneous transformation of the form

$$Z^* = (Z^{\tau'}B + D)^{-1}(Z^{\tau'}A + C), \tag{36}$$

where τ is an anti-automorphism of the sfield. Among such transformations the most essential one is

$$Z^* = Z^{\tau'}. \tag{37}$$

If homogeneous coordinates are again introduced, we have

$$Y^{*-1}X^* = X^{\tau'}Y^{\tau'-1},$$

that is

$$X^* Y^{\tau'} = Y^* X^{\tau'},$$

we write this relation as

$$(X^*, Y^*)\mathfrak{F}(X, Y)^{\tau'} = 0, \tag{38}$$

where
$$\mathfrak{F} = \begin{pmatrix} O & I \\ -I & O \end{pmatrix}. \tag{39}$$

Now we are going to prove that (38) defines a one-to-one mapping carrying the projective space of matrices onto itself. Let (X_1, Y_1) be another point which satisfies
$$(X_1, Y_1)\mathfrak{F}(X, Y)^{\tau'} = 0.$$

Then, we have
$$\begin{pmatrix} X^* & Y^* \\ X_1 & Y_1 \end{pmatrix} \mathfrak{F}(X, Y)^{\tau'} = 0.$$

By Sylvester's law of nullity, we deduce that
$$\begin{pmatrix} X^* & Y^* \\ X_1 & Y_1 \end{pmatrix}$$
is of rank $\leqslant n$. Since (X^*, Y^*) is of rank n, we have a non-singular Q such that
$$(X^*, Y^*) = Q(X_1, Y_1).$$

More generally, let $\mathfrak{R}$ be any $2n$-rowed non-singular matrix. Then
$$(X^*, Y^*)\mathfrak{R}(X, Y)^{\tau'} = 0 \tag{40}$$
defines a mapping from (X, Y) into (X^*, Y^*).

The product of two such relations is a transformation of the form (1). In fact, let $\sigma = \tau\tau_1$ which is an automorphism of the field, then from
$$(X_1, Y_1)\mathfrak{R}_1(X, Y)^{\tau'} = 0$$
and
$$(X_2, Y_2)\mathfrak{R}_2(X_1, Y_1)^{\tau_1'} = 0,$$
we deduce
$$\begin{pmatrix} (X, Y)^\sigma \mathfrak{R}_1^{\tau_1'} \\ (X_2, Y_2)\mathfrak{R}_2 \end{pmatrix} (X_1, Y_1)^{\tau_1'} = 0.$$

By Sylvester's law of nullity, we have
$$(X_2, Y_2)\mathfrak{R}_2 = Q(X, Y)^\sigma \mathfrak{R}_1^{\tau_1'},$$
i.e.,
$$(X_2, Y_2) = Q(X, Y)^\sigma \mathfrak{R}_1^{\tau_1'} \mathfrak{R}_2, \tag{41}$$

which is a transformation of the form (1).

This establishes also the following.

Theorem 11 *The enlarged group generated by all the transformations of* (1) *and all the relations* (40) *is simply generated by* (1) *and one of relations of* (40), *say for example* (38).

Now we are going to state the fact that the coherence also remains unaltered under the enlarged group.

Theorem 12 *The mapping* (40) *keeps the arithmetic distance invariant. More precisely, let*

$$(X^*, Y^*)\mathfrak{R}(X, Y)^{\tau'} = 0$$

and

$$(X_1^*, Y_1^*)\mathfrak{R}(X_1, Y_1)^{\tau'} = 0.$$

The two matrices

$$\begin{pmatrix} X & Y \\ X_1 & Y_1 \end{pmatrix} \text{ and } \begin{pmatrix} X^* & Y^* \\ X_1^* & Y_1^* \end{pmatrix}$$

have the same rank.

Proof We have a transformation (1) carrying (X, Y) and (X_1, Y_1) into (O, I) and (L_r, I), where

$$L_r = \begin{pmatrix} I^{(r)} & O \\ O & O \end{pmatrix};$$

let the transformation be

$$(X, Y) = Q(U, V)T.$$

Let

$$(X^*, Y^*)\mathfrak{R}T^{\tau'} = (U^*, V^*),$$

then, we have

$$(U^*, V^*)(O, I)^{\tau'} = 0$$

and

$$(U_1^*, V_1^*)(L_r, I)^{\tau'} = 0.$$

Consequently

$$V^* = 0, \quad U_1^* L_r + V_1^* = 0.$$

The rank of

$$\begin{pmatrix} U^* & V^* \\ U_1^* & V_1^* \end{pmatrix} = \begin{pmatrix} U^* & O \\ U_1^* & -U_1^* L_r \end{pmatrix}$$

is equal to the rank of $U_1^* L_r$, i.e. $\leqslant n+r$. Therefore we have

$$\text{rank of } \begin{pmatrix} X & Y \\ X_1 & Y_1 \end{pmatrix} = \text{rank of } \begin{pmatrix} U & V \\ U_1 & V_1 \end{pmatrix}$$

$$\geqslant \text{rank of } \begin{pmatrix} U^* & V^* \\ U_1^* & V_1^* \end{pmatrix} = \text{rank of } \begin{pmatrix} U^* & V^* \\ U_1^* & V_1^* \end{pmatrix}.$$

By symmetry, we have the theorem.

From (41), we deduce easily that (40) is involutory, if and only if $\sigma = 1$ and $\mathfrak{R} = \varrho \mathfrak{R}^{\tau'}$ where ϱ belongs to the center of Φ. Consequently $\mathfrak{R} = \varrho(\varrho \mathfrak{R}^{\tau'})' = \varrho \varrho^\tau \mathfrak{R}$, that is $\varrho \varrho^\tau = 1$. Such an involutorial correspondence (40) is called a duality. The fixed elements (if exist) of a duality, i.e. those points (X, Y) satisfying

$$(X,\ Y) \mathfrak{R} (X,\ Y)^{\tau'} = 0$$

form a set which is called a *complex*. The coherence of a pair of points (X, Y) and (X_1, Y_1) both on the complex is equivalent to the statement that the rank of the matrix

$$(X,\ Y) \mathfrak{R}(X_1, Y_1)^{\tau'}$$

is one. In fact, the rank of

$$\begin{pmatrix} X & Y \\ X_1 & Y_1 \end{pmatrix} \mathfrak{R}(X_1, Y_1)^{\tau'} = \begin{pmatrix} (X,\ Y)\mathfrak{R}(X_1, Y_1)^{\tau'} \\ 0 \end{pmatrix}$$

is equal to 1. By Sylvester's law of nullity, we have

$$\begin{pmatrix} X & Y \\ X_1 & Y_1 \end{pmatrix}$$

is of rank $\leqslant n+1$. The assertion follows.

In case the complex is not an empty set and Φ is of characteristic $\neq 2$, $\mathfrak{R}$ can be transformed into a particularly convenient normal form. We have two n-rowed matrices S and T such that

$$\mathfrak{F} = \begin{pmatrix} X & Y \\ S & T \end{pmatrix}$$

is non-singular. Then

$$\mathfrak{F} \mathfrak{R} \mathfrak{F}^{\tau'} = \begin{pmatrix} O & B \\ \varrho^\tau B^{\tau'} & H \end{pmatrix},$$

where B is non-singular and $H^{\tau'} = \varrho H$. Since

$$\begin{pmatrix} B^{-1} & O \\ O & I \end{pmatrix} \begin{pmatrix} O & B \\ \varrho^\tau B^{\tau'} & H \end{pmatrix} \begin{pmatrix} (B^{\tau'})^{-1} & O \\ O & I \end{pmatrix} = \begin{pmatrix} O & I \\ \varrho^\tau I & H \end{pmatrix}$$

and

$$\begin{pmatrix} I & O \\ -\frac{1}{2}H & I \end{pmatrix} \begin{pmatrix} O & I \\ \varrho^\tau I & H \end{pmatrix} \begin{pmatrix} I & -\frac{1}{2}H^{\tau'} \\ O & I \end{pmatrix} = \begin{pmatrix} O & I \\ \varrho^\tau I & O \end{pmatrix},$$

we have a matrix $\mathfrak{Y}$ such that

$$\mathfrak{Y}\mathfrak{R}\mathfrak{Y}^{\tau'} = \begin{pmatrix} O & I \\ \varrho^\tau I & O \end{pmatrix}.$$

Corresponding to this normal form, we have the transformation, in non-homogeneous coordinates

$$Z^* = -\varrho Z^{\tau'}.$$

It is an interesting problem to characterize the geometry of matrices on a complex by means of the property of coherence. In a previous occation, the author established a particular case with Φ commutative and $\tau = 1$ and $\varrho = -1$.

Remark In the non-homogeneous coordinates the transformation takes the form

$$Z^* = (ZB + D)^{-1}(ZA + C)$$

and

$$Z^* = (Z^{\tau'}B + D)^{-1}(Z^{\tau'}A + C).$$

Apparently, it seems to be difficult to construct the product of two transformations. This difficulty can be avoided by means of the following identity: for any $Z (= Z^{(n)})$,

$$(ZB + D)^{-1}(ZA + C) = (SZ - R)(-QZ + P)^{-1},$$

where

$$\begin{pmatrix} A & B \\ C & D \end{pmatrix} \begin{pmatrix} P & Q \\ R & S \end{pmatrix} = \begin{pmatrix} I & O \\ O & I \end{pmatrix}.$$

11. Maximal set

We define similarly that

A *maximal set* (of rank 1) is a set of points in the projective space of rectangular matrices, any pair of which are coherent, if there is no other point of the space can be added to the set such that the set still have the same property.

Similar to Theorem 3, we have

Theorem 13 *Under the group* (32), *every maximal set (of rank* 1) *is equivalent either to the set constituted by the finite points*

$$\begin{pmatrix} x_1 & \cdots & x_m & 1 & 0 & \cdots & 0 \\ 0 & \cdots & 0 & 0 & 1 & \cdots & 0 \\ \vdots & & \vdots & \vdots & \vdots & & \vdots \\ 0 & \cdots & 0 & 0 & 0 & \cdots & 1 \end{pmatrix} \qquad (42)$$

and the infinite points

$$\begin{pmatrix} x_1 & \cdots & x_m & 0 & 0 & \cdots & 0 \\ 0 & \cdots & 0 & 0 & 1 & \cdots & 0 \\ \vdots & & \vdots & \vdots & \vdots & & \vdots \\ 0 & \cdots & 0 & 0 & 0 & \cdots & 1 \end{pmatrix} \qquad (43)$$

or to the set constituted by the finite points

$$\begin{pmatrix} y_1 & 0 & \cdots & 0 & 1 & 0 & \cdots & 0 \\ y_2 & 0 & \cdots & 0 & 0 & 1 & \cdots & 0 \\ \vdots & \vdots & & \vdots & \vdots & \vdots & & \vdots \\ y_n & 0 & \cdots & 0 & 0 & 0 & \cdots & 1 \end{pmatrix} \qquad (44)$$

and the infinite points

$$\begin{pmatrix} y_1 & 0 & \cdots & 0 & 0 & 0 & \cdots & 0 \\ y_2 & 0 & \cdots & 0 & 0 & 1 & \cdots & 0 \\ \vdots & \vdots & & \vdots & \vdots & \vdots & & \vdots \\ y_n & 0 & \cdots & 0 & 0 & 0 & \cdots & 1 \end{pmatrix}. \qquad (45)$$

The proof is similar to that of Theorem 3 with some slight modifications.

Analogously, we define "lines" as the intersections of two distinct maximal sets which contain more than one point in common, and the arithmetic distance between two points as we did in §6. By these concepts, we construct linear subspaces (cf. §5). The linear subspaces of dimension $m-n$ are of primary importance. It can be proved without difficulty that every subspace of dimension $m-n$ can be carried into the following normal form: the set S is formed by the points

$$\begin{pmatrix} x_{n+1} & \cdots & x_m & 1 & 0 & \cdots & 0 & 0 & \cdots & 0 \\ 0 & \cdots & 0 & 0 & 1 & \cdots & 0 & 0 & \cdots & 0 \\ \vdots & & \vdots & \vdots & \vdots & & \vdots & \vdots & & \vdots \\ 0 & \cdots & 0 & 0 & 0 & \cdots & 1 & 0 & \cdots & 0 \end{pmatrix}$$

and

$$\begin{pmatrix} y_{n+1} & \cdots & y_m & 0 & 0 & \cdots & 0 & 0 & \cdots & 0 \\ 0 & \cdots & 0 & 0 & 1 & \cdots & 0 & 0 & \cdots & 0 \\ \vdots & & \vdots & \vdots & \vdots & & \vdots & \vdots & & \vdots \\ 0 & \cdots & 0 & 0 & 0 & \cdots & 1 & 0 & \cdots & 0 \end{pmatrix},$$

where at least one of the y's is not zero.

12. Fundamental theorem in the projective geometry of rectangular matrices

Theorem 14 *Any one-to-one mapping carrying the projective space of rectangular matrices onto itself is of the form* (37). *In case* $m = n$, *we have an extra type as described in* §8.

Proof Since the linear subspaces of dimension $m - n$ are transitive, we may assume, without loss of generality that the mapping A under consideration leaves the manifold S invariant.

It is easy to verify that a point P is a point at infinity, if and only if, there is a point on S, of which the arithmetic distance from P is $< n$.

Thus the mapping A carries finite points into finite points and infinite points into infinite points. By Theorem 1, we may assume that all the finite points are left fixed elementwisely by A. Since any $m - n$ dimensional linear subspace may be legarded as a manifold S, applying the same argument as above to new S's and comparing the common finite points arising from different manifolds S we know that all points, both finite and infinite are left fixed elementwisely by A. The theorem is thus proved.

Part III. Applications to Algebra

13. Module

Now we consider the set of $n \times m$ matrices as a module, for it is closed with respect to addition. A one-to-one mapping $(A \to A^\sigma)$ of a module onto itself is called a module automorphism, if it keeps the additive relation invariant, that is,

$$(A + B)^\sigma = A^\sigma + B^\sigma. \tag{46}$$

An easy consequence of our Theorem 1 is the following

Theorem 15 *Every module automorphism of the set of $n \times m$ matrices carrying matrices of rank one into matrices of rank one is of the form*

$$Z_1 = PZ^\sigma Q, \tag{47}$$

where $P\,(=P^{(n)})$ and $Q\,(=Q^{(m)})$ are non-singular and σ is an automorphism of the sfield Φ. When $m = n$, in addition to (47), we have also

$$Z_1 = PZ'^\tau Q, \tag{48}$$

where τ is an anti-automorphism of the sfield.

In fact, let $\Gamma(Z) = Z_1$ be the mapping under consideration. From (46), we deduce immediately $\Gamma(0) = 0$, $\Gamma(-A) = -\Gamma(A)$. If $P-Q$ is of rank 1, then $\Gamma(P-Q) = \Gamma(P) - \Gamma(Q)$ is also of rank 1. Thus the theorem follows immediately from Theorem 1.

Now we restrict ourselves to the case $m = n$. The set of all n-rowed matrices form a ring which is known as a total matrix ring. A semi-automorphism $(A \to A^\sigma)$ of a ring is a one to one mapping of the ring onto itself satisfying

$$(A+B)^\sigma = A^\sigma + B^\sigma, \tag{49}$$

$$(ABA)^\sigma = A^\sigma B^\sigma A^\sigma \tag{50}$$

and

$$I^\sigma = I. \tag{51}$$

Evidently, automorphisms and anti-automorphisms are semi-automorphisms. From (50) and (49), we dequce

$$\begin{aligned}(XY+YX)^\sigma &= [(X+Y)^2 - X^2 - Y^2]^\sigma \\ &= [(X+Y)^\sigma]^2 - (X^\sigma)^2 - (Y^\sigma)^2 = X^\sigma Y^\sigma + Y^\sigma X^\sigma.\end{aligned} \tag{52}$$

It is easy to deduce (50) and (51) from (52) and (49), provided that the characteristic of the sfield is different from two. In fact, it follows from the identity

$$2(XYX) = (XY+YX)X + X(XY+YX) - (X^2Y+YX^2).$$

Theorem 16 *Every semi-automorphism of the total matrix ring over a sfield is either of the form*

$$Z_1 = PZ^\tau P^{-1}, \tag{53}$$

where τ is an automorphism of the sfield, or of the form

$$Z_1 = PZ'^\tau P^{-1}, \tag{54}$$

where τ is an anti-automorphism of the sfield.

Proof 1) By Theorem 15, it is enough to prove that a semi-automorphism carries matrices of rank 1 into matrices of rank 1. Let $A \to A^\sigma$ be the mapping under consideration.

2) An element A satisfying $A^2 = A$ is called an idempotent matrix. From (50) with $B = 1$, we have $(A^\sigma)^2 = A^\sigma$. That is, a semi-automorphism carries idempotent elements into idempotent elements.

3) Let B be an element orthogonal to an idempotent element A, that is, $AB = BA = O$. Then, by (52), we have

$$O = (AB + BA)^\sigma = A^\sigma B^\sigma + B^\sigma A^\sigma$$

and by (50),

$$O = (ABA)^\sigma = A^\sigma B^\sigma A^\sigma = -(A^\sigma)^2 B^\sigma = -A^\sigma B^\sigma.$$

Therefore

$$A^\sigma B^\sigma = B^\sigma A^\sigma = 0.$$

4) An idempotent element A is called irreducible, if $A = B+C$, $B^2 = B$, $C^2 = C$, $BC = CB = O$ implies either $B = O$ or $C = O$. This property is evidently invariant under a semi-automorphism.

5) The theorem follows from Theorem 1 and the following

Lemma A matrix M $(\neq 0)$ is of rank 1 if and only if there exists an irreducible idempotent matrix E such that

$$(I - E)M(I - E) = 0$$

and

$$(M - EME)^2 = 0.$$

Notice that both relations are invariant under a semi-automorphism.

6) *Proof of the lemma.* If M is of rank 1, there is a matrix P such that

$$PMP^{-1} = \begin{pmatrix} u \\ 0 \\ \vdots \\ 0 \end{pmatrix},$$

where u is a row vector. Taking $E = E_1 = (a_{ij})$, $a_{ij} = 0$ except $a_{11} = 1$, we can verify both equations. Conversely, without loss of generality, we take $E = E_1$ Then from the first equation, we have

$$M = \begin{pmatrix} a & b \\ c & o \end{pmatrix}, \quad b = b^{(1,n-1)}, \quad c = c^{(n-1,1)}.$$

From the second equation

$$(M - EME)^2 = \begin{pmatrix} o & b \\ c & o \end{pmatrix}^2 = 0,$$

it follows that either $b = o$ or $c = o$. Therefore M is of rank 1.

Remark It is known[9] that a simple ring with descending chain condition for left idelas is simply a total matrix ring over a sfield. Therefore Theorem 16 can be "dignified" as follows:

Theorem 17 *A semi-automorphism of a simple ring with descending chain condition for left ideals is either an automorphism or an anti-automorphism.*

Theorem 18 *An automorphism (or anti-automorphism) of a simple ring with descending chain condition for left ideals is a product of an inner automorphism and an automorphism (or an anti-automorphism) induced by an automorphism (or an anti-automorphism) of the sfield.*

The method of the proof can be easily extended to establish a corresponding theorem for semi-simple rings.

Let R be a semi-simple ring with descending chain condition for left ideals. It is known that it is isomorphic to the ring formed by all the matrices of the form

$$X_1^{(m_1)} + X_2^{(m_2)} + \cdots + X_r^{(m_r)},$$

where $X_i^{(m_i)}$ runs through all (m_i, m_i) matrices with elements in a sfield Φ_i.

The steps 1)~8) are the same except that the E of 5) may be one of that form in a certain block. Let E belong to the i-th component, and let its image belong to the j-th component. The same method 7) asserts that the mapping carries all the elements of rank 1 in the i-th component into all the elements of rank 1 in the j-th component. Therefore our mapping carries the i-th component into the j-th component. The i-th component is matrices in Φ_i and contains m_i mutually orthogonal irreducible idempotent elements, and we deduce then that $m_i = m_j$ and Φ_i and Φ_j are isomorphic or anti-isomorphic. If we arrange the decomposition into a proper order, the semi-automorphism induces a semi-automorphism on each component. This is a generalization of a result of Kaplansky[4].

Certainly, we can apply the previous method to the problem of isomorphism of different Jordan rings.

14. Jordan and Lie automorphisms

A one-to-one mapping of a ring onto itself satisfying (49) and (52) is called a Jordan automorphism of the ring. Since (49) and (52) imply (50) and (51), in case that the basis sfield is of characteristic different from 2, we can restate Theorem 15 as

Theorem 19 *A Jordan automorphism of a simple ring with descending chain condition for left ideals is either an automorphism or an anti-automorphism, provided that the characteristic of the sfield is different from* 2.

An analogous, but much more difficult, problem is about the Lie automorphism.

A Lie automorphism $(A \to A^\sigma)$ is a one-to-one mapping of a ring onto itself satisfying the properties that
$$(A+B)^\sigma = A^\sigma + B^\sigma$$
and
$$(AB - BA)^\sigma = A^\sigma B^\sigma - B^\sigma A^\sigma. \tag{55}$$
For brevity, we introduce the notation
$$[A, B] = AB - BA. \tag{56}$$
Therefore (55) can be written as
$$[A, B]^\sigma = [A^\sigma, B^\sigma].$$

The module generated by the elements of the form $[A,B]$ is called the derived Lie module of the ring, or simply the derived module. We use L to denote the derived Lie module of the sfield Φ.

Now we consider again the total matrix ring R over a sfield Φ. The elements of the form λI is isomorphic to the ground sfield, we identify these elements with those of the ground sfield. The results of Lie automorphism is not so perfact as that of Jordan automorphism; we have to impose several conditions in order to obtain a similar conclusion. Before we are going to prove the theorem, we need quite a number of lemmas.

15. Properties of the derived Lie module

Theorem 20 *A necessary and sufficient condition for a matrix belonging to the derived Lie module of R is that its trace belongs to the derided Lie module L of Φ.*

Proof 1) Since
$$[A_1 + A_2, B] = [A_1, B] + [A_2, B],$$
it is enough to verify that, for $A = aE_{ij}$, the trace of $[A, B]$ belongs to the derived Lie module L of Φ, where E_{ij} is the matrix with zero element everywhere except 1 at the (i,j)-position. Now we have
$$\text{tr}[aE_{ij}, B] = \text{tr}(aE_{ij}B - BE_{ij}a) = ab_{ji} - b_{ji}a,$$
which evidently belongs to L.

2) we can readily verify that
$$[aE_{ij}, E_{jj}] = aE_{ij}, \quad \text{for } i \neq j$$
and
$$[aE_{1i}, E_{i1}] = aE_{11} - aE_{ii}, \quad \text{for } i \neq 1.$$
Any matrix with trace belonging to L is a sum of the previous elements and
$$bE_{11},$$
where b belongs to L. Since
$$b = \sum_i (a_i b_i - b_i a_i),$$
we deduce that
$$bE_{11} = \sum_i (a_i E_{11} b_i E_{11} - b_i E_{11} a_i E_{11}),$$
the theorem is now proved.

Consequently, the property that the trace belongs to the derived module of K is invariant under similarity transformation, since
$$P[S, T]P^{-1} = [PSP^{-1}, PTP^{-1}].$$
Incidentally, we obtain the following

Theorem 21 *If S and T are similar to each other, then the trace of S is congruent to that of T, mod L, where L is the derived Lie module of Φ.*

Proof Let
$$PSP^{-1} = T.$$
It is sufficient to prove successively that the theorem is true for $n = 2$ and for
$$P = \begin{pmatrix} 0 & 1 \\ 1 & 0 \end{pmatrix}, \quad \begin{pmatrix} \lambda & 0 \\ 0 & 1 \end{pmatrix} \text{ and } \begin{pmatrix} 1 & \mu \\ 0 & 1 \end{pmatrix},$$
since the general linear group is generated by these elements. Since
$$\lambda a \lambda^{-1} - a = (\lambda a)\lambda^{-1} - \lambda^{-1}(\lambda a)$$
belongs to L, our theorem follows immediately.

16. Lemmas

Now we assume that the characteristic of the sfield diffes from 2 and 3.

A matrix A is calld an *I-matrix*, if it can be expressed as

$$A = \alpha I + E, \tag{57}$$

where α belongs to the center of Φ and E is an idempotent matrix, i.e., $E^2 = E$ and $E \neq 0$.

Theorem 22 *A matrix A is an I-matrix, if and only if*

$$[A, [A, [A, B]]] = [A, B] \tag{58}$$

for all B.

Proof We have

$$[A, [A, [A, B]]] = A^3 B - 3A^2 BA + 3ABA^2 - BA^3.$$

Since $[A, B] = [A - \alpha I, B]$, evidently (57) implies (58).

Now we suppose that

$$A^3 B - 3A^2 BA + 3ABA^2 - BA^3 = AB - BA \tag{59}$$

for all B. We write

$$A = (a_{ij}), \quad A^l = (a_{ij}^{(l)}), \quad B = (b_{ij}).$$

Then, for all B, we have

$$\sum_{j=1}^{n}(a_{ij}^{(3)}b_{jk} - b_{ij}a_{jk}^{(3)}) - 3\sum_{j,l=1}^{n}(a_{ij}^{(2)}b_{jl}a_{lk} - a_{ij}b_{jl}a_{lk}^{(2)}) = \sum_{j=1}^{n}(a_{ij}b_{jk} - b_{ij}a_{jk}).$$

Putting $B = xE_{jl}$, we have

$$a_{ij}^{(2)} x a_{lk} = a_{ij} x a_{lk}^{(2)}, \quad \text{for } j \neq i, k \neq l, \tag{60}$$

$$a_{ij}^{(3)}x - a_{ij}x - 3a_{ij}^{(2)}xa_{kk} + 3a_{ij}xa_{kk}^{(2)} = 0, \quad \text{for } j \neq i, k = l, \tag{61}$$

$$xa_{lk}^{(3)} - xa_{lk} + 3a_{ii}^{(2)}xa_{lk} - 3a_{ii}xa_{lk}^{(2)} = 0, \quad \text{for } j = i, k \neq l \tag{62}$$

and

$$a_{ii}^{(3)}x - xa_{kk}^{(3)} - 3a_{ii}^{(2)}xa_{kk} + 3a_{ii}xa_{kk}^{(2)} = a_{ii}x - xa_{kk} \tag{63}$$

for $j = i, k = l$. (60)–(63) hold for all x.

If A is a central element, i.e., $A = cI$, where c is a cnetral element of K, then $A = (c-1)I + I$ and A is an I-matrix. Thus there remains to consider the case that

A is non-central. If A does not belong to the center, we have an invertible matrix P such that the element at (1,2)-position of PAP^{-1} is different from zero. In fact, if there is a non-diagonal element of A different from zero, we have a permutation matrix P such that PAP^{-1} satisfies our requirement. Next, if A is a diagonal matrix, the statement follows from the following formula

$$\begin{pmatrix} 1 & t \\ 0 & 1 \end{pmatrix} \begin{pmatrix} a & 0 \\ 0 & b \end{pmatrix} \begin{pmatrix} 1 & t \\ 0 & 1 \end{pmatrix}^{-1} = \begin{pmatrix} a & tb - at \\ 0 & b \end{pmatrix}.$$

Since the relation (58) and the property of a matrix being an I-matrix are both invariant under similarity transformations, without loss of generality we can assume that $a_{12} \neq 0$.

From (60), we have

$$a_{12}^{-1} a_{12}^{(2)} x a_{lk} = x a_{lk}^{(2)},$$

for $k \neq l$ and all x. If $a_{12}^{(2)} \neq 0$ and $a_{lk} \neq 0$, then

$$a_{12}^{-1} a_{12}^{(2)} x = x a_{lk}^{(2)} a_{lk}^{-1}$$

for all x. Consequently, we have

$$a_{12}^{-1} a_{12}^{(2)} = a_{lk}^{(2)} a_{lk}^{-1} = \varrho,$$

where ϱ belongs to the center. Thus

$$a_{lk}^{(2)} = \varrho a_{lk} \qquad (64)$$

for $k \neq l$ when $a_{12}^{(2)} \neq 0$ and $a_{lk} \neq 0$. The element ϱ is independent of l and k. If $a_{12}^{(2)} \neq 0$ but $a_{lk} = 0$, then $a_{lk}^{(2)} = 0$, (64) is also true. Finally, if $a_{12}^{(2)} = 0$, then $a_{lk}^{(2)} = 0$; in this case (64) also holds for $\varrho = 0$. Therefore (64) holds for all $k \neq l$. From (61) with $x = 1$, we have

$$a_{12}^{(3)} - a_{12} = 3 a_{12}^{(2)} a_{11} - 3 a_{12} a_{11}^{(2)} = 3 a_{12}^{(2)} a_{kk} - 3 a_{12} a_{kk}^{(2)},$$

then, since the characteristic of the sfield is $\neq 3$,

$$a_{12}^{(2)}(a_{11} - a_{kk}) = a_{12}(a_{11}^{(2)} - a_{kk}^{(2)}).$$

From (64), we deduce

$$a_{kk}^{(2)} - a_{11}^{(2)} = \varrho(a_{kk} - a_{11}). \qquad (65)$$

From (64) and (65), we have

$$A^2 - a_{11}^{(2)} I = \varrho(A - a_{11} I).$$

That is
$$A^2 = \varrho A + \lambda I, \tag{66}$$
where ϱ belongs to the center. We deduce immediately that λI is commutative with A. Repeating (66), we have
$$A^3 = \varrho A^2 + \lambda A = (\varrho^2 + \lambda)A + \varrho\lambda I. \tag{67}$$
Substituting (66) and (67) into (59), we have
$$[(\varrho^2 + \lambda)A + \varrho\lambda I]B - 3(\varrho A + \lambda I)BA + 3AB(\varrho A + \lambda I)$$
$$-3B[(\varrho^2 + \lambda)A + \varrho\lambda I] = AB - BA$$
for all B, and then
$$(\varrho^2 + 4\lambda - 1)(AB - BA) = 0$$
for all matrix B which are commutative with λ. Since A does not belong to the center, we can choose a B which is commutative with λ but not with A. In fact, we can choose B as a matrix with elements in the prime field and $AB \neq BA$. Therefore we have
$$\varrho^2 + 4\lambda - 1 = 0.$$
Then, from (66),
$$\left(A - \frac{\varrho - 1}{2} - I\right)^2 = A - \frac{\varrho - 1}{2}I,$$
that is, $A - \dfrac{\varrho - 1}{2}I$ is an idempotent element.

Remark The theorem is not true for fields of characteristic 3. In fact $A^3 = A + \tau I$ where τ belongs to the center, is a solution, but A is not an I-matrix,
$$A = \begin{pmatrix} 0 & 1 & 0 \\ 0 & 0 & 1 \\ \tau & 1 & 0 \end{pmatrix}$$
is such a matrix. The theorem does not hold for fields of characteristic 2, since
$$A = \begin{pmatrix} 0 & 1 \\ 1 & 0 \end{pmatrix}$$
is a "Gegenbeispiel".

A matrix A is called an N-matrix, if it can be expressed as
$$A = \alpha I + N, \tag{68}$$
where α belongs to the center of Φ and $N^2 = 0$.

Theorem 23 *A matrix A is an N-matrix, if and only if*
$$[A,[A,[A,B]]] = 0 \tag{69}$$
for all B.

Proof Instead of (59), we now have
$$A^3B - 3A^2BA + 3ABA^2 - BA^3 = 0. \tag{70}$$
Proceeding as in the proof of Theorem 22, we prove that
$$A^2 = \varrho A + \lambda I,$$
where $\varrho^2 + 4\lambda = 0$. Consequently
$$\left(A - \frac{\varrho}{2}I\right)^2 = 0.$$

17. Continuation

Theorem 24 *Suppose that $A^2 = A$, $AC = CA$, $AD = DA$ and $A \neq 0$ and $\neq I$. Then*
$$[D,[A,B]] + [C[A,[C,[A,B]]]] = 0 \tag{71}$$
for all B if and only if either
$$C = \lambda I + F, \quad D = \mu I + F^2, \tag{72}$$
where λ and μ belong to the center and $AF = FA = 0$, or
$$C = \lambda I - G, \quad D = \mu I - G^2, \tag{73}$$
where λ and μ belong to the center, $AG = GA = G$.

Proof We expand (71) as
$$D(AB - BA) - (AB - BA)D + AC^2B + C^2BA + ABC^2 + BAC^2$$
$$-2(AC^2BA + ABAC^2 + ACBC + CBAC) + 4ACBAC = 0,$$
which can be written as
$$(C^2 + D)AB + BA(C^2 + D) + (C^2 - D - 2C^2A)BA$$
$$+AB(C^2 - D - 2C^2A) + 2C(2ABA - AB - BA)C = 0. \tag{74}$$

In order to verify that (72) implies (71), we need only to prove that $C = F$ and $D = F^2$ satisfy (74). This is evident. In order to verify that (73) implies (71), we need only to prove that $C = -G$, $D = -G^2$ satisfy (74). This is also evident.

Conversely, without loss of generality, we may assume that

$$A = \begin{pmatrix} I^{(r)} & O \\ O & O \end{pmatrix},$$

since for every idempotent A, we have an invertible matrix P such that PAP^{-1} is of the prescribed form. We take

$$B = \begin{pmatrix} O^{(r)} & O \\ B_{21} & O \end{pmatrix}, \quad B_{21} = B_{21}^{(n-r,r)}.$$

Then $AB = 0$ and

$$BA = \begin{pmatrix} O & O \\ B_{21} & O \end{pmatrix} = B.$$

Substituting into (74), we have

$$\begin{pmatrix} O & O \\ B_{21} & O \end{pmatrix} \left(\begin{pmatrix} C_{11}^* & C_{12}^* \\ C_{21}^* & C_{22}^* \end{pmatrix} + \begin{pmatrix} D_{11} & D_{12} \\ D_{21} & D_{22} \end{pmatrix} \right)$$

$$+ \left(\begin{pmatrix} C_{11}^* & C_{12}^* \\ C_{21}^* & C_{22}^* \end{pmatrix} - \begin{pmatrix} D_{11} & D_{12} \\ D_{21} & D_{22} \end{pmatrix} \right) \begin{pmatrix} O & O \\ B_{21} & O \end{pmatrix}$$

$$-2 \begin{pmatrix} C_{11} & C_{12} \\ C_{21} & C_{22} \end{pmatrix} \begin{pmatrix} O & O \\ B_{21} & O \end{pmatrix} \begin{pmatrix} C_{11} & C_{12} \\ C_{21} & C_{22} \end{pmatrix} = 0,$$

where

$$C^2 = \begin{pmatrix} C_{11}^* & C_{12}^* \\ C_{21}^* & C_{22}^* \end{pmatrix}, \quad D = \begin{pmatrix} D_{11} & D_{12} \\ D_{21} & D_{22} \end{pmatrix}, \text{ etc.}$$

Then, for all B_{21}, we have

(i) $(C_{12}^* - D_{12})B_{21} = 2C_{12}B_{21}C_{11}$;

(ii) $C_{12}B_{21}C_{12} = 0$;

(iii) $B_{21}(C_{11}^* + D_{11}) + (C_{22}^* - D_{22})B_{21} = 2C_{22}B_{21}C_{11}$

and

(iv) $B_{21}(C_{12}^* + D_{12}) = 2C_{22}B_{21}C_{12}$.

From (ii), we deduce that $C_{12} = 0$ ① and from (i) and (iv) we deduce $D_{12} = C_{12}^* = 0$.

Putting

$$B = \begin{pmatrix} O & B_{12} \\ O & O \end{pmatrix}, \quad B_{12} = B_{12}^{(r,n-r)},$$

---

① In fact, it is evident that if $AXB = 0$ for all possible X, then we have either $A = 0$ or $B = 0$.

we deduce similarly $D_{21} = C_{12}^* = C_{21} = 0$ and

(v) $(C_{11}^* + D_{11})B_{12} + B_{12}(C_{22}^* - D_{22}) = 2C_{11}B_{12}C_{22}$.

Since $C_{22}^* = C_{22}^2$, $C_{11}^* = C_{11}^2$, we write

$$C_{22}^2 - D_{22} = (\alpha_{ij}), \quad C_{11}^2 + D_{11} = (\beta_{ij}),$$

$$C_{22} = (\gamma_{ij}), \quad C_{11} = (\delta_{ij})$$

and $B_{21} = (b_{ij})$, then, from (iii), we have

$$\sum_{j=1}^{r} b_{ij}\beta_{jk} + \sum_{j=1}^{n-r} \alpha_{ij}b_{jk} = 2\sum_{s=1}^{n-r}\sum_{t=1}^{r} \gamma_{is}b_{st}\delta_{tk}$$

for all b_{ij}. Consequently, we have

(vi) $\gamma_{is}\delta_{tk} = 0$ for $t \neq k, i \neq s$;
(vii) $2\gamma_{ii}b\delta_{tk} = b\beta_{tk}$ for $t \neq k$;
(viii) $2\gamma_{is}b\delta_{kk} = \alpha_{is}b$ for $i \neq s$;
(ix) $2\gamma_{ij}b\delta_{kk} = \alpha_{ii}b + b\beta_{kk}$

for all b.

If there is a non-diagonal element of C_{22} different from zero, say $\gamma_{is} \neq 0$ and $i \neq s$, then, from (vi) we have $\delta_{tk} = 0$ for all $t \neq k$. Then $\beta_{tk} = 0$ for all $t \neq k$, by (vii). Further, from (viii), we deduce that

$$\frac{1}{2}\gamma_{is}^{-1}\alpha_{is} = \delta$$

belongs to the center and $\delta_{kk} = \delta$. That is, $C_{11} = \delta I$.

From (ix), we have $\beta_{kk} = \beta$ for all k and

$$2\gamma_{ii}\delta b = \alpha_{ii}b + b\beta$$

for all b. In particular $b = 1$, then $\alpha_{ii} = 2\gamma_{ii}\delta - \beta$. Substituting into the previous equation, we have $b\beta - \beta b = 0$ for all b. Therefore β belongs to the center. Then

(x) $C_{11}^2 + D_{11} = \beta I$.

From (iii) and (ix), we have

$$C_{22}^2 - D_{22} = (\alpha_{ij}) = 2(\gamma_{ij})\delta - \beta I = 2C_{22}\delta - \beta I.$$

That is

$$D_{22} = C_{22}^2 - 2C_{22}^{ij}\delta + \beta I = (C_{22} - \delta I)^2 - (\delta^2 - \beta)I.$$

Therefore

$$C = \begin{pmatrix} \delta I^{(r)} & O \\ O & C_{22} \end{pmatrix} = \delta I^{(n)} + \begin{pmatrix} O & O \\ O & C_{22} - \delta I \end{pmatrix}$$

and
$$D = \begin{pmatrix} D_{11} & O \\ O & D_{22} \end{pmatrix} = \begin{pmatrix} \beta I - C_{11}^2 & O \\ O & (C_{22} - \delta I)^2 - (\delta^2 - \beta)I \end{pmatrix}$$
$$= (\beta - \delta^2)I + \begin{pmatrix} O & O \\ O & (C_{22} - \delta I)^2 \end{pmatrix}.$$

This proves the theorem.

By a slightly different method, we establish our theorem for the case C_{11} is not diagonal but notice that now we obtain the second conclusion instead of (8).

Suppose that C_{11} and C_{22} are both diagonal, then, from (vii) and (viii), both D_{11} and D_{22} are diagonal, and

(xi) $2\gamma_i b \delta_k = \alpha_i b + b \beta_k$

for all b, we write γ_i for γ_{ii} etc. If there is one of the differences $\gamma_i - \gamma_j$ different from zero, then we have

$$2(\gamma_i - \gamma_j) b \delta_k = (\alpha_i - \alpha_j) b$$

for all b. Then δ_k belongs to the center and

$$\delta_k = \delta = \frac{1}{2}(\gamma_i - \gamma_j)^{-1}(\alpha_i - \alpha_j),$$

which is independent of k. From (xi), we have

$$(2\gamma_i \delta - \alpha_i) b = b \beta_k$$

for all b. It follows that

$$\beta_k = \beta = 2\gamma_i \delta - \alpha_i$$

is a central element for all k. Therefor we have

$$C_{11}^2 + D_{11} = \beta I, \quad C_{11} = \delta I.$$

Then
$$\begin{pmatrix} C_{11} & O \\ O & C_{22} \end{pmatrix} = \begin{pmatrix} \delta I & O \\ O & C_{22} \end{pmatrix} = \delta I + \begin{pmatrix} O & O \\ O & C_{22} - \delta I \end{pmatrix}$$

and
$$\begin{pmatrix} D_{11} & O \\ O & D_{22} \end{pmatrix} = \begin{pmatrix} (\beta - \delta^2)I & O \\ O & C_{22}^2 - 2\delta C_{22} + \beta I \end{pmatrix}$$
$$= (\beta - \delta)^2 I + \begin{pmatrix} O & O \\ O & (C_{22} - \delta I)^2 \end{pmatrix}.$$

The theorem follows.

In case there is one of the difference $\delta_i - \delta_j$ different from zero, then we have
$$\gamma_i = \gamma, \quad \alpha_i = \alpha,$$
which belong to the center, and
$$2\gamma\delta_k = \alpha + \beta_k.$$
Then
$$C_{22}^2 - D_{22} = \alpha I, \quad C_{22} = \gamma I$$
and
$$C_{11}^2 + D_{11} = -\alpha I + 2\gamma C_{11}.$$
Then
$$C = \begin{pmatrix} C_{11} & O \\ O & \gamma I \end{pmatrix} = \gamma I + \begin{pmatrix} C_{11} - \gamma I & O \\ O & O \end{pmatrix}$$
and
$$D = \begin{pmatrix} D_{11} & O \\ O & D_{22} \end{pmatrix} = (\gamma^2 - \alpha)I + \begin{pmatrix} -(C_{11} - \gamma I)^2 & O \\ O & O \end{pmatrix}.$$
That is
$$C = \lambda I - G, \quad D = \mu I - G^2,$$
where
$$G = \begin{pmatrix} -(C_{11} - \gamma I) & O \\ O & O \end{pmatrix}$$
satisfies evidently $GA = AG = G$.

Theorem 25 Suppose that $A^2 = A \, (\neq 0, 1)$, $C^2 = C$, $AC = CA$. Then
$$[C, [A, B]] + [C[A[C, [A, B]]]] = 0 \tag{75}$$
for all B, if and only if
$$AC = 0 \quad \text{or} \quad (I - A)(I - C) = 0.$$

Proof The theorem is an easy consequence of Theorem 24, by putting $D = C^2 = C$. From (72), we have
$$C^2 = (\lambda I + F)^2 = \mu I + F^2,$$
then $\lambda^2 = \mu$, $2\lambda = 0$. Consequently $C = F$ and $AC = CA = 0$.

From (73) and $C^2 = D = C$, we have

$$C^2 = (\lambda I - G)^2 = \mu I - G^2 = \lambda I - G,$$

then $\lambda = 1$ and

$$(I - A)(I - C) = (I - A)G = G - G = 0.$$

Remark Theorem 25 can be proved very simply by the direct expansion of (75). In fact (75) is equivalent to

$$CA(CAB - CBA - ABC + BAC)$$
$$- C(CAB - CBA - ABC + BAC)A$$
$$- A(CAB - CBA - ABC + BAC)C$$
$$+ (CAB - CBA - ABC + BAC)AC$$
$$= -(CAB - CBA - ABC + BAC). \tag{76}$$

Evidently $CA = O$ or $(I - C)(I - A) = O$ implies (76). Conversely multiplying (76) by AC on the left, we have

$$ACB(I - A)(I - C) = 0$$

for all B. This implies either AC or $(I - A)(I - C) = O$. Therefore we have Theorem 25.

18. A theorem on Lie automorphism

The situation about Lie automorphism of the matrix ring R differs greatly from that of Jordan automorphism.

Definition A mapping $A \to A^\sigma$ of R into itself is called a Lie representation, if it satisfies

$$(A + B)^\sigma = A^\sigma + B^\sigma, \quad [A, B]^\sigma = [A^\sigma, B^\sigma].$$

The representation is called central, if it maps R into its center, that is

$$A \to \lambda(A)I. \tag{77}$$

where $\lambda(A)$ belongs to the center of the basic sfield Φ.

Theorem 26 *Every central representation can be expressed as*

$$\lambda(A) = \mu(\mathrm{tr}A),$$

where $\mathrm{tr}A$ *denotes the trace of* A *and* $\mu(a)$ *is a central Lie representation of the sfield* Φ.

Theorem 27 *A central Lie representation of a sfield is a representation of the factor group of the additive group of* Φ *over its derived Lie module into the center of* Φ.

Owing to their simplicity, we omit the proofs of both theorems.

If $A \to A^\sigma$ is a Lie representation, so is

$$A \to \lambda(A)I + A^\sigma. \tag{78}$$

This very fact makes some trouble. In fact, sometimes, (78) ceases to be one-to-one.

First of all, let us investigate the condition under which (78) is one-to-one. That is, to find the condition under which the equality

$$\lambda(A)I + A^\sigma = 0 \tag{79}$$

implies $A = O$.

Evidently, if A belongs to the center, A^σ belongs to the center. Conversely, suppose that A^σ belongs to the center, but A does not. There exist a matrix B such that $C = [A, B] \neq 0$. Then C is mapped into

$$[\lambda(A)I + A^\sigma, \lambda(B)I + B^\sigma] = [A^\sigma, B^\sigma] = 0,$$

which contradicts the assumption "one to one". Therefore, if A^σ belongs to the center, so does A.

Foom (79), we therefore deduce that A belongs to the center, say $A = \alpha I$, where α belongs to the center of Φ. From (79), we deduce the condition

$$n\mu(\alpha) + \alpha^\sigma = 0.$$

Therefore (78) is one to one, if and only if

$$n\mu(\alpha) + \alpha^\sigma \neq 0 \tag{80}$$

for all $\alpha (\neq 0)$ belonging to the center of Φ.

Theorem 28 *Let Φ be a sfield with characteristic different from 2 and 3. Every Lie automorphism of R ($n > 2$) is either of the form*

$$A \to \mu(\text{tr} A)I + PA^\tau P^{-1}, \tag{81}$$

where $\mu(d)$ is a central Lie representation of Φ and τ is an automorphism of Φ and $n\mu(\alpha) - \alpha^\tau \neq 0$ for all $\alpha (\neq 0)$ belonging to the center of Φ; or of the form

$$A \to \mu(\text{tr} A)I - PA^{\tau'} P^{-1}, \tag{82}$$

where τ is an anti-automorphism of Φ and $n\mu(\alpha) - \alpha^\tau \neq 0$ for all $\alpha (\neq 0)$ belonging to the center of Φ.

Proof 1) Idempotent elements. Let $E_1, \cdots, E_n$ be n elements different from zero such that

$$E_i^2 = E_i \quad \text{for} \quad 1 \leqslant i \leqslant n$$

and
$$E_i E_j = E_j E_i = 0 \quad \text{for} \quad i \neq j.$$

By Theorem 22, E_i's are carried into
$$\lambda_i I + E_i^*, \quad E_i^{*2} = E_i^*,$$

where λ_j belongs to the center. From Theorem 25, we have either
$$E_i^* E_j^* = E_j^* E_i^* = 0$$

or
$$(I - E_i^*)(I - E_j^*) = (I - E_j^*)(I - E_i^*) = 0.$$

Suppose that we have
$$E_1^* E_2^* = E_1^* E_3^* = 0,$$

but
$$(I - E_2^*)(I - E_3^*) = 0.$$

Then, we have
$$E_1^* = E_1^*(I - E_2^*) = E_1^*(I - E_2^*)(I - E_3^*) = 0,$$

which is impossible. Therefore we can only have either

(i) $E_i^* E_j^* = E_j^* E_i^* = 0 \quad \text{for all} \quad i \neq j$,

or
$$(I - E_i^*)(I - E_j^*) = (I - E_j^*)(I - E_i^*) = 0 \quad \text{for all} \quad i \neq j.$$

For the second case, we replace $I - E_i^*$ by E_i^* we have then

(ii) $E_i \to \lambda_i I - E_i^*$,

where
$$E_i^{*2} = E_i^*, \quad E_i^* E_j^* = E_j^* E_i^* = 0 \quad \text{for} \quad i \neq j.$$

Without loss of generality, after subjecting the given Lie automorphism to an inner automorphism, we may assume that
$$E_i = E_i^* = (\alpha_{st}),$$

where $\alpha_{st} = 0$ except $\alpha_{ii} = 1$.

2) We have
$$[E_i, [E_j, P]] = pE_{ij} + qE_{ji} \quad \text{for} \quad i \neq j, \tag{83}$$

where pE_{ij} denotes a matrix (α_{st}) with $\alpha_{st} = 0$ except $\alpha_{ij} = p$. Consider elements of the form (83) whose squares are zero. Then $pq = 0$. Therefore the matrix pE_{ij} is mapped into
$$\lambda I + N, \quad N^2 = 0, \quad \lambda \in \text{center}$$
by Theorem 23. Moreover, by 1), (83) is mapped into
$$p^* E_{ij} + q^* E_{ij} = \lambda I + N.$$
We deduce
$$O = N^2 = (p^* E_{ij} + q^* E_{ji} - \lambda I)^2,$$
that is
$$p^* q^* E_i + q^* p^* E_j - 2(p^* E_{ij} + q^* E_{ji})\lambda + \lambda^2 I = 0.$$
If $\lambda \neq 0$, then $p^* = q^* = 0$, which is impossible. For $\lambda = 0$, we have $p^* q^* = q^* p^* = 0$. Consequently, a matrix of the form pE_{ij} is mapped either into $p^* E_{ij}$ or into $q^* E_{ji}$.

Next, from
$$\alpha E_{12} \to \alpha^* E_{12}, \quad \beta E_{13} \to \beta^* E_{31},$$
we deduce
$$O = [\alpha E_{12}, \beta E_{13}] \to [\alpha^* E_{12}, \beta^* E_{31}] = -\beta^* \alpha^* E_{32},$$
which is absurd. Therefore we have either
$$\alpha E_{ij} \to \alpha^* E_{ij} \quad \text{for all } i \neq j \text{ and } \alpha \in \Phi \tag{84}$$
or
$$\alpha E_{ij} \to \alpha^* E_{ji} \quad \text{for all } i \neq j \text{ and } \alpha \in \Phi. \tag{85}$$

Suppose that we have (84), then
$$E_{1i} \to \gamma_i E_{1i}.$$
The mapping
$$[1, \gamma_2, \cdots, \gamma_n] X [1, \gamma_2, \cdots, \gamma_n]^{-1}$$
leaves E_i's invariant and carries $\gamma_i E_{1i}$ into E_{1i}, therefore after subjecting our Lie automorphism to an inner automorphism we can assume
$$E_{1i} \to E_{1i}, \quad i = 1, \cdots, n.$$
Further, let $E_{ij} \to \gamma E_{ij}$ then
$$[E_{1j}, E_{ij}] = E_{1j} \quad (j \neq 1)$$

is mapped into
$$[E_{1i}, \gamma E_{ij}] = \gamma E_{1j},$$
thus $\gamma = 1$. Further, for $j = 1$, the element
$$[E_{1i}, E_{i1}] = E_1 - E_i$$
is mapped into
$$[E_{1i}, \gamma E_{i1}] = (\lambda_1 - \lambda_i)I + E_1 - E_i$$
in case (i). Consequently
$$(\gamma - 1)(E_1 - E_i) = (\lambda_1 - \lambda_i)I.$$
Then $\lambda_1 = \lambda_i$ and $\gamma = 1$. Consequently, we have

(i) $E_i \to \lambda I + E_i, E_{ij} \to E_{ij}$.

In case (ii), it is mapped into
$$[E_{1i}, \gamma E_{i1}] = (\lambda_1 - \lambda_i)I - (E_1 - E_i),$$
then
$$(\gamma + 1)(E_1 - E_i) = (\lambda_1 - \lambda_i)I.$$
Then $\lambda_1 = \lambda_i$ and $\gamma = -1$. Consequently, $E_{i1} \to -E_{i1}$ for $i \neq 1$. For $i \neq 1$, we have
$$[E_{i1}, E_1] = E_{i1},$$
which is mapped into
$$[-E_{i1}, \lambda_1 I - E_1] = E_{i1} = -E_{i1},$$
which is impossible.

Similarly, with (85), we have
$$E_i \to \lambda I - E_i, \quad E_{1i} \to E_{i1}, \quad E_{j1} \to E_{1j},$$
$$E_{ij} \to -E_{ji} \quad \text{for} \quad i, j \neq 1.$$
Notice that the combination of (85) and (i) is impossible. The inner automorphism $[-1, 1, \cdots, 1]A[-1, 1, \cdots, 1]$ carries the automorphism into one satisfying

(ii) $E_i \to \lambda I - E_i, E_{ij} \to -E_{ji}$.

3) Principal matrices of rank 1.

Let $P (\neq 0)$ be a matrix satisfying
$$PE_1 = E_1 P = P, \quad PE_i = E_i P = 0 \quad \text{for} \quad 2 \leqslant i \leqslant n.$$

For $PE_i = E_iP = 0$ or P, we deduce from Theorem 24 with $D = C^2$ and $C = P$ that P is mapped into $\mu I + P^*$ such that

$$P^*E_i = E_iP^* = 0 \quad \text{or} \quad P^*.$$

If $P^*E_i = E_iP^* = 0$ for all i, then $P^* = 0$ which is impossible. From $P^*E_i = E_iP^* = P^*$, we deduce $P^* = aE_i$. Therefore we have

$$\alpha E_1 \to \lambda I + \alpha^* E_i, \quad \lambda \in \text{center}.$$

If $i \neq 1$, we have

$$[\alpha E_1, E_{1k}] = \alpha E_{1k}, \quad k \neq 1, \quad k \neq i,$$

but

$$[a^* E_1, E_{1k}] = 0 \quad \text{and} \quad [\alpha^* E_i, -E_{k1}] = 0,$$

both are impossible. Therefore, since $n \geq 3$, we have

$$\alpha E_1 \to \lambda I + \alpha^* E_1.$$

Therefore, we may assume that

$$\alpha E_i \to \lambda_i I + \beta_i E_i. \tag{86}$$

4) In conclusion, we established that either

(i) $\alpha E_i \to \mu_i(\alpha) J + \alpha_{ii}^* E_i$, $\alpha E_{ij} \to \alpha_{ij}^* E_{ij}$, $1_{ij}^* = 1$

or

(ii) $\alpha E_i \to \mu_i(\alpha) I = \alpha_{ii}^* E_i$, $\alpha E_{ij} \to -\alpha_{ij}^* E_{ij}$, $1_{ij}^* = 1$.

For (i), the element

$$\alpha E_i - \alpha E_j = [\alpha E_{ij}, E_{ji}]$$

is mapped into

$$(\mu_i(\alpha) - \mu_j(\alpha))I + \alpha_{ii}^* E_i - \alpha_{jj}^* E_j = [\alpha_{ij}^* E_{ij}, E_{ji}] = \alpha_{ij}^* E_i - \alpha_{ij}^* E_j.$$

We have then, since $n \geq 3$,

$$\mu_i(\alpha) = \mu_j(\alpha), \quad \alpha_{ii}^* = \alpha_{ij}^* = \alpha_{jj}^*,$$

that is, we have

(i) $\alpha E_i \to \mu(\alpha) I + \alpha^* E_i, \alpha E_{ij} \to \alpha^* E_{ij}, 1^* = 1$.

Similarly for case (ii), we have

(ii) $\alpha E_i \to \mu(\alpha) I - \alpha^* E_i, \alpha E_{ij} \to -\alpha^* E_{ij}, 1^* = 1$.

Notice that the mapping $\alpha \to \alpha^*$ is one-to-one, for otherwise from $\alpha \ne \alpha'$ and
$$\alpha E_i \to \mu(\alpha) \pm \alpha^* E_i, \quad \alpha' E_i \to \mu(\alpha')I \pm \alpha^* E_i,$$
we deduce
$$(\alpha - \alpha')E_i \to (\mu(\alpha) - \mu(\alpha'))I,$$
which is impossible, since the right hand side belongs to the center but the left does not. Now the mapping (i) (or (ii)) is decomposed into two parts:
$$A \to \mu(\mathrm{tr} A) \tag{87}$$
and
$$A = (\alpha_{ij}) \to A^* = (\alpha^*_{ij}). \tag{88}$$
The second mapping is a one-to-one mapping carrying R onto itself. It carries matrices of rank one into matrices of rank one, since the first mapping is independent of the choice of $E_1, \cdots, E_n$.

By Theorem 1, we have consequenently either
$$A^* = P A^\tau Q, \tag{89}$$
where τ is an automorphism of Φ or
$$A^* = P A^{\tau'} Q, \tag{90}$$
where τ is an anti-automorphism. Since the mapping brings central elements into central elements, we have $Q = \beta P^{-1}$ where β belongs to the center. From (89), we deduce $A^* = \beta P A^\tau P^{-1}$, then
$$[A,B]^* = \beta P[A,B]P^{-1} = [A^*, B^*] = \beta^* P[A,B]P^{-1}.$$
Consequently $\beta^2 = \beta$, i.e., $\beta = 0$ or 1. From (90), we deduce $\beta = 0$ or -1. Therefore we have either
$$A^* = P A^\tau P^{-1}$$
or
$$A^* = P A^{\tau'} P^{-1},$$
the theorem is now proved.

Remark For $n = 2$, the theorem is not true. For example, for the field of all complex numbers, we have a mapping
$$\begin{pmatrix} a_1 + ia_2 & b_1 + ib_2 \\ c_1 + ic_2 & d_1 + id_2 \end{pmatrix} \to \begin{pmatrix} a - id_2 & b_1 + ib_2 \\ c_1 + ic_2 & d_1 - ia_2 \end{pmatrix},$$
which is not of the type described in the theorem.

References

[1] Hua. *Annals of Math.*, 1949, **50**: 8-31.
[2] Albert. *Structure of Algebras.*
[3] Ancochea. *Annals of Math.*, 1947, **48**: 147-153.
[4] Kaplansky. *Duke Math. Jour.*, 1947, **14**: 521-535.
[5] Ancochea. *Jour. für Math.*, 1942, **184**: 192-198.
[6] Hua. *Proc. of Nat. Aca. of Sc.*, U. S. A., 1949, **35**: 386-389
[7] Jacobson. *Duke Math. Jour.*, 1938, **5**: 534-551 and *Amer. Jour. of Math*, 1941, **63**: 481-515.
[8] Veblen and Whitehead, *Foundations of Differential Geometry.* Cambridge Tracts, 1932, **29**: 12.
[9] Artin and Whaples. *Amer. Jour. of Math.*, 1943, **65**: 87-107.
[10] Hua. *Science Reports of Nat. Tsinghua Univ.* Series A, 1948, **5**: 150-181.

Supplement to the paper of Dieudonné on the automorphisms of classical groups*

1. Introduction

In a previous paper, the author[1] determined the group of automorphisms of the symplectic group $\text{Sp}_n(K)$ over a field K of characteristic different from two. The method used there applies equally well to the general linear group $\text{GL}_n(K)$ and after some complicated modifications to the orthogonal group $O_n(K, f)$, with a quadratic form f of index $\nu \geqslant 1$. Professor Dieudonné[2] used an entirely different approach and comprehensively worked out the group of automorphisms of classical groups with several exceptions (see the last section of his paper). It is the aim of this paper to give solutions of some of the problems which he left open.

The first difficult problem cited in his paper is the determination of the group of automorphisms of $\text{GL}_2(K)$, where K is a sfield. For K of characteristic different from two he fails to characterize the transvections. Even if he were to succeed in characterizing the transvections, as he does when K is of characteristic two, the notion of semi-automorphisms of a sfield is still an obscure point of the final result. As a by-product of the present investigation, the author[3] proves that a semi-automorphism (in the sense of Ancochea[4] and Kaplansky[5]) of any sfield is either an automorphism or an anti-automorphism. Combining this with the result of Dieudonné, we can immediately solve the problem of automorphisms of $\text{GL}_2(K)$ (and $\text{PGL}_2(K)$), for K of characteristic 2. In this paper we shall solve the more difficult problem concerning the automorphisms of $\text{GL}_2(K)$ for K a sfield of characteristic different from two.

The second problem which he left unsolved concerns the automorphisms of the special linear groups $\text{SL}_4(K)$ and $\text{PSL}_4(K)$. Since the dimension is higher, the problem is much easier than the previous one. The group of automorphisms is determined in §6.

The only problem left unsettled in the family of orthogonal groups with indefinite fundamental form is the determination of the group of automorphisms of $O_4^+(K, f)$ (and $PO_4^+(K, f)$), for the case where K is of characteristic different from two and the quadratic form f has index 2. This is closely related to the author's study of geom-

---

* Reprinted from Memoirs of the *American Mathemaiical Society*, Number 2, 96-122

etry of skew-symmetric matrices. Some new types of automorphisms appear; these surprising phenomena throw a new light on the study of the commutator subgroup $\Omega_n(K, f)$ of the orthogonal group, which will be a subject of our later discussion.

It seems worthwhile to mention that the nature of such problems show that the lower the dimension the harder the problem. Dieudonné adopted a method which worked smoothly for large n, and treated individually the cases with small n. As the author mentioned before, the difficulty increases as n diminishes; Dieudonné's method becomes very clumsy for smaller n, and sometimes he is unable to solve the case for smallest n. On the other hand the author's method starts with the least possible n, which is usually the most difficult case. Therefore the reader will have little difficulty in extending the special results of this paper to the general case by means of the inductive method used in [1]. Moreover, in contrast with that of Dieudonné, the author's method uses only the calculus of matrices.

I. Linear groups

2. Automorphisms of the general linear group $\mathrm{GL}_2(K)$

Let K be a sfield of characteristic $\neq 2$. An element M of $\mathrm{GL}_2(K)$ is said to be an involution if $M^2 = I$. Since $-I$ is the only involution commutative with every element of $\mathrm{GL}_2(K)$, any automorphism of $\mathrm{GL}_2(K)$ must leave $-I$ invariant.

Let A and B be two anti-commutative involutions, that is

$$A^2 = B^2 = I, \quad AB = -BA. \tag{1}$$

These are invariant relations under any automorphism. Since A and B can be simultaneously carried into

$$\begin{pmatrix} 1 & 0 \\ 0 & -1 \end{pmatrix} \quad \text{and} \quad \begin{pmatrix} 0 & 1 \\ 1 & 0 \end{pmatrix} \tag{2}$$

by a similarity relation[①], we may assume that the automorphism under consideration leaves both matrices of (2) unaltered.

It is easy to verify that a matrix commutative with both matrices of (2) takes the form

$$\begin{pmatrix} \lambda & 0 \\ 0 & \lambda \end{pmatrix}, \tag{3}$$

---

① If there is a non singular P such that $PXP^{-1} = Y$, then we say that X and Y are similar.

where λ runs over all non-zero elements of the sfield K. Therefore the automorphism under consideration carries the group constituted by all the matrices of the form (3) onto itself.

Let K_0 be the center of the sfield K. Every matrix commutative with all the elements of (3) belongs to $\text{GL}_2(K_0)$ and conversely. Therefore our automorphism induces an automorphism on $\text{GL}_2(K_0)$ which leaves both matrices of (2) unaltered.

The parabolic elements [1] are defined in the same way as before ([1], p.756); namely, if the characteristic p of the sfield differs from 0, a matrix A is called parabolic if $A^{2p} = I$; and in case $p = 0$, if there are infinitely many matrices similar to A and commutative with A. Since $\text{SL}_2(K_0)$ is generated by all the parabolic elements of $\text{GL}_2(K_0)$, the automorphism under consideration induces an automorphism on $\text{SL}_2(K_0)$. It was proved in [1] that every automorphism of $\text{SL}_2(K_0)$ is of the form

$$A \to PA^\sigma P^{-1}, \tag{4}$$

where P is a non-singular matrix and σ is an automorphism of the field K_0. This is defined for A belonging to $\text{SL}_2(K_0)$.

Since

$$\begin{pmatrix} 1 & 0 \\ 0 & -1 \end{pmatrix} A \begin{pmatrix} 1 & 0 \\ 0 & -1 \end{pmatrix}, \quad \begin{pmatrix} 0 & 1 \\ 1 & 0 \end{pmatrix} A \begin{pmatrix} 0 & 1 \\ 1 & 0 \end{pmatrix}, \quad \begin{pmatrix} 0 & 1 \\ 1 & 0 \end{pmatrix} A \begin{pmatrix} 1 & 0 \\ 0 & -1 \end{pmatrix}$$

all belong to $\text{SL}_2(K_0)$ for A belonging to $\text{SL}_2(K_0)$, and our automorphism leaves (2) unaltered, we deduce immediately that P is either λI or $\mu \begin{pmatrix} 0 & 1 \\ -1 & 0 \end{pmatrix}$. That is, we have only two possible cases of (4), namely,

$$A \to A^\sigma \tag{5}$$

and

$$A \to \begin{pmatrix} 0 & 1 \\ -1 & 0 \end{pmatrix} A^\sigma \begin{pmatrix} 0 & 1 \\ -1 & 0 \end{pmatrix}^{-1}. \tag{6}$$

We are going to discuss both cases separately. For the first case we have, in particular,

$$\begin{pmatrix} 1 & 0 \\ x & 1 \end{pmatrix} \to \begin{pmatrix} 1 & 0 \\ x^\sigma & 1 \end{pmatrix}, \quad \text{for } x \in K_0, \tag{7}$$

---

[1] Notice that the definition does not coincide with the usual definition of parabolic element.

$$\begin{pmatrix} a & 0 \\ 0 & a^{-1} \end{pmatrix} \to \begin{pmatrix} a^\sigma & 0 \\ 0 & (a^\sigma)^{-1} \end{pmatrix}, \quad \text{for } a \, (\neq 0) \in K_0 \tag{8}$$

and

$$\begin{pmatrix} 0 & 1 \\ -1 & 0 \end{pmatrix} \to \begin{pmatrix} 0 & 1 \\ -1 & 0 \end{pmatrix}. \tag{9}$$

Notice that

$$\begin{pmatrix} 0 & 1 \\ -1 & 0 \end{pmatrix} \begin{pmatrix} 1 & 0 \\ x & 1 \end{pmatrix} \begin{pmatrix} 0 & -1 \\ 1 & 0 \end{pmatrix} = \begin{pmatrix} 1 & -x \\ 0 & 1 \end{pmatrix} \to \begin{pmatrix} 1 & -x^\sigma \\ 0 & 1 \end{pmatrix}.$$

Next we are going to consider those parabolic elements of $\mathrm{GL}_2(K)$ which are commutative with (7). By commutativity they are of the form

$$\begin{pmatrix} a & 0 \\ y & a \end{pmatrix},$$

since they are parabolic, we have $a = \pm 1$. Squaring all those elements, we deduce that our automorphism induces the mapping

$$\begin{pmatrix} 1 & 0 \\ x & 1 \end{pmatrix} \to \begin{pmatrix} 1 & 0 \\ x^\sigma & 1 \end{pmatrix}, \tag{10}$$

for all x belonging to K, noticing that the definition of x^σ is now extended to any element x of K. Evidently, we have

$$(x+y)^\sigma = x^\sigma + y^\sigma \tag{11}$$

for all x and y of K.

Since

$$\begin{pmatrix} x & 0 \\ 0 & x^{-1} \end{pmatrix}, \quad x \, (\neq 0) \in K$$

is commutative with (8), and

$$\begin{pmatrix} x & 0 \\ 0 & x^{-1} \end{pmatrix} = \begin{pmatrix} 1 & -x \\ 0 & 1 \end{pmatrix} \begin{pmatrix} 1 & 0 \\ x^{-1} & 1 \end{pmatrix} \begin{pmatrix} 1 & 0 \\ -1 & 1 \end{pmatrix} \begin{pmatrix} 1 & 1 \\ 0 & 1 \end{pmatrix}$$
$$\times \begin{pmatrix} 1 & 0 \\ -1 & 1 \end{pmatrix} \begin{pmatrix} 1 & 0 \\ x & 1 \end{pmatrix}, \tag{12}$$

it is mapped by our automorphism into the following form

$$\begin{pmatrix} \lambda & 0 \\ 0 & \mu \end{pmatrix} = \begin{pmatrix} 1 & -x^\sigma \\ 0 & 1 \end{pmatrix} \begin{pmatrix} 1 & 0 \\ x^* & 1 \end{pmatrix} \begin{pmatrix} 1 & 0 \\ -1 & 1 \end{pmatrix} \begin{pmatrix} 1 & 1 \\ 0 & 1 \end{pmatrix}$$
$$\times \begin{pmatrix} 1 & 0 \\ -1 & 1 \end{pmatrix} \begin{pmatrix} 1 & 0 \\ x^\sigma & 1 \end{pmatrix}.$$

Comparing the elements in the (1, 2) position, we have

$$x^\sigma x^* = 1,$$

that is, $(x^{-1})^\sigma = (x^\sigma)^{-1}$, and

$$\begin{pmatrix} x & 0 \\ 0 & x^{-1} \end{pmatrix} \to \begin{pmatrix} x^\sigma & 0 \\ 0 & x^{-\sigma} \end{pmatrix}.$$

for all $x \in K$. Considering the element

$$\begin{pmatrix} xyx & 0 \\ 0 & (xyx)^{-1} \end{pmatrix} = \begin{pmatrix} x & 0 \\ 0 & x^{-1} \end{pmatrix} \begin{pmatrix} y & 0 \\ 0 & y^{-1} \end{pmatrix} \begin{pmatrix} x & 0 \\ 0 & x^{-1} \end{pmatrix},$$

we deduce that

$$(xyx)^\sigma = x^\sigma y^\sigma x^\sigma.$$

Evidently $1^\sigma = 1$; therefore, by a theorem of the author[3], we deduce that the mapping $x \to x^\sigma$ is either an automorphism or an anti-automorphism.

Since $\mathrm{SL}_2(K)$ is generated by the elements of (9) and (10) (Dieudonné[6]), we have the following two possibilities:

1) If $x \to x^\sigma$ is an automorphism, the automorphism of $\mathrm{SL}_2(K)$ under consideration is

$$A \to A^\sigma. \tag{13}$$

2) If $x \to x^\sigma$ is an anti-automorphism, since

$$\begin{pmatrix} 1 & x \\ 0 & 1 \end{pmatrix} \to \begin{pmatrix} 1 & x^\sigma \\ 0 & 1 \end{pmatrix} = \begin{pmatrix} 0 & 1 \\ -1 & 0 \end{pmatrix} \begin{pmatrix} 1 & 0 \\ x^\sigma & 1 \end{pmatrix}^{-1} \begin{pmatrix} 0 & 1 \\ -1 & 0 \end{pmatrix}^{-1},$$

$$\begin{pmatrix} x & 0 \\ 0 & x^{-1} \end{pmatrix} \to \begin{pmatrix} x & 0 \\ 0 & x^{-\sigma} \end{pmatrix} = \begin{pmatrix} 0 & 1 \\ -1 & 0 \end{pmatrix} \begin{pmatrix} x & 0 \\ 0 & x^{-\sigma} \end{pmatrix}^{-1} \begin{pmatrix} 0 & 1 \\ -1 & 0 \end{pmatrix}^{-1},$$

$$\begin{pmatrix} 0 & 1 \\ -1 & 0 \end{pmatrix} \to \begin{pmatrix} 0 & 1 \\ -1 & 0 \end{pmatrix} = \begin{pmatrix} 0 & 1 \\ -1 & 0 \end{pmatrix} \begin{pmatrix} 0 & -1 \\ 1 & 0 \end{pmatrix}^{-1} \begin{pmatrix} 0 & -1 \\ 1 & 0 \end{pmatrix},$$

we have the automorphism, after an inner automorphism,

$$A \to (A'^\sigma)^{-1}, \tag{14}$$

where A' denotes the transpose of A.

By similar methods applied to (6), we obtain the same results (13) and (14). Form now on we assume that the automorphism of $GL_2(K)$ under consideration satisfies (13) or (14) for A belonging to $SL_2(K)$. We now consider $GL_2(K)$; certainly the most essential part is to find the mapping of those elements of the form

$$\begin{pmatrix} 1 & 0 \\ 0 & \lambda \end{pmatrix}, \quad \lambda \in K. \tag{15}$$

Since such an element permutes with any element (12), with $x \in K_0$, it is mapped into an element of the form

$$\begin{pmatrix} a_\lambda & 0 \\ 0 & d_\lambda \end{pmatrix}.$$

Since

$$\begin{pmatrix} 1 & 0 \\ 0 & \lambda \end{pmatrix} \begin{pmatrix} 1 & 0 \\ x & 1 \end{pmatrix} \begin{pmatrix} 1 & 0 \\ 0 & \lambda \end{pmatrix}^{-1} = \begin{pmatrix} 1 & 0 \\ \lambda x & 1 \end{pmatrix}$$

is mapped into

$$\begin{pmatrix} 1 & 0 \\ (\lambda x)^\sigma & 1 \end{pmatrix},$$

we deduce

$$d_\lambda x^\sigma a_\lambda^{-1} = (\lambda x)^\sigma \tag{16}$$

for all x. In particular setting $x = 1$, we have

$$d_\lambda = \lambda^\sigma a_\lambda, \tag{17}$$

that is, (15) is mapped into

$$\begin{pmatrix} 1 & 0 \\ 0 & \lambda^\sigma \end{pmatrix} a_\lambda = d_\lambda \begin{pmatrix} \lambda^{-\sigma} & 0 \\ 0 & 1 \end{pmatrix}. \tag{18}$$

Combining (16) and (17), we have

$$\lambda^\sigma a_\lambda x^\sigma = (\lambda x)^\sigma a_\lambda.$$

In case σ is an automorphism, we have

$$a_\lambda x^\sigma = x^\sigma a_\lambda,$$

that is, a_λ belongs to the center, and (15) is mapped into

$$a_\lambda \begin{pmatrix} 1 & 0 \\ 0 & \lambda^\sigma \end{pmatrix},$$

where a_λ is a representation of the multiplicative group into the center.

In case σ is an anti-automorphism, we have

$$(\lambda^\sigma a_\lambda) x^\sigma = x^\sigma (\lambda^\sigma a_\lambda),$$

that is, $d_\lambda = \lambda^\sigma a_\lambda$ belongs to the center, and (15) is mapped into

$$d_\lambda \begin{pmatrix} \lambda^{-\sigma} & 0 \\ 0 & 1 \end{pmatrix}.$$

Therefore we have

Theorem 1 *Every automorphism of* $\mathrm{GL}_2(K)$ *is a product of an inner automorphism and one of the following two types:* (i)

$$A \to \chi(A) A^\sigma, \tag{19}$$

where σ is an automorphism of K and $\chi(A)$ is a representation of $\mathrm{GL}_2(K)$ *into the multiplicative group of* K_0; *and* (ii)

$$A \to \chi(A) (A'^{-1})^\sigma, \tag{20}$$

where σ is an anti-automorphism of K.

Remark The theorem also holds for sfield of characteristic 2.

3. Automorphisms of $\mathrm{SL}_2^\pm(K)$

We now consider a group $\mathrm{SL}_2^\pm(K)$ which differs slightly from $\mathrm{SL}_2(K)$. The projective geometry on a line over a sfield suggests that the group $\mathrm{SL}_2^\pm(K)$, obtained by successive application of harmonic separation, is of primary interest (Ancochea[7]). As we know, in the group $\mathrm{SL}_2(K)$, involutions do not always exist. More precisely, if -1 does not belong to the commutator subgroup of the group of multiplication, there does not exist an involution (a relation obtained by harmonic conjugacy).

By an easy investigation, we find that if -1 belongs to the commutator subgroup of the multiplicative group of K, then $\mathrm{SL}_2^{\pm}(K) = \mathrm{SL}_2(K)$; otherwise, $\mathrm{SL}_2^{\pm}(K)$ is obtained from $\mathrm{SL}_2(K)$ by adjoining an involution, such as

$$\begin{pmatrix} 1 & 0 \\ 0 & -1 \end{pmatrix}.$$

The automorphisms of $\mathrm{SL}_2(K)$ are still to be determined when -1 is not in the commutator subgroup.

The theorem which we shall prove in this section is the following:

Theorem 2 *Every automorphism of $\mathrm{SL}_2^{\pm}(K)$ is either of the form*

$$\chi(A) P A^\sigma P^{-1}, \tag{21}$$

where P is a non singular matrix, σ is an automorphism of K, and $\chi(A)$ is a representation of $\mathrm{SL}_2^{\pm}(K)$ into the multiplicative group of the center K_0, or of the form

$$\chi(A) P A'^{\sigma-1} p^{-1}, \tag{22}$$

where σ is an anti-automorphism.

Remark Concerning $\chi(A)$: in case -1 belongs to the commutator subgroup of the multiplicative group of K, then we have always $\chi(A) = 1$. Otherwise we have another possibility, namely

$$\chi(A) = \begin{cases} 1, & \text{if } A \text{ belongs to } \mathrm{SL}_2(K), \\ -1, & \text{otherwise.} \end{cases}$$

Proof of Theorem 2 Let us start with a pair of anti-commutative involutions A and B, that is

$$A^2 = B^2 = I, \quad AB = -BA.$$

There exists a matrix P, which may not belong to $\mathrm{SL}_2^{\pm}(K)$ such that

$$PAP^{-1} = \begin{pmatrix} 1 & 0 \\ 0 & -1 \end{pmatrix}, \quad PBP^{-1} = \begin{pmatrix} 0 & 1 \\ 1 & 0 \end{pmatrix}.$$

An element of $\mathrm{SL}_2^{\pm}(K)$ commuting with both matrices of (2) is of the form

$$\begin{pmatrix} \lambda & 0 \\ 0 & \lambda \end{pmatrix}, \tag{23}$$

where either λ^2 or $-\lambda^2$ belongs to the commutator subgroup of the group of multiplication of K. By Theorem 4 of Hua[8], every element commutative with all λ belongs to the center K_0. Therefore a matrix commutative with all elements of (23) is of the form

$$\begin{pmatrix} a & b \\ c & d \end{pmatrix}, \tag{24}$$

where a, b, c, d belong to K_0, and the determinant of (24) (or its negative) belongs to the intersection of the center and the commutator subgroup of the multiplicative group. The remaining part of the proof is similar to that of Theorem 1.

4. Automorphisms of $\mathrm{PGL}_2(K)$

We identify all the elements of the form γM of $\mathrm{GL}_2(K)$ as a single element of $\mathrm{PGL}_2(K)$, where $\gamma\,(\neq 0)$ runs over all the elements of the center. A matrix M represents an involution if it satisfies

$$M^2 = \gamma I, \quad \gamma \in K_0 \tag{25}$$

and if M does not represent the identity of $\mathrm{PGL}_2(K)$. γ is called the scalar of the matrix M.

Since we have a matrix Q such that

$$QMQ^{-1} = \begin{pmatrix} 0 & 1 \\ c & d \end{pmatrix}, \tag{26}$$

therefore an involution satisfying (25) is similar to the normal form

$$\begin{pmatrix} 0 & 1 \\ \gamma & 0 \end{pmatrix}. \tag{27}$$

If $\begin{pmatrix} 0 & 1 \\ \gamma & 0 \end{pmatrix}$ and $\begin{pmatrix} 0 & 1 \\ \delta & 0 \end{pmatrix}$ are similar in $\mathrm{PGL}_2(K)$, then from

$$\begin{pmatrix} 0 & 1 \\ \gamma & 0 \end{pmatrix} \begin{pmatrix} a & b \\ c & d \end{pmatrix} = \varepsilon \begin{pmatrix} a & b \\ c & d \end{pmatrix} \begin{pmatrix} 0 & 1 \\ \delta & 0 \end{pmatrix},$$

where ε belongs to the center, we deduce

$$c = \varepsilon\delta b, \quad d = \varepsilon a, \quad \gamma a = \varepsilon\delta b, \quad \gamma b = \varepsilon c;$$

consequently

$$(\gamma - \varepsilon^2\delta)b = (\gamma - \varepsilon^2\delta)a = 0.$$

Since a and b cannot be both zero, we have $\gamma = \varepsilon^2\delta$. Conversely, we have

$$\begin{pmatrix} 0 & 1 \\ \gamma & 0 \end{pmatrix} = \varepsilon \begin{pmatrix} 1 & 0 \\ 0 & \varepsilon \end{pmatrix} \begin{pmatrix} 0 & 1 \\ \delta & 0 \end{pmatrix} \begin{pmatrix} 1 & 0 \\ 0 & \varepsilon^{-1} \end{pmatrix}.$$

Therefore two involutions, represented by matrices with scalar γ and δ, are similar in $\mathrm{PGL}_2(K)$ if and only if $\gamma\delta^{-1}$ is the square of a central element.

Therefore, all the involutions can be classified into conjugate sets Γ_α, where α runs over all the elements of the factor group of the multiplicative group of the center by its subgroup of all square elements.

We are now going to prove that the automorphism under consideration leaves Γ_1 invariant. Γ_1 is a conjugate set containing two distince commuting elements, namely

$$\begin{pmatrix} 1 & 0 \\ 0 & -1 \end{pmatrix} \quad \text{and} \quad \begin{pmatrix} 0 & 1 \\ 1 & 0 \end{pmatrix}.$$

It is enough to distinguish Γ_1 from those Γ_α containing two commutative involutions. Let Γ_α be a set containing two commuting involutions A and B with scalars α and $\alpha\delta^2$. From $AB = \varepsilon BA$, ε belonging to the center, we deduce

$$BA^2 = B\alpha = \alpha B = A^2 B = \varepsilon^2 BA^2, \tag{28}$$

that is, $\varepsilon = \pm 1$, Therefore $(AB)^2 = \pm \alpha^2 \delta^2 I$, so that AB belongs either to Γ_1 and Γ_{-1}. Therefore our automorphism either leaves both Γ_1 and Γ_{-1} invariant, or permutes them.

1°. If -1 is the square of a central element, then Γ_{-1} coincides with Γ_1.

2°. Suppose that -1 is not a square of a central element but is a square of an element a in K. The product of any two distinct commuting elements of Γ_1 belongs to Γ_{-1}. In fact, let us fix one of them as $A = \begin{pmatrix} 1 & 0 \\ 0 & -1 \end{pmatrix}$; then, from $AB = BA$, we deduce that B is either the identity or γA, γ belonging to center. Thus, we always have $AB = -BA$ (the existence of B is shown by (2)), and $(AB)^2 = -I$. In Γ_{-1} we have two commuting elements

$$\begin{pmatrix} a & 0 \\ 0 & -a \end{pmatrix}, \quad \begin{pmatrix} 0 & a \\ a & 0 \end{pmatrix},$$

whose product is

$$a^2 \begin{pmatrix} 0 & 0 \\ -1 & 0 \end{pmatrix} = -\begin{pmatrix} 0 & 1 \\ -1 & 0 \end{pmatrix},$$

which belongs to Γ_{-1}. Therefore our automorphism cannot interchange Γ_1 and Γ_{-1}.

3°. Next we suppose that -1 is not the square of an element in K. Let A and B be involutions of Γ_1 and Γ_{-1} respectively, and let them commute. We may fix A to be $\begin{pmatrix} 1 & 0 \\ 0 & -1 \end{pmatrix}$. Then B takes one of the forms: $\begin{pmatrix} \lambda & 0 \\ 0 & \mu \end{pmatrix}$ or $\begin{pmatrix} 0 & \lambda \\ \mu & 0 \end{pmatrix}$. However, the first is impossible, since $\lambda^2 = -1$ is insoluble. Therefore we have

$$B = \begin{pmatrix} 0 & \lambda \\ -\lambda^{-1} & 0 \end{pmatrix}.$$

The product AB belongs to Γ_1. Therefore the automorphism under consideration cannot interchange Γ_1 and Γ_{-1}.

Take a pair of commuting involutions of Γ_1. Without loss of generality, we assume that the automorphism leaves (2) invariant. The elements commuting with both matrices of (2) are of the form

$$\begin{pmatrix} a & 0 \\ 0 & a \end{pmatrix}, \begin{pmatrix} a & 0 \\ 0 & -a \end{pmatrix}, \begin{pmatrix} 0 & b \\ b & 0 \end{pmatrix}, \begin{pmatrix} 0 & b \\ -b & 0 \end{pmatrix}, \quad (29)$$

by (28). Squaring all these elements, we obtain

$$\begin{pmatrix} a^2 & 0 \\ 0 & a^2 \end{pmatrix}. \quad (30)$$

Consider the set of elements commuting with all these elements, that is

$$\begin{pmatrix} p & q \\ r & s \end{pmatrix} \begin{pmatrix} a^2 & 0 \\ 0 & a^2 \end{pmatrix} = \rho \begin{pmatrix} a^2 & 0 \\ 0 & a^2 \end{pmatrix} \begin{pmatrix} p & q \\ r & s \end{pmatrix}, \quad (31)$$

where ρ belongs to the center and may depend on a. Suppose $p \neq 0$; then from

$$pa^2 = \rho a^2 p,$$

we have

$$\begin{pmatrix} 1 & qp^{-1} \\ rp^{-1} & sp^{-1} \end{pmatrix} \begin{pmatrix} a^2 & 0 \\ 0 & a^2 \end{pmatrix} p = \frac{1}{\rho} \begin{pmatrix} p & q \\ r & s \end{pmatrix} \begin{pmatrix} a^2 & 0 \\ 0 & a^2 \end{pmatrix}$$
$$= \begin{pmatrix} a^2 & 0 \\ 0 & a^2 \end{pmatrix} \begin{pmatrix} p & q \\ r & s \end{pmatrix},$$

that is
$$\begin{pmatrix} 1 & qp^{-1} \\ rp^{-1} & sp^{-1} \end{pmatrix} \begin{pmatrix} a^2 & 0 \\ 0 & a^2 \end{pmatrix} = \begin{pmatrix} a^2 & 0 \\ 0 & a^2 \end{pmatrix} \begin{pmatrix} 1 & qp^{-1} \\ rp^{-1} & sp^{-1} \end{pmatrix}.$$

By Theorem 6 of Hua[8], qp^{-1}, rp^{-1} and sp^{-1} all belong to the center. By similarly treating the case $p = 0, q \neq 0$, we reach the conclusion that

$$\begin{pmatrix} p & q \\ r & s \end{pmatrix} = \lambda M,$$

where M belongs to $\mathrm{PGL}_2(K_0)$, and $\lambda a^2 \lambda^{-1} a^{-2}$ belongs to the center for all a, by (31).

Consider those elements of the form

$$(\lambda M)(\mu N)^2 (\lambda M)^{-1} (\mu N)^{-2} = \xi M N^2 M^{-1} N^{-2}, \quad \xi \in K_0,$$

where M and N belong to $\mathrm{PGL}_2(K_0)$. The elements $MN^2 M^{-1} N^{-2}$ generate the group $\mathrm{PSL}_2(K_0)$; in fact, they generate a normal subgroup of $\mathrm{PSL}_2(K_0)$, but since $\mathrm{PGL}_2(K_0)$ is simple①, we have the assertion. Therefore the automorphism of $\mathrm{PSL}_2(K)$ induces an automorphism on $\mathrm{PSL}_2(K_0)$.

Now we use non-homogeneous representations: an element of $\mathrm{PGL}_2(K)$, represented by a matrix $\begin{pmatrix} a & b \\ c & d \end{pmatrix}$, can be realized as a transformation

$$z_1 = (az + b)(cz + d)^{-1} \tag{32}$$

on a projective line over the sfield K. It is evident that the realization is one-to-one.

We define a parabolic element t by the following properties: if the characteristic p of our sfield is not zero, a transformation t is called parabolic, if t^p equals the identity of $\mathrm{PGL}_2(K)$; and, for $p = 0$, a transformation t is called parabolic, if there are infinitely many elements similar to t and commutative with t. We shall now prove that every parabolic element of $\mathrm{PGL}_2(K_0)$ can be carried into the form

$$z_1 = z + a \quad \text{(where } a \text{ belongs to } K_0\text{)}.$$

In fact, suppose $p = 0$, and that t is represented by a matrix T belonging to $\mathrm{SL}_2(K_0)$, where

$$T^p = \alpha I.$$

---

① Notice that if K_0 has only 3 elements, the characteristic of K is equal to 2, the case which was excluded at the beginning.

Since T^p belongs to $\mathrm{SL}_2(K_0)$, we have $\alpha = \pm 1$. As in [1], we have our assertion. In case $p = 0$, the argument of [1] can still be applied.

After we have taken care of the parabolic elements, we can follow either the method of Schreier-van der Waerden[9] or the method of the author[1] to obtain the automorphism of $\mathrm{PSL}_2(K_0)$. More precisely, we may assume that our automorphism of $\mathrm{PGL}_2(K)$, induces an automorphism of $\mathrm{PSL}_2(K)$ which carries the transformations

$$z_1 = z + t \quad \text{to} \quad z_1 = z + t^\sigma, \quad t \in K_0 \tag{33}$$

and

$$z_1 = -1/z \quad \text{to} \quad z_1 = -1/z, \tag{34}$$

where t^σ is an automorphism of the field K_0.

Consider all the parabolic elements commuting with the translation group (33). It is easy to see that they are of the form

$$z_1 = z + t, \quad t \in K. \tag{35}$$

Therefore, our automorphism carries (35) into

$$z_1 = z + t^\sigma,$$

where t^σ is defined for all t belonging to K. From the identity (12), we deduce by the same argument following (12), that t^σ is either an automorphism or an anti-automorphism of the sfield, and

Theorem 3 *Every automorphism of* $\mathrm{PGL}_2(K)$ *is induced by an automorphism of* $\mathrm{GL}_2(K)$.

Remark The theorem is still true for characteristic two.

5. Automorphisms of $\mathrm{PSL}_2^\pm(K)$

As in §4, we classify the involutions into conjugate sets; the situation now becomes very much more complicated, however, since neither can the scalar be arbitrary, nor dose the scalar determine a unique conjugate set. It is not always possible to find the matrix Q, belonging to $\mathrm{SL}_2^\pm(K)$, such that (26) is true. A matrix of scalar γ is similar to a matrix of the form

$$\begin{pmatrix} 0 & q^{-1} \\ \gamma q & 0 \end{pmatrix} \tag{36}$$

under $SL_2(K)$. A conjugate set Γ_α of §4 may now break up into several conjugate sets. Nevertheless, the same argument gives us the result that the pair of sets Γ_1 and Γ_{-1} is invariant under automorphisms.

Fortunately Γ_1 does not split into several conjugate sets; in fact,

$$\begin{pmatrix} 0 & q^{-1} \\ q & 0 \end{pmatrix} \begin{pmatrix} 1 & -\frac{1}{2}q^{-1} \\ q & \frac{1}{2} \end{pmatrix} = \begin{pmatrix} 1 & -\frac{1}{2}q^{-1} \\ q & \frac{1}{2} \end{pmatrix} \begin{pmatrix} 1 & 0 \\ 0 & -1 \end{pmatrix}$$

and

$$\begin{pmatrix} 1 & -\frac{1}{2}q^{-1} \\ q & \frac{1}{2} \end{pmatrix} = \begin{pmatrix} 1 & 0 \\ q & 1 \end{pmatrix} \begin{pmatrix} 1 & -\frac{1}{2}q^{-1} \\ 0 & 1 \end{pmatrix},$$

which belongs to $SL_2(K)$.

As in §4, we divide our treatment into three cases: 1° and 3° can be treated analogously. The only difficult case is where $(2°) - 1$ is a square of a non-central element a. Now Γ_{-1} splits into several conjugate sets. We shall now prove that every involution (namely (36) with $\gamma = -1$) of Γ_{-1} is similar to

$$\begin{pmatrix} a & 0 \\ 0 & -qaq^{-1} \end{pmatrix} \tag{37}$$

under the group $SL_2(K)$. In fact this is a consequence of the following simple calculation:

$$\begin{pmatrix} 1 & -\frac{1}{2}a \\ 0 & 1 \end{pmatrix} \begin{pmatrix} 1 & 0 \\ -a & 1 \end{pmatrix} \begin{pmatrix} 0 & 1 \\ -1 & 0 \end{pmatrix} \begin{pmatrix} 1 & 0 \\ a & 1 \end{pmatrix} \begin{pmatrix} 1 & \frac{1}{2}a \\ 0 & 1 \end{pmatrix} = \begin{pmatrix} a & 0 \\ 0 & -a \end{pmatrix}. \tag{38}$$

Then (36) with $\gamma = -1$ is similar to (37) under $SL_2(k)$, since

$$\begin{pmatrix} 1 & 0 \\ 0 & q \end{pmatrix} \begin{pmatrix} 1 & -\frac{1}{2}a \\ 0 & 1 \end{pmatrix} \begin{pmatrix} 1 & 0 \\ -a & 1 \end{pmatrix} \begin{pmatrix} 1 & 0 \\ 0 & q \end{pmatrix}^{-1}$$

belongs to $SL_2(K)$. Putting qa instead of q in (36), we find that

$$\begin{pmatrix} a & 0 \\ 0 & -qaq^{-1} \end{pmatrix} \text{ and } \begin{pmatrix} 0 & (qa)^{-1} \\ -qa & 0 \end{pmatrix} \tag{39}$$

are similar under the group $SL_2(K)$. Their product is

$$\begin{pmatrix} 0 & q^{-1} \\ -q & 0 \end{pmatrix},$$

which is similar to (39) too. Therefore any set γ contains three mutually commuting elements.

As we proved beofre (§4), the product of two commuting elements of Γ_1 is not an element of Γ_1.

Therefore the automorphism carries Γ_1 onto itself. By an argument similar to that of §4, we may assume that our automorphism leaves

$$\begin{pmatrix} 1 & 0 \\ 0 & -1 \end{pmatrix}, \begin{pmatrix} 0 & 1 \\ 1 & 0 \end{pmatrix}$$

invariant.

The elements of $\mathrm{PSL}_2(K)$ commuting with both these elements are of the form

$$\begin{pmatrix} a & 0 \\ 0 & a \end{pmatrix}, \begin{pmatrix} a & 0 \\ 0 & -a \end{pmatrix}, \begin{pmatrix} 0 & b \\ b & 0 \end{pmatrix}, \begin{pmatrix} 0 & b \\ -b & 0 \end{pmatrix}.$$

Squaring all of these elements, we obtain

$$\begin{pmatrix} a^2 & 0 \\ 0 & a^2 \end{pmatrix}.$$

Since they belong to $\mathrm{PSL}_2(K)$, then a^2 or $-a^2$ belongs to the commutator subgroup.

Using the lemma below, we proceed as in §4, and obtain

Theorem 4 *Every automorphism of* $\mathrm{PSL}_2^\pm(K)$ *is induced by an automorphism of* $\mathrm{SL}_2^\pm(K)$.

Lemma *Let C be the commutator subgroup of the multiplicative group of K and let C^2 be the group generated by all the square elements of C. Then either K is commutative, or the sfield K is generated by all the elements of C^2. Consequently, if an element commutes with all the elements of C^2, then it belongs to center.*

Proof If C^2 is not contained in the center, the lemma follows from Theorem 1 of Hua[8].

Now we suppose C^2 is contained in the center K_0 and K is not commutative. Let

$$1, a_1, a_2, \cdots, \quad a_i \in C$$

be the representative system of the factor group of C over C^2. Since C generates the sfield, by Theorems 4 and 1 of Hua[8], we have therefore

$$K = K_0(a_1, a_2, \cdots),$$

where
$$a_i^2 = \gamma_i, \quad (a_i a_j)^2 = \gamma_i \gamma_j \delta_{ij}$$

and γ's and δ's belong to the field K_0.

Since K is not a field, there is a $\delta_{ij} \neq 1$. we may suppose that $\delta_{12} = \delta \neq 1$.

From
$$a_1(a_2 + 1) = \delta(a_2 + 1)a_1 + a_1(1 - \delta),$$

we deduce
$$a_1(a_2 + 1)a_1^{-1}(a_2 + 1)^{-1} = \delta + (a_2 + 1)^{-1}(1 - \delta).$$

Suppose that the squares of commutators belong to the center. Then
$$(\delta + (a_2 + 1)^{-1}(1 - \delta))^2 = \beta,$$

where β belongs to K_0. Consequently
$$(\delta a_2 + 1)^2 = \beta(a_2 + 1)^2,$$

we deduce immediately
$$\beta = \delta, \quad \delta \gamma_2 = 1.$$

Consider
$$(a_2 + 1)a_1 = \frac{1}{\delta} a_1(a_2 + 1) + a_1\left(1 - \frac{1}{\delta}\right);$$

we deduce similarly
$$\frac{1}{\delta} \gamma_2 = 1.$$

Consequently $\delta = -1$ and $\gamma_2 = -1$. Similarly $\gamma_1 = -1$.

Suppose that K_0 has more than three elements (it is easy to prove that if K_0 contains less than 4 elements, there does not exits such a K), Take x of K_0 different form 0 and ± 1. Then
$$a_1(1 + xa_2)a_1^{-1}(1 + xa_2)^{-1} = (1 - xa_2)(1 + xa_2)^{-1} = \frac{1 - x^2}{1 + x^2} - \frac{2x}{1 + x^2} a_2,$$

its square is not in the center. This is a contradiction. We have therefore the lemma.

6. Automorphism of $SL_4(K)$

The case left open by Dieudonné concerning the automorphism of $SL_4(K)$ is that where K is not commutative, K has characteristic $\neq 2$ and -1 is not contained in the commutator subgroup of the multiplicative group.

Any two commuting involutions of $SL_4(K)$ can be brought into the form

$$J_1 = \begin{pmatrix} 1 & 0 & 0 & 0 \\ 0 & 1 & 0 & 0 \\ 0 & 0 & -1 & 0 \\ 0 & 0 & 0 & -1 \end{pmatrix}, \quad J_2 = \begin{pmatrix} 1 & 0 & 0 & 0 \\ 0 & -1 & 0 & 0 \\ 0 & 0 & 1 & 0 \\ 0 & 0 & 0 & -1 \end{pmatrix}. \tag{40}$$

Let

$$J_3 = J_1 J_2 = \begin{pmatrix} 1 & 0 & 0 & 0 \\ 0 & -1 & 0 & 0 \\ 0 & 0 & -1 & 0 \\ 0 & 0 & 0 & 1 \end{pmatrix}. \tag{41}$$

We may assume, that after a transformation, our automorphism leaves J_1, J_2 and J_3 invariant.

We define K_1 by means of the equation

$$K_1 J_1 = J_1 K_1, \quad K_1 J_2 = -J_3 K_1, \quad K_1^2 = -J_1. \tag{42}$$

Solving (42), we obtain

$$K_1 = \begin{pmatrix} 0 & a & 0 & 0 \\ -a^{-1} & 0 & 0 & 0 \\ 0 & 0 & \pm 1 & 0 \\ 0 & 0 & 0 & \pm 1 \end{pmatrix},$$

where the signs $\pm$ are either both $+$ or both $-$. By a transformation, we may assume without loss of generality that our automorphism leaves J_1, J_2, J_3 invariant, and either leaves

$$K_1^+ = \begin{pmatrix} 0 & 1 & 0 & 0 \\ -1 & 0 & 0 & 0 \\ 0 & 0 & 1 & 0 \\ 0 & 0 & 0 & 1 \end{pmatrix}, \quad K_1^- = \begin{pmatrix} 0 & 1 & 0 & 0 \\ -1 & 0 & 0 & 0 \\ 0 & 0 & -1 & 0 \\ 0 & 0 & 0 & -1 \end{pmatrix} \tag{43}$$

invariant, or permutes them. By similar considerations, we may assume that our automorphism leaves the following two pairs pair-wise invariant:

$$K_2^+ = \begin{pmatrix} 0 & 0 & 1 & 0 \\ 0 & 1 & 0 & 0 \\ -1 & 0 & 0 & 0 \\ 0 & 0 & 0 & 1 \end{pmatrix}, \quad K_2^- = \begin{pmatrix} 0 & 0 & 1 & 0 \\ 0 & -1 & 0 & 0 \\ -1 & 0 & 0 & 0 \\ 0 & 0 & 0 & -1 \end{pmatrix} \quad (44)$$

and

$$K_3^+ = \begin{pmatrix} 0 & 0 & 0 & 1 \\ 0 & 1 & 0 & 0 \\ 0 & 0 & 1 & 0 \\ -1 & 0 & 0 & 0 \end{pmatrix}, \quad K_3^- = \begin{pmatrix} 0 & 0 & 0 & 1 \\ 0 & -1 & 0 & 0 \\ 0 & 0 & -1 & 0 \\ -1 & 0 & 0 & 0 \end{pmatrix} \quad (45)$$

($K_2 = K_2^\pm$ satisfies

$$K_2 J_2 = J_2 K_2, \quad K_2 J_3 = -J_1 K_2, \quad K_2^2 = -J_2$$

and $K_3 = K_3^\pm$ satisfies

$$K_3 J_3 = J_3 K_3, \quad K_3 J_1 = -J_2 K_3, \quad K_3^2 = -J_3).$$

The subgroup commuting with J_1 consists of all elements of the form

$$T = \begin{pmatrix} A & 0 \\ 0 & B \end{pmatrix}, \quad (46)$$

where A and B are two-rowed matrices. Consider the group G formed by

$$T K_1^\pm T^{-1} = \begin{pmatrix} A \begin{pmatrix} 0 & 1 \\ -1 & 0 \end{pmatrix} A^{-1} & 0 \\ 0 & \pm I^{(2)} \end{pmatrix}. \quad (47)$$

Since all of the elements of the form $A \begin{pmatrix} 0 & 1 \\ -1 & 0 \end{pmatrix} A^{-1}$ generate (Dieudonné[6]) the group $\mathrm{SL}_2(K)$, all the elements of the form (47) generate a group formed by all elements of the form

$$\begin{pmatrix} A & 0 \\ 0 & \pm I \end{pmatrix}, \quad (48)$$

where A runs over all elements of $\mathrm{SL}_2(K)$. Squaring all of the elements of (48), we obtain

$$\begin{pmatrix} A^2 & 0 \\ 0 & I \end{pmatrix}.$$

Since $SL_2(K)$ is also generated by all of its square elements (Dieudonné[6]) it follows that our automorphism induces an automorphism on the subgroup of elements

$$\begin{pmatrix} A & 0 \\ 0 & I \end{pmatrix},$$

where A runs over all the elements of $SL_2(K)$.

If we also use J_3, our automorphism induces an automorphism on

$$\begin{pmatrix} A & 0 \\ 0 & \begin{pmatrix} 1 & 0 \\ 0 & -1 \end{pmatrix}^l \end{pmatrix},$$

where A runs over all the elements of $SL_2^{\pm}(K)$. Notice that the integer l is uniquely determined by A. More precisely, in case -1 belongs to the commutator subgroup of the multiplicative group, then l is always even, otherwise l is even for $A \in SL_2(K)$ and l is odd for $A \notin SL_2(K)$.

By Theorem 2, our automorphism gives us the following mappings:

$$\begin{pmatrix} A & 0 \\ 0 & I \end{pmatrix} \to \begin{pmatrix} PA^\sigma P^{-1} & 0 \\ 0 & I \end{pmatrix}, \tag{49}$$

where σ is an automorphism of K or

$$\begin{pmatrix} A & 0 \\ 0 & I \end{pmatrix} \to \begin{pmatrix} PA'^{\sigma-1}P^{-1} & 0 \\ 0 & I \end{pmatrix}, \tag{50}$$

where σ is an anti-automorphism of K, and where A runs over all the elements of $SL_2(K)$. Consequently our automorphism cannot permute K_1^+ and K_1^{-1}, and must keep K_1^+ invariant. By an inner automorphism, keeping K_1^+ invariant, we may assume that our automorphism induces either a mapping

$$\begin{pmatrix} A & 0 \\ 0 & I \end{pmatrix} \to \begin{pmatrix} A^\sigma & 0 \\ 0 & I \end{pmatrix} \tag{51}$$

or

$$\begin{pmatrix} A & 0 \\ 0 & I \end{pmatrix} \to \begin{pmatrix} A'^{\sigma-1} & 0 \\ 0 & I \end{pmatrix}. \tag{52}$$

Using J_2 and K_2 instead of J_1 and K_1, we can prove that our automorphism keeps K_2^+ invariant, and similarly K_3^+. Since $SL_4(K)$ is generated by K_1^+, K_2^+, K_3^+

and
$$\begin{pmatrix} 1 & x & 0 & 0 \\ 0 & 1 & 0 & 0 \\ 0 & 0 & 1 & 0 \\ 0 & 0 & 0 & 1 \end{pmatrix},$$
we have the following theorem:

Theorem 5 *Every automorphism of* $\mathrm{SL}_4(K)$ *is the restriction of an automorphism of* $\mathrm{GL}_4(K)$.

Since there is no essential difficulty in the case of $\mathrm{PSL}_4(K)$, the author will not discuss it here.

II. Orthogonal groups

The case of the orthogonal group left open by Dieudonné is $O_4^+(K, f)$, where K is a field of characteristic $\neq 2$ and f is a quadratic form of index 2.

7. Preliminaries

Theorem 6 (Witt[10]) *Every quaternary quadratic form of index 2 is equivalent to*
$$f = x_1 x_3 + x_2 x_4,$$
whose matrix is
$$\mathfrak{F} = \begin{pmatrix} 0 & I \\ I & 0 \end{pmatrix}, \tag{53}$$
where I and 0 are two-rowed identity and zero matrices.

From now on, we assume that our fundamental quadratic form has the matrix $\mathfrak{F}$.

A four-rowed matrix $\mathfrak{T}$ satisfying
$$\mathfrak{T}\mathfrak{F}\mathfrak{T}' = \mathfrak{F} \tag{54}$$
with determinant of $\mathfrak{T}$ equal to 1, is an element of $O_4^+(K)$. If we put
$$\mathfrak{T} = \begin{pmatrix} A & B \\ C & D \end{pmatrix}, \quad A = A^{(2)}, \quad \text{etc.}, \tag{55}$$
then we have
$$AB' + BA' = 0, \quad CD' + DC' = 0, \quad AD' + BC' = I. \tag{56}$$

Since $\mathfrak{T}'$ is also orthogonal, we deduce

$$A'C + C'A = 0, \quad B'D + D'B = 0, \quad A'D + C'B = I. \tag{57}$$

Later we shall need another group containing O_4^+ as its subgroup and which is defined by the matrices $\mathfrak{P}$ satisfying

$$\mathfrak{P}\mathfrak{F}\mathfrak{P}' = a\mathfrak{F}, \tag{58}$$

where a is an element ($\neq 0$) of the field. This group is denoted by $GO_4 = GO_4(K, f)$.

Theorem 7 If $\mathfrak{T}$ belongs to O_4^+, then in the expression (55), A is either zero or non-singular.

Proof Suppose that A is of rank 1, we have two matrices $P(=P^{(2)})$ and $Q(=Q^{(2)})$ such that

$$PAQ = \begin{pmatrix} 1 & 0 \\ 0 & 0 \end{pmatrix}.$$

Since both

$$\begin{pmatrix} P & 0 \\ 0 & P'^{-1} \end{pmatrix} \quad \text{and} \quad \begin{pmatrix} Q & 0 \\ 0 & Q'^{-1} \end{pmatrix}$$

belongs to O_4^+, without loss of generality we may consider those T with $A = \begin{pmatrix} 1 & 0 \\ 0 & 0 \end{pmatrix}$. From (56), AB' is skew symmetric so we may deduce that

$$B = \begin{pmatrix} 0 & b_2 \\ 0 & b_4 \end{pmatrix}, \quad b_4 \neq 0$$

and from (57)

$$C = \begin{pmatrix} 0 & 0 \\ c_3 & c_4 \end{pmatrix}, \quad c_4 \neq 0.$$

Again from (56)

$$D = \begin{pmatrix} 1 & d_2 \\ -b_2 c_4 & d_4 \end{pmatrix}, \quad b_4 c_4 = 1;$$

in conclusion, we have

$$\mathfrak{T} = \begin{pmatrix} 1 & 0 & 0 & b_2 \\ 0 & 0 & 0 & b_4 \\ 0 & 0 & 1 & d_2 \\ c_3 & c_4 & -b_2 c_4 & d_4 \end{pmatrix}.$$

Its determinant is $-b_4 c_4 = -1$. This proves Theorem 2.

Remark The converse of the theorem is also true, i.e., any $\mathfrak{T}$ satisfying (54) with non-singular A must belong to O_4^+, This is an easy consequence of Theorem 8.

Theorem 8 Write $K = \begin{pmatrix} 0 & 1 \\ -1 & 0 \end{pmatrix}$. Every element of O_4^+ can be expressed uniquely as one of the following forms:

$$\begin{pmatrix} I & 0 \\ xK & I \end{pmatrix} \begin{pmatrix} A & 0 \\ 0 & A'^{-1} \end{pmatrix} \begin{pmatrix} I & yK \\ 0 & I \end{pmatrix} = \begin{pmatrix} A & yAK \\ xKA & (1 - xyd(A))A'^{-1} \end{pmatrix}^{①} \quad (59)$$

and

$$\begin{pmatrix} 0 & I \\ I & 0 \end{pmatrix} \begin{pmatrix} A & 0 \\ 0 & A'^{-1} \end{pmatrix} \begin{pmatrix} I & zK \\ 0 & I \end{pmatrix} = \begin{pmatrix} 0 & A'^{-1} \\ A & zAK \end{pmatrix}, \quad (60)$$

where x, y and z are elements of K and A is a non-singular two-rowed matrix.

Proof In case A is non-singular, from (56) and (57) we deduce

$$B = yAK, \quad C = xKA$$

and

$$D = xyKAK + A'^{-1} = (1 - xyd(A))A'^{-1},$$

since $AKA' = d(A)K$. The case $A = 0$ can easily be expressed as (60).

The uniqueness can also be proved without any difficulty.

Definition We use

$$(x, A, y), \quad (\infty, A, z)$$

to denote the matrices of (59) and (60) respectively, which may be called the coordinates of an element of O_4^+.

8. An extra type of automorphisms

From our previous experience, we might expect that the only automorphism is

$$\mathfrak{P} \mathfrak{U} \mathfrak{P}^{-1},$$

where $\mathfrak{P}$ belongs to $GO_4(K)$. But this is entirely false; there are many automorphisms of $O_4^+(K)$ which are not those of $O_4(K)$.

---

① We use $d(A)$ to denote the determinant of A.

Let $a \to a^\tau$ be an automorphism of the field. Let $\varphi(a)$ be a representation of the multiplicative group of the field into itself, and let it satisfy

$$\varphi^2(a) = \frac{a^\tau}{a}. \tag{61}$$

This is called a representation induced by the automorphism.

Example 1 Let K be a rational field, The identity automorphism is the only one. We write a rational number as

$$r = \pm 2^{l_1} 3^{l_2} \cdots p_n^{l_n} \cdots,$$

where p_n denotes the n-th prime, and the l_n are integers. We fix a prime p_n. Then we define

$$\varphi(r) = (-1)^{l_n}.$$

This evidently satisfies (61). There is exactly one other induced representation

$$\varphi(r) = \begin{cases} 1, & \text{if } r > 0, \\ -1, & \text{if } r < 0. \end{cases}$$

Example 2 Let K be a real field. The induced representation

$$\varphi(a) = \begin{cases} 1, & \text{if } a > 0, \\ -1, & \text{if } a < 0 \end{cases}$$

is the only one.

Example 3 Let K be a complex field, and a^τ be the conjugate of a. Then

$$\varphi(a) = |a|/a$$

is an induced mapping. Since

$$\frac{a^\tau}{a} = \frac{|a|^2}{a^2} = (\varphi(a))^2,$$

this does not exhaust all of the induced representations of the complex number field, which contains infinitely many automorphisms.

Now we express the previous notion in matrix form. We define

$$\chi(A) = \varphi(\det A). \tag{62}$$

Theorem 9 *The mapping*

$$(x, A, y) \to (x^\tau, \chi(A)A, y^\tau) \tag{63}$$

defines an automorphism of O_4^+, where $\chi(A)$ is related to τ by (62) and (61).

Proof The theorem follows from Theorem 8, if we can prove that (63) carries the following identities into their corresponding ones:

$$(0, I, y)(x_1, I, 0) = (x_1(1 - yx_1)^{-1}, (1 - yx_1)I, y(1 - yx_1)^{-1}), \quad \text{if } x_1 y = 1, \quad (64)$$

$$(0, I, x_1^{-1})(x_1, I, 0) = (\infty, x_1 K, -x_1^{-1}), \quad (65)$$

$$(0, I, x)(\infty, I, 0) = (-x^{-1}, xK, -x^{-1}) \quad (66)$$

and

$$(0, A, 0)(\infty, I, 0) = (\infty, {A'}^{-1}, 0). \quad (67)$$

To verify (64), we need to prove that

$$(y_1^\tau (1 - y^\tau x_1^\tau)^{-1}, (1 - y^\tau x_1^\tau)I, y^\tau (1 - y^\tau x_1^\tau)^{-1})$$
$$= ((x_1(1 - y_1 x_1)^{-1})^\tau, \chi((1 - yx_1)I)(1 - yx_1)I, (y(1 - yx_1)^{-1})^\tau).$$

The first and the last terms are evidently equal to the corresponding ones, by the definition of an automorphism. The middle terms are equal, since

$$\chi((1 - yx_1)I) = \varphi((1 - yx_1)^2) = (\varphi(1 - yx_1))^2 = (1 - y^\tau x_1^\tau)/(1 - yx_1).$$

To verify (65), we need only establish that

$$x_1^\tau K = \chi(x_1 K) x_1 K.$$

This is also true, since

$$\chi(x_1 K) = \varphi(x_1^2) = x_1^\tau / x_1.$$

The other two can be proved similarly.

9. Structure of the group $O_4^+(K, f)$

Theorem 10 *The commutator subgroup $\Omega_4^+(K, f)$ of $O_4^+(K, f)$ consists of those elements*

$$\begin{pmatrix} A & B \\ C & D \end{pmatrix}, \quad (68)$$

where the determinants of A, B, C, D are square elements of the field.

Proof Notice that from the expressions (59) and (60) we deduce easily that if one of the determinants of A, B, C, D is a square element different from zero, then all the others are squares. By means of (59) and (60), we can easily verify that all the elements of (68) form a normal subgroup.

The factor group of $O_4^+(K,f)$ by (68) is generated by

$$\begin{pmatrix} 1 & 0 & 0 & 0 \\ 0 & \lambda & 0 & 0 \\ 0 & 0 & 1 & 0 \\ 0 & 0 & 0 & \lambda^{-1} \end{pmatrix} \text{ and } \begin{pmatrix} 0 & I \\ I & 0 \end{pmatrix},$$

and it is abelian, since

$$\begin{pmatrix} 1 & 0 & 0 & 0 \\ 0 & \lambda & 0 & 0 \\ 0 & 0 & 1 & 0 \\ 0 & 0 & 0 & \lambda^{-1} \end{pmatrix} \begin{pmatrix} 0 & I \\ I & 0 \end{pmatrix} \begin{pmatrix} 1 & 0 & 0 & 0 \\ 0 & \lambda & 0 & 0 \\ 0 & 0 & 1 & 0 \\ 0 & 0 & 0 & \lambda^{-1} \end{pmatrix}^{-1} \begin{pmatrix} 0 & I \\ I & 0 \end{pmatrix}^{-1}$$

belongs to (68). Therefore $\Omega_4^+(K,f)$ is contained in (68), by the property of commuator subgroup.

Next we are going to prove that the subgroup (68) is generated by

$$\begin{pmatrix} I & xK \\ 0 & I \end{pmatrix}, \begin{pmatrix} 0 & I \\ I & 0 \end{pmatrix} \tag{69}$$

and

$$\begin{pmatrix} A & 0 \\ I & A'^{-1} \end{pmatrix}, \tag{70}$$

where A belongs to $SL_2(K)$. In fact, the group generated by (69) and (70) evidently contains

$$\begin{pmatrix} I & 0 \\ yK & I \end{pmatrix} = \begin{pmatrix} 0 & I \\ I & 0 \end{pmatrix} \begin{pmatrix} I & yK \\ 0 & I \end{pmatrix} \begin{pmatrix} 0 & I \\ I & 0 \end{pmatrix} \tag{71}$$

and consequently it contains

$$\begin{pmatrix} I & (1-a^{-1})K \\ 0 & I \end{pmatrix} \begin{pmatrix} I & 0 \\ K & I \end{pmatrix} \begin{pmatrix} I & (1-a)K \\ 0 & I \end{pmatrix} \begin{pmatrix} K & 0 \\ -a^{-1}K & I \end{pmatrix}$$
$$= \begin{pmatrix} a^{-1}I & 0 \\ 0 & aI \end{pmatrix}.$$

Since a matrix Q of determinant q^2 can be expressed as $\dfrac{1}{q}Q\, qI$, the group generated by (69) and (70) also contains

$$\begin{pmatrix} Q & 0 \\ 0 & Q'^{-1} \end{pmatrix} = \begin{pmatrix} qI & 0 \\ 0 & q^{-1}I \end{pmatrix} \begin{pmatrix} \frac{1}{q}Q & 0 \\ 0 & qQ'^{-1} \end{pmatrix}. \tag{72}$$

Evidently (69), (71) and (72) generate the group formed by (68). Therefore (69) and (70) generate the group (68).

The group $\Omega_4^+(K, f)$ contains

$$\begin{pmatrix} A & 0 \\ 0 & A'^{-1} \end{pmatrix} \begin{pmatrix} B & 0 \\ 0 & B'^{-1} \end{pmatrix} \begin{pmatrix} A & 0 \\ 0 & A'^{-1} \end{pmatrix}^{-1} \begin{pmatrix} B & 0 \\ 0 & B'^{-1} \end{pmatrix}^{-1},$$

so it contains (70). Since

$$\begin{pmatrix} rI & 0 \\ 0 & r^{-1}I \end{pmatrix} \begin{pmatrix} I & sK \\ 0 & I \end{pmatrix} \begin{pmatrix} rI & 0 \\ 0 & r^{-1}I \end{pmatrix}^{-1} \begin{pmatrix} I & sK \\ 0 & I \end{pmatrix}^{-1} = \begin{pmatrix} I & s(r^2-1)K \\ 0 & I \end{pmatrix},$$

$\Omega_4^+(K, f)$ contains the first term of (69); similarly it contains $\begin{pmatrix} I & 0 \\ yK & I \end{pmatrix}$. Since

$$\begin{pmatrix} I & K \\ 0 & I \end{pmatrix} \begin{pmatrix} I & 0 \\ K & I \end{pmatrix} \begin{pmatrix} -K & 0 \\ 0 & -K \end{pmatrix} \begin{pmatrix} I & K \\ 0 & I \end{pmatrix} = \begin{pmatrix} 0 & I \\ I & 0 \end{pmatrix},$$

therefore $\Omega_4^+(K, f)$ contains (69) and (70), and consequently if contains the group (68). The theorem follows.

Theorem 11 *The group $P\Omega_4^+(K, f)$ is a direct product of two groups each of which is isomorphic to $\mathrm{PSL}_2(K)$. More precisely, $\Omega_4^+(K, f)$ can be expressed as a direct product of three irreducible components: one formed by the identity and its negation, and two being isomorphic to $\mathrm{PSL}_2(K)$.*

Proof The two subgroups are the one formed by (70) and the one generated by (69). More precisely, we may write the second as

$$\begin{pmatrix} qI & rK \\ -sK & tI \end{pmatrix}, \quad qt - rs = 1. \tag{73}$$

The correspondence of (73) and $\begin{pmatrix} q & r \\ -s & t \end{pmatrix}$ established the isomorphism of the group (73) and $\mathrm{SL}_2(K)$.

The intersection of the groups (69) and (73) is given by

$$\pm \begin{pmatrix} I & 0 \\ 0 & I \end{pmatrix},$$

which is the identity of $P\Omega_4^+(K,f)$. Moreover, for determinant of A equal to q^2, we have

$$\begin{pmatrix} A & xAK \\ yKA & (1-xyd(A))A'^{-1} \end{pmatrix} = \begin{pmatrix} \frac{1}{q}A & 0 \\ 0 & qA'^{-1} \end{pmatrix} \begin{pmatrix} qI & qxK \\ qyK & (1-xyq^2)q^{-1}I \end{pmatrix}$$

and

$$\begin{pmatrix} 0 & A \\ A'^{-1} & zA'^{-1}K \end{pmatrix} = \begin{pmatrix} \frac{1}{q}AK & 0 \\ 0 & q(AK)'^{-1} \end{pmatrix} \begin{pmatrix} 0 & -qK \\ -\frac{1}{q}K & \frac{z}{q}I \end{pmatrix},$$

Theorem 11 follows from the definition of the direct product.

Theorem 12 *Every automorphism of $O_4^+(K,f)$ is a combination of*

$$\mathfrak{T} \to \mathfrak{D}\mathfrak{T}^\sigma \mathfrak{D}^{-1}, \tag{74}$$

where $\mathfrak{D}$ belongs to GO_4 and σ is an automorphism and that given by (63). Those of $PO_4^+(K,f)$ are induced from the automorphisms of $O_4^+(K,f)$.

Proof Since $\Omega_4^+(K,f)$ (or $P\Omega_4^+(K,f)$ is a characteristic invariant subgroup, an automorphism of $O^+(K,f)$ (or PO_4^+) induces an automorphism on $\Omega_4^+(K,f)$ (or $P\Omega_4^+$). Since $\Omega_4^+(K,f)$ is a direct product of three groups G_0, G_1 and G_2, where G_1 and G_2 denote the subgroups of $PO_4^+(K,f)$ represented by (70) and (73), and since these are isomorphic to $\mathrm{PSL}_2(K)$, it follows that the automorphism of $\Omega_4^+(K,f)$ can be obtained by those of G_1 and G_2 and by adjoining one which permutes G_1 and G_2 (since G_0 has only identity automorphism).

Since we have, by putting $A = \begin{pmatrix} a & b \\ c & d \end{pmatrix}$,

$$\begin{pmatrix} 1 & 0 & 0 & 0 \\ 0 & 0 & 0 & 1 \\ 0 & 0 & 1 & 0 \\ 0 & 1 & 0 & 0 \end{pmatrix} \begin{pmatrix} A & B \\ C & D \end{pmatrix} \begin{pmatrix} 1 & 0 & 0 & 0 \\ 0 & 0 & 0 & 1 \\ 0 & 0 & 1 & 0 \\ 0 & 1 & 0 & 0 \end{pmatrix} = \begin{pmatrix} aI & bK \\ -cK & dI \end{pmatrix},$$

therefore, the permutation of the two components of G_1 and G_2 is merely an automorphism induced by an element of $O_4(K,f)$. Hence after an automorphism of the type (74) if necessary, we may assume that the automorphism of $\Omega_4^+(K, f)$ induces automorphisms on G_1 and G_2, say

$$\begin{pmatrix} A & 0 \\ 0 & A'^{-1} \end{pmatrix} \to \pm \begin{pmatrix} Q & 0 \\ 0 & Q'^{-1} \end{pmatrix} \begin{pmatrix} A & 0 \\ 0 & A'^{-1} \end{pmatrix}^\sigma \begin{pmatrix} Q & 0 \\ 0 & Q'^{-1} \end{pmatrix}^{-1} \tag{75}$$

and

$$\begin{pmatrix} qI & rK \\ -sK & tI \end{pmatrix} \to \pm \begin{pmatrix} q_1 I & r_1 K \\ -s_1 K & t_1 I \end{pmatrix} \begin{pmatrix} qI & rK \\ -sK & tI \end{pmatrix}^\tau \begin{pmatrix} q_1 I & r_1 K \\ -s_1 K & t_1 I \end{pmatrix}^{-1}, \tag{76}$$

where σ and τ are two automorphsims of the field. Squaring the elements on both sides of (74) and (76), and since $\mathrm{SL}_2(K)$ is generated by all its square elements, we may omit the $\pm$ signs in (75) and (76).

We shall now construct a matrix $\mathfrak{T}$ belonging to $GO_4(K, f)$ which effects both (74) and (75) at the same time. In fact, let the determinant of Q be λ, and write $R = \begin{pmatrix} 1 & 0 \\ 0 & \lambda \end{pmatrix}^{-1} Q$. Then

$$\mathfrak{T} = \begin{pmatrix} I & 0 \\ 0 & \lambda I \end{pmatrix} \begin{pmatrix} 1 & 0 & 0 & 0 \\ 0 & \lambda & 0 & 0 \\ 0 & 0 & 1 & 0 \\ 0 & 0 & 0 & \lambda^{-1} \end{pmatrix} \begin{pmatrix} R & 0 \\ 0 & R'^{-1} \end{pmatrix} \begin{pmatrix} q_1 I & r_1 K \\ -s_1 K & t_1 I \end{pmatrix},$$

which evidently belongs to $GO_4(K, f)$, and

$$\mathfrak{T} \begin{pmatrix} A & 0 \\ 0 & A'^{-1} \end{pmatrix} \mathfrak{T}^{-1} = \begin{pmatrix} I & 0 \\ 0 & \lambda I \end{pmatrix} \begin{pmatrix} Q & 0 \\ 0 & Q'^{-1} \end{pmatrix} \begin{pmatrix} A & 0 \\ 0 & A'^{-1} \end{pmatrix}$$

$$\times \begin{pmatrix} Q & 0 \\ 0 & Q'^{-1} \end{pmatrix}^{-1} \begin{pmatrix} I & 0 \\ 0 & \lambda I \end{pmatrix}^{-1}$$

$$= \begin{pmatrix} Q & 0 \\ 0 & Q'^{-1} \end{pmatrix} \begin{pmatrix} A & 0 \\ 0 & A'^{-1} \end{pmatrix} \begin{pmatrix} Q & 0 \\ 0 & Q'^{-1} \end{pmatrix}^{-1},$$

since

$$\begin{pmatrix} A & 0 \\ 0 & A'^{-1} \end{pmatrix} \begin{pmatrix} q_1 I & r_1 K \\ -s_1 K & t_1 I \end{pmatrix} = \begin{pmatrix} q_1 I & r_1 K \\ -s_1 K & t_1 I \end{pmatrix} \begin{pmatrix} A & 0 \\ 0 & A'^{-1} \end{pmatrix}$$

and

$$\mathfrak{T}\begin{pmatrix} qI & rK \\ -sK & tI \end{pmatrix}\mathfrak{T}^{-1} = \begin{pmatrix} 1 & 0 & 0 & 0 \\ 0 & \lambda & 0 & 0 \\ 0 & 0 & \lambda & 0 \\ 0 & 0 & 0 & 1 \end{pmatrix}\begin{pmatrix} q_1I & r_1K \\ -s_1K & t_1I \end{pmatrix}$$

$$\times \begin{pmatrix} R & 0 \\ 0 & R'^{-1} \end{pmatrix}\begin{pmatrix} qI & rK \\ -sK & tI \end{pmatrix}$$

$$\times \begin{pmatrix} R & 0 \\ 0 & R'^{-1} \end{pmatrix}^{-1}\begin{pmatrix} q_1I & r_1K \\ -s_1K & t_1I \end{pmatrix}^{-1}\begin{pmatrix} 1 & 0 & 0 & 0 \\ 0 & \lambda & 0 & 0 \\ 0 & 0 & \lambda & 0 \\ 0 & 0 & 0 & 1 \end{pmatrix}^{-1}$$

$$= \begin{pmatrix} q_1I & r_1K \\ -s_1K & t_1I \end{pmatrix}\begin{pmatrix} qI & rK \\ -sK & tI \end{pmatrix}\begin{pmatrix} q_1I & r_1K \\ -s_1K & t_1I \end{pmatrix}^{-1},$$

since

$$\begin{pmatrix} 1 & 0 & 0 & 0 \\ 0 & \lambda & 0 & 0 \\ 0 & 0 & \lambda & 0 \\ 0 & 0 & 0 & 1 \end{pmatrix} \text{ and } \begin{pmatrix} q_1 & 0 & 0 & r_1 \\ 0 & q_1 & -r_1 & 0 \\ 0 & -s_1 & t_1 & 0 \\ s_1 & 0 & 0 & t_1 \end{pmatrix}$$

evidently commute. Therefore, after an automorphism of the type (74), we may assume that our automorphism under consideration induces the mapping

$$\begin{pmatrix} A & 0 \\ 0 & A'^{-1} \end{pmatrix} \to \begin{pmatrix} A & 0 \\ 0 & A'^{-1} \end{pmatrix} \tag{77}$$

and

$$\begin{pmatrix} qI & rK \\ -sK & tI \end{pmatrix} \to \begin{pmatrix} q^\tau I & r^\tau K \\ -s^\tau K & t^\tau I \end{pmatrix}. \tag{78}$$

If K contains more than three elements, we have the following particular case of (78):

$$\begin{pmatrix} aI & 0 \\ 0 & a^{-1}I \end{pmatrix} \to \begin{pmatrix} a^\tau I & 0 \\ 0 & a^{-\tau}I \end{pmatrix}, \quad a^\tau \neq \pm 1. \tag{79}$$

Thus elements which commute with (79) are of the form

$$\begin{pmatrix} P & 0 \\ 0 & P'^{-1} \end{pmatrix}; \tag{80}$$

by Theorem 1, we have

$$\begin{pmatrix} P & 0 \\ 0 & P'^{-1} \end{pmatrix} \rightarrow \begin{pmatrix} \chi(P)P^\mu & 0 \\ 0 & \chi(P)^{-1}P^{\mu'-1} \end{pmatrix},$$

from (77), we have $\mu = 1$, and from (78),

$$\chi(qI) = q^\tau/q.$$

Consequently

$$\left(\chi\begin{pmatrix} 1 & 0 \\ 0 & q \end{pmatrix}\right)^2 = \chi\begin{pmatrix} 1 & 0 \\ 0 & q \end{pmatrix} \times \begin{pmatrix} q & 0 \\ 0 & 1 \end{pmatrix} = \chi(qI) = q^\tau/q$$

and

$$(\chi(A))^2 = \det A^\tau / \det A.$$

Our theorem is proved for K containing more than three elements (certainly a little detail is needed to complete the proofs of O_4^+ and PO_4^+). In case K contains only three elements, the factor group O_4^+/Ω_4^+ is of order two, and it is represented by

$$\begin{pmatrix} I & 0 \\ 0 & I \end{pmatrix}, \quad \mathfrak{U} = \begin{pmatrix} U & 0 \\ 0 & U^{-1} \end{pmatrix}, \quad U = \begin{pmatrix} 1 & 0 \\ 0 & -1 \end{pmatrix}.$$

Therefore we need only find the mapping of $\mathfrak{U}$. $\mathfrak{U}$ is mapped to $\mathfrak{U}\mathfrak{P}$, where $\mathfrak{P}$ belongs to Ω_4. There are two ways to map the element

$$\mathfrak{U}\begin{pmatrix} A & 0 \\ 0 & A'^{-1} \end{pmatrix}\begin{pmatrix} qI & rK \\ -sK & tI \end{pmatrix}\mathfrak{U}^{-1} = \begin{pmatrix} UAU^{-1} & 0 \\ 0 & (UAU^{-1})'^{-1} \end{pmatrix}\begin{pmatrix} qI & -rK \\ sK & tI \end{pmatrix},$$

they give us the identity

$$\mathfrak{U}\mathfrak{P}\begin{pmatrix} A & 0 \\ 0 & A'^{-1} \end{pmatrix}\begin{pmatrix} q^\tau I & r^\tau K \\ -s^\tau K & t^\tau I \end{pmatrix}\mathfrak{P}^{-1}\mathfrak{U}^{-1} = \begin{pmatrix} UAU^{-1} & 0 \\ 0 & (UAU^{-1})'^{-1} \end{pmatrix}$$

$$\times \begin{pmatrix} q^\tau I & -r^\tau K \\ s^\tau K & t^\tau I \end{pmatrix}$$

for all possible A, q, r, s and t. It follows that $\mathfrak{P}$ is either an identity or its negation, and we obtain the same conclusion.

References

[1] Hua L K. On the automorphisms of the symplectic group over any field. *Annals of Math.*, 1948, **49**: 739-759.

[2] Dieudonné J. On the automorphism of classical groups. This volume.

[3] Hua L K. On the automorphisms of a sfield. *Proc. Nat. Acad. Sci.*, 1949, **35**: 386-389.

[4] Ancochea G. On semi-automorphisms of division algebras. *Annals of Math.*, 1947, **48**: 147-153.

[5] Kaplansky I. Semi-automorphisms of rings. *Duke Math. Jour.*, 1947, **14**: 521-525.

[6] Dieudonné J. Les déterminants sur un corps non commutatif. *Bull. Soc. Math. Fr.*, 1943, **79**: 27-45.

[6*] Ibid., p.41, Theorem 2. Dieudonné. Complénebts à trois articles antérieurs. *Bull. Soc. Math. Fr.*, 1946, **74**: 59-68.

[7] Ancochea G. Le théorème de van Staudt en géométrie projective quaternionne. *Jour. für die reine und angewandte Math.*, 1942, **184**: 192-198.

[8] Hua L K. Some properties of a sfield. *Proc. of Nat. Aca. Sci.*, 1949, **35**: 533-537.

[9] Schreier O and van der Waerden B L. Die Automorphismen der projektiven gruppen. *Hamburg Univ, Math. Seminar, Abn.*, 1928, **6**: 303-322, cf. also appendix of (1).

[10] Witt E. Theorie der quadratische Formen in beliebigen Körpern. *Jour. für Math.*, 1937, **176**: 21-44.

Automorphisms of the unimodular group*

Notation Let $\mathfrak{M}_n$ denote the group of $n \times n$ integral matrices of determinant ± 1 (the unimodular group). By $\mathfrak{M}_n^+$ we denote that subset of $\mathfrak{M}_n$ where the deferminant is $+1$; $\mathfrak{M}_n^-$ is correspondingly defined. Let $I^{(n)}$ (or briefly I) be the identity matrix in $\mathfrak{M}_n$, and let X' represent the transpose of X. The direct sum of the matrices A and B will be represented by $A \dotplus B$;

$$A \stackrel{s}{=} B$$

will mean that A is similar to B. In this paper, we shall find explicitly the generators of the group $\mathfrak{U}_n$ of all automorphisms of $\mathfrak{M}_n$.

1. The commutator subgroup of $\mathfrak{M}_n$

The following result is useful, and is of independent interest.

Theorem 1 *Let $\mathfrak{R}_n$ be the commutator subgroup of $\mathfrak{M}_n$. Then trivially $\mathfrak{R}_n \subset \mathfrak{M}_n^+$. For $n = 2$, $\mathfrak{R}_n$ is of index 2 in $\mathfrak{M}_n^+$, while for $n > 2$, $\mathfrak{R}_n = \mathfrak{M}_n^+$.*

Proof Consider first the case where $n = 2$. Define

$$S = \begin{pmatrix} 0 & 1 \\ -1 & 0 \end{pmatrix}, \quad T = \begin{pmatrix} 1 & 1 \\ 0 & 1 \end{pmatrix}. \tag{1}$$

It is well known that S and T generate $\mathfrak{M}_2^+$. An element X of $\mathfrak{M}_2^+$ is called *even* if, when X is expressed as a product of powers of S and T, the sum of the exponents is even; otherwise, X is called *odd*. Since all relations satisfied by S and T are consequences of

$$S^2 = -I, \quad (ST)^3 = I,$$

it follows that the parity of $X \in \mathfrak{M}_2^+$ depends only on X, and not on the manner in which X is expressed as a product of powers of S and T. Let $\mathfrak{G}$ be the subgroup of $\mathfrak{M}_2^+$ consisting of all even elements; then clearly $\mathfrak{G}$ is of index 2 in $\mathfrak{M}_2^+$. It suffices to prove that $\mathfrak{G} = \mathfrak{R}_2$.

---

* Presented to the Society, December 29, 1950; received by the editors January 8, 1951.
Reprinted from *Transactions of The American Mathematical Society*, 1951, **71**(3): 331–348.
By Hua L K and Reiner I.

We prove first that $\mathfrak{R}_2 \subset \mathfrak{G}$. Since the commutator subgroup of a group is always generated by squares, it suffices to show that $A \in \mathfrak{M}_2$ implies $A^2 \in \mathfrak{G}$. For $A \in \mathfrak{M}_2^+$, this is clear. If $A \in \mathfrak{M}_2^-$, set $A = XJ = JY$, where

$$J = \begin{pmatrix} 1 & 0 \\ 0 & -1 \end{pmatrix}, \tag{2}$$

and X and $Y \in \mathfrak{M}_2^+$. Then $A^2 = XY = XJ^{-1}XJ$. Hence we need only prove that if $X \in \mathfrak{M}_2^+$, X and $J^{-1}XJ$ are of the same parity. This is easily verified for $X = S$ or T; since S and T generate $\mathfrak{M}_2^+$, and $J^{-1}X_1X_2J = J^{-1}X_1J \cdot J^{-1}X_2J$, the result follows.

On the other hand we can show that $\mathfrak{G} \subset \mathfrak{R}_2$. For, $\mathfrak{G}$ is generated by T^2 and ST, since $TS = (ST \cdot T^{-2})^2$. However, $T^2 = TJT^{-1}J^{-1} \in \mathfrak{R}_2$, and therefore also $(T')^{-2} \in \mathfrak{R}_2$. Furthermore, $ST = TST^{-1}S^{-1}(T')^{-2}T^2 \in \mathfrak{R}_2$. This completes the proof for $n = 2$.

Suppose now that $n > 2$, and define

$$R = \begin{pmatrix} 0 & \cdots & 0 & (-1)^{n-1} \\ 1 & \cdots & 0 & 0 \\ \vdots & & \vdots & \vdots \\ 0 & \cdots & 1 & 0 \end{pmatrix} \in \mathfrak{M}_n^+, \quad S = \begin{pmatrix} 0 & 1 \\ -1 & 0 \end{pmatrix} + I^{(n-2)}, \tag{3}$$

$$T = \begin{pmatrix} 1 & 1 \\ 0 & 1 \end{pmatrix} + I^{(n-2)}$$

(the symbols S and T defined here are the analogues in $\mathfrak{M}_n^+$ of those defined by (1). It will be clear from the context which are meant). For $n > 2$, we have[①]

$$T' = [R^{-1}(TR)^{-(n-2)}R(TR)^{n-2}](TR)^{-1}[R(TR)^{-(n-2)}R^{-1}(TR)^{n-2}](TR) \in \mathfrak{R}_n.$$

Further $S = TST^{-1}S^{-1}(T')^{-2}T \in \mathfrak{R}_n$. Finally, for odd n there exists a permutation matrix P such that $R^2 = P^{-1}RP$, whence $R = R^{-1}P^{-1}RP \in \mathfrak{R}_n$. For even n, R represents the monomial transformation

$$\begin{pmatrix} x_1 & x_2 & \cdots & x_{n-1} & x_n \\ x_2 & x_3 & \cdots & x_n & -x_1 \end{pmatrix},$$

which is a product of

[①] Hua L K and Reiner I. Trans. Amer. Math, Soc., 1949, **65**: 423.

$$\begin{pmatrix} x_1 & x_2 & x_3 & \cdots & x_{n-1} & x_n \\ x_2 & -x_1 & x_3 & \cdots & x_{n-1} & x_n \end{pmatrix}, \begin{pmatrix} x_1 & x_2 & x_3 & x_4 & \cdots & x_n \\ -x_3 & x_2 & x_1 & x_4 & \cdots & x_n \end{pmatrix},$$

$$\begin{pmatrix} x_1 & x_2 & x_3 & x_4 & \cdots & x_n \\ x_4 & x_2 & x_3 & -x_1 & \cdots & x_n \end{pmatrix}, \cdots, \begin{pmatrix} x_1 & x_2 & \cdots & x_{n-1} & x_n \\ x_n & x_2 & \cdots & x_{n-1} & -x_1 \end{pmatrix},$$

each factor of which is similar to S (and hence is in $\mathfrak{R}_n$). Since T and R generate $\mathfrak{M}_n^+$, the theorem is proved.

Corollary 1 *In any automorphism of $\mathfrak{M}_n$, always $\mathfrak{M}_n^+ \to \mathfrak{M}_n^+$.*

Proof For $n > 2$ this is an immediate corollary, since the commutator subgroup goes into itself in any automorphism. For $n = 2$, let $S \to S_1$ and $T \to T_1$. Then $ST \in \mathfrak{R}_2$ implies $S_1 T_1 \in \mathfrak{R}_2$, so $\det(S_1 T_1) = 1$. Further, $S^2 = -I$ implies $S_1^2 = -I$, so $\det S_1 = 1$, since the minimum function of S_1 is $x^2 + 1$, and the characteristic function must therefore be a power of $x^2 + 1$. This completes the proof when $n = 2$.

2. Automorphisms of $\mathfrak{M}_2^+$

We wish to determine the automorphisms of $\mathfrak{M}_2$. Since every automorphism of $\mathfrak{M}_2$ takes $\mathfrak{M}_2^+$ into itself, we shall first determine all automorphisms of $\mathfrak{M}_2^+$. For $X \in \mathfrak{M}_2^+$ define $\varepsilon(X) = +1$ or -1, according as X is even or odd.

Theorem 2 *Every automorphism of $\mathfrak{M}_2^+$ is of one of the forms*

(I) $X \in \mathfrak{M}_2^+ \to AXA^{-1}, A \in \mathfrak{M}_2$

or

(II) $X \in \mathfrak{M}_2^+ \to \varepsilon(X) \cdot AXA^{-1}, A \in \mathfrak{M}_2$.

That is, the automorphism group of $\mathfrak{M}_2^+$ is generated by the set of "inner" automorphisms $X \to AXA^{-1}$ ($A \in \mathfrak{M}_2$) and the automorphism $X \to \varepsilon(X) \cdot X$.

Proof Let τ be an automorphism of $\mathfrak{M}_2^+$; it certainly leaves $I^{(2)}$ and $-I^{(2)}$ individually unaltered. Let S and T (as given by (1)) be mapped into S^τ and T^τ. Then $(S^\tau)^2 = -I$. Since all second order fixed points are equivalent, there exists a matrix $B \in \mathfrak{M}_2$ such that $BS^\tau B^{-1} = S$. Instead of τ, consider the automorphism $\tau' : X \to BX^\tau B^{-1}$, which leaves S unaltered. Assume hereafter that τ leaves S invariant (it is this sort of replacement of τ by τ' which we shall mean when we refer to some property holding "after a suitable inner automorphism"). Set

$$T^\tau = \begin{pmatrix} a & b \\ c & d \end{pmatrix}.$$

From $(ST)^3 = I$ we obtain $(ST^\tau)^3 = I$, whence $b - c = 1$. Since $\det T^\tau = 1$, we get

$$ad = 1 + bc = c^2 + c + 1 > 0.$$

Set $N = |a+d|$. If $N \geqslant 3$, consider the elements generated by S and T^τ (mod N). Since $a + d \equiv 0 \pmod{N}$, we find that $(T^\tau)^2 \equiv I \pmod{N}$. Furthermore $(ST^\tau)^3 \equiv I$ (mod N); therefore S and T^τ generate (mod N) at most the 12 elements

$$\pm I, \quad \pm S, \quad \pm T^\tau, \quad \pm ST^\tau, \quad \pm T^\tau S, \quad \pm ST^\tau S.$$

But if τ is an automorphism, S and T^τ generate $\mathfrak{M}_2^+$, which has more than 12 elements (mod N) for $N \geqslant 3$.

Therefore $N \leqslant 2$. Since $ad > 0$, either $a = d = 1$ or $a = d = -1$, and thence $b = 1, c = 0$ or $b = 0, c = -1$. There are 4 possibilities for T^τ:

$$T^\tau = \begin{cases} T_0 = \begin{pmatrix} 1 & 1 \\ 0 & 1 \end{pmatrix}, & T_2 = \begin{pmatrix} -1 & 1 \\ 0 & -1 \end{pmatrix}, \\ T_1 = \begin{pmatrix} 1 & 0 \\ -1 & 1 \end{pmatrix}, & T_3 = \begin{pmatrix} -1 & 0 \\ -1 & -1 \end{pmatrix}. \end{cases}$$

Since S and T generate $\mathfrak{M}_2^+$, to determine τ it is sufficient to specify S^τ and T^τ. Thus every automorphism of $\mathfrak{M}_2^+$ is of the form $S \to BSB^{-1}$, $T \to BT_iB^{-1}$ (for some i, $i = 0, 1, 2, 3$), where $B \in \mathfrak{M}_2$. If J is given by (2), we have

$$T_0 = T, \quad T_1 = STS^{-1}, \quad T_2 = -JTJ^{-1}, \quad T_3 = -SJTJ^{-1}S^{-1},$$

and also $S = -JSJ^{-1}$. The possible automorphisms are:

$i = 0:$ $S \to BSB^{-1}, \quad T \to BTB^{-1}$;

$i = 1:$ $S \to BS \cdot S \cdot S^{-1}B^{-1}, \quad T \to BS \cdot T \cdot S^{-1}B^{-1}$;

$i = 2:$ $S \to -BJ \cdot S \cdot J^{-1}B^{-1}, \quad T \to -BJ \cdot T \cdot J^{-1}B^{-1}$;

$i = 3:$ $S \to -BSJ \cdot S \cdot J^{-1}S^{-1}B^{-1}, \quad T \to -BSJ \cdot T \cdot J^{-1}S^{-1}B^{-1}$.

These automorphisms are of two types: for $i = 0$ and 1, $S \to ASA^{-1}$, $T \to ATA^{-1}$, which imply that $X \in \mathfrak{M}_2^+ \to AXA^{-1}$; for $i = 2$ and 3, $S \to -ASA^{-1}$, $T \to -ATA^{-1}$, which imply that $X \in \mathfrak{M}_2^+ \to \varepsilon(X) \cdot AXA^{-1}$. This completes the proof.

3. Automorphisms of $\mathfrak{M}_n^+$ and $\mathfrak{M}_n$

We are now faced with the problem of determining the automorphisms of $\mathfrak{M}_2$ from those of $\mathfrak{M}_2^+$. We shall have the same problem for $\mathfrak{M}_n$ and $\mathfrak{M}_n^+$. As we shall see, the passage from $\mathfrak{M}_n^+$ to $\mathfrak{M}_n$ is trivial, and most of the difficulty lies in determining the automorphisms of $\mathfrak{M}_n^+$. In this paper we shall prove the following results:

Theorem 3 *For $n > 2$, the group of those automorphisms of $\mathfrak{M}_n^+$ which are induced by automorphisms of $\mathfrak{M}_n$ is generated by*

(i) *the set of all "inner" automorphisms*

$$X \in \mathfrak{M}_n^+ \to AXA^{-1} \quad (A \in \mathfrak{M}_n)$$

and

(ii) *the automorphism*

$$X \in \mathfrak{M}_n^+ \to X'^{-1}.$$

Remark When $n = 2$, the automorphism (ii) is the same as $X \to SXS^{-1}$, hence is included in (i). The automorphism $X \to \varepsilon(X) \cdot X$ occurs only for $n = 2$. Furthermore, for odd n all automorphisms of $\mathfrak{M}_n^+$ are induced by automorphisms of $\mathfrak{M}_n$.

Theorem 4 *The generators of $\mathfrak{U}_n$ are*
(i) *the set of all inner automorphisms*

$$X \in \mathfrak{M}_n \to AXA^{-1} \quad (A \in \mathfrak{M}_n);$$

(ii) *the automorphism* $X \in \mathfrak{M}_n \to X'^{-1}$;
(iii) *for even n only, the automorphism*

$$X \in \mathfrak{M}_n \to (\det X) \cdot X$$

and

(iv) *for $n = 2$ only, the automorphism*

$$X \in \mathfrak{M}_2^+ \to \varepsilon(X) \cdot X, \quad X \in \mathfrak{M}_2^- \to \varepsilon(JX) \cdot X,$$

where J is given by (2).

Further, when $n = 2$, the automorphism (ii) *may be omitted from this list.*

Let us show that Theorem 4 is a simple consequence of Theorem 3. Let τ be any automorphism of $\mathfrak{M}_n$. By Corollary 1, τ induces an automorphism on $\mathfrak{M}_n^+$ which, by Theorems 2 and 3, can be written as:

$$X \in \mathfrak{M}_n^+ \to \alpha(X) \cdot AX^*A^{-1},$$

where $A \in \mathfrak{M}_n, \alpha(X) = 1$ for all X or $\alpha(X) = \varepsilon(X)$ for all X (this can occur only when $n = 2$), and where either $X^* = X$ for all X or $X^* = X'^{-1}$ for all X.

Let Y and $Z \in \mathfrak{M}_n^-$; then

$$Y^\tau Z^\tau = (YZ)^\tau = \alpha(YZ) \cdot A(YZ)^* A^{-1},$$

whence

$$Y^\tau = \alpha(YZ) \cdot AY^* Z^* A^{-1} (Z^\tau)^{-1}.$$

Let $Z \in \mathfrak{M}_n^-$ be fixed; then

$$Y^\tau = \alpha(YZ) \cdot AY^* B \quad \text{for all } Y \in \mathfrak{M}_n^-,$$

where A and B are independent of Y. But then

$$AY^*B \cdot AY^*B = (Y^\tau)^2 = (Y^2)^\tau = \alpha(Y^2)A(Y^2)^*A^{-1},$$

so that

$$(BA)Y^*(BA) = \alpha(Y^2)Y^*.$$

Since this is valid for all $Y \in \mathfrak{M}_n^-$, we see that of necessity $\alpha(Y^2) = 1$ for all Y, and $BA = \pm I$. This shows that either $Y^\tau = \alpha(YZ) \cdot AY^*A^{-1}$ for all $Y \in \mathfrak{M}_n^-$, or $Y^\tau = -\alpha(YZ) \cdot AY^*A^{-1}$ for all $Y \in \mathfrak{M}_n^-$. If $n = 2$ and $\alpha(YZ) = \varepsilon(YZ)$, it is trivial to verify that either $\varepsilon(YZ) = \varepsilon(JY)$ for all $Y \in \mathfrak{M}_2^-$ or $\varepsilon(YZ) = -\varepsilon(JY)$ for all $Y \in \mathfrak{M}_2^-$.

The remainder of the paper will be concerned with proving Theorem 3.

4. Canonical forms for involutions

In the proof of Theorem 3 we shall use certain canonical forms of involutions under similarity transformations.

Lemma 1 *Under a similarity transformation, every involution $X \in \mathfrak{M}_n$ such that $X^2 = I^{(n)}$ can be brought into the form*

$$W(x,y,z) = \underbrace{L \dotplus \cdots \dotplus L}_{(x \text{ xterms})} \dotplus (-I)^{(y)} \dotplus I^{(z)}, \tag{4}$$

where $2x + y + z = n$ and

$$L = \begin{pmatrix} 1 & 0 \\ 1 & -1 \end{pmatrix}.$$

Proof We prove first, by induction on n, that every $X \in \mathfrak{M}_n$ satisfying $X^2 = I$ is similar to a matrix of the form

$$\begin{pmatrix} I^{(l)} & 0 \\ M & -I^{(n-l)} \end{pmatrix}. \tag{5}$$

For $n = 1$ and 2, this is trivial. Let the theorem be proved for n, and assume that $X^2 = I^{(n+1)}$, where $n \geqslant 2$. Then $X^2 - I = 0$, or $(X-I)(X+I) = 0$. If $X - I$ is nonsingular, then $X = -I$ and the result is obvious. Hence, supposing that $X - I$ is singular (so that $\lambda = 1$ is a characteristic root of X), there exists a primitive column vector $t = (t_1, \cdots, t_{n+1})'$ with integral elements such that $t'X = t'$. Choose $P \in \mathfrak{M}_{n+1}$ with first row t'. Then

$$PXP^{-1} = \begin{pmatrix} 1 & \mathfrak{n}' \\ \mathfrak{x} & X_1 \end{pmatrix},$$

where $\mathfrak{n}$ denotes a vector whose components are 0; thus

$$X \stackrel{s}{=} \begin{pmatrix} 1 & \mathfrak{n}' \\ \mathfrak{x} & X_1 \end{pmatrix}.$$

But

$$I^{(n+1)} = X^2 \stackrel{s}{=} \begin{pmatrix} 1 & \mathfrak{n}' \\ (I+X_1)\mathfrak{x} & X_1^2 \end{pmatrix}$$

shows that $X_1^2 = I^{(n)}$ and $(I+X_1)\mathfrak{x} = \mathfrak{n}$. By the induction hypothesis,

$$X_1 \stackrel{s}{=} \begin{pmatrix} I^{(m)} & 0 \\ M & -I^{(n-m)} \end{pmatrix}$$

and, after making the similarity transformation, we have (as a consequence of $(I+X_1)\mathfrak{x} = \mathfrak{n}$)

$$\begin{pmatrix} 2I^{(m)} & 0 \\ M & 0 \end{pmatrix} \mathfrak{x} = \mathfrak{n}.$$

Therefore

$$\mathfrak{x} = (0, \cdots, 0, *, \cdots, *)',$$
$$\phantom{\mathfrak{x} = (}(m \text{ terms}) (n-m \text{ terms})$$

where $*$ denotes an arbitrary element. Thus

$$X \stackrel{s}{=} \begin{pmatrix} 1 & & \mathfrak{n}' & & \\ 0 & & & & \\ \vdots & I^{(m)} & & 0 & \\ 0 & & & & \\ * & & & & \\ \vdots & M & & -I^{(n-m)} & \\ * & & & & \end{pmatrix} = \begin{pmatrix} I^{(m+1)} & 0 \\ M & -I^{(n-m)} \end{pmatrix}.$$

This completes the first part of the proof.

Suppose we now subject (5) to a further similarity transformation by

$$\begin{pmatrix} A^{(l)} & 0 \\ C & D^{(n-l)} \end{pmatrix} \in \mathfrak{M}_n.$$

A simple calculation shows that we obtain a matrix given by (5) with M replaced by $\overline{M}$, where $\overline{M} = 2CA^{-1} + DMA^{-1}$. Choosing firstly $C = 0$, A and D unimodular, we find that $\overline{M} = DMA^{-1}$, and by proper choice of A and D we can make $\overline{M}$ diagonal.

Supposing this done, secondly put $A = I, D = I$; we find that $\overline{M} = M + 2C$. Since C is arbitrary, we can bring $\overline{M}$ into the form

$$\begin{pmatrix} I^{(k)} & 0 \\ 0 & 0 \end{pmatrix},$$

where k is the rank of M. Since we can interchange two rows and simultaneously interchange the corresponding columns by means of a similarity transformation, the lemma follows.

It is easily seen that

$$W(x, y, z) \stackrel{s}{=} W(\bar{x}, \bar{y}, \bar{z})$$

only when $x = \bar{x}$, $y = \bar{y}$, and $z = \bar{z}$. Furthermore, changing the order of terms in the direct summation does not alter the similarity class. The number A_n of nonsimilar involutions in $\mathfrak{M}_n$ is therefore equal to the number of solutions of $2x + y + z = n$, $x \geqslant 0$, $y \geqslant 0$, $z \geqslant 0$. This gives

$$A_n = \begin{cases} \left(\dfrac{n+2}{2}\right)^2, & n \text{ even,} \\ \dfrac{(n+1)(n+3)}{4}, & n \text{ odd.} \end{cases} \tag{6}$$

Let B_n be the number of nonsimilar involutions in $\mathfrak{M}_n^+$, where the similarity factors are in $\mathfrak{M}_n$. One easily obtains

$$B_n = \begin{cases} (A_n - 1)/2, & \text{if } n \equiv 0 \,(\text{mod } 4), \\ A_n/2, & \text{otherwise.} \end{cases} \tag{7}$$

5. Automorphisms of $\mathfrak{M}_3^+$

We shall now prove Theorem 3 for $n = 3$. Let

$$I_1 = \begin{pmatrix} -1 & 0 & 0 \\ 0 & -1 & 0 \\ 0 & 0 & 1 \end{pmatrix}, \quad I_2 = \begin{pmatrix} 1 & 0 & 0 \\ 1 & -1 & 0 \\ 0 & 0 & -1 \end{pmatrix} \in \mathfrak{M}_3^+.$$

Then $I_1^2 = I^{(3)}$. Let τ be any automorphism of $\mathfrak{M}_3^+$ and let $X = I_1^\tau$; then $X^2 = I^{(3)}$. By Lemma 1, the matrices I_1, I_2 and $I^{(3)}$ form a complete system of nonsimilar involutions in $\mathfrak{M}_3^+$. Therefore

$$X \stackrel{s}{=} I_1 \text{ or } I_2.$$

After a suitable inner automorphism, we may assume that either $I_1 \to I_1$ or $I_1 \to I_2$. We shall show that this latter case is impossible by considering the normalizer groups of I_1 and I_2. The normalizer group of I_1, that is, the group of matrices $\in \mathfrak{M}_3^+$ which commute with I_1 consists of all elements of $\mathfrak{M}_3^+$ of the form

$$\begin{pmatrix} a & b & 0 \\ c & d & 0 \\ 0 & 0 & e \end{pmatrix}$$

and is isomorphic to $\mathfrak{M}_2$. That of I_2 consists of all elements of $\mathfrak{M}_3^+$ of the form

$$\begin{pmatrix} a & 0 & 0 \\ (a-e)/2 & e & f \\ -h/2 & h & i \end{pmatrix}$$

and is isomorphic to that subgroup $\mathfrak{G}$ of $\mathfrak{M}_2$ consisting of the elements

$$\begin{pmatrix} e & f \\ h & i \end{pmatrix} \in \mathfrak{M}_2, \quad \text{where} \quad \left.\begin{matrix} e \equiv 1 \\ h \equiv 0 \\ i \equiv 1 \end{matrix}\right\} \pmod{2}.$$

Since e and i are both odd, $\mathfrak{G}$ contains no element of order 3, and hence is not isomorphic to $\mathfrak{M}_2$. But then $I_1 \to I_2$ is impossible.

We may assume thus that after a suitable inner automorphism, I_1 is invariant. Thence elements of $\mathfrak{M}_3^+$ which commute with I_1 map into elements of the same kind, so that

$$\begin{pmatrix} X & \mathfrak{n}' \\ \mathfrak{n} & \pm 1 \end{pmatrix} \in \mathfrak{M}_3^+ \to \begin{pmatrix} X^\tau & \mathfrak{n}' \\ \mathfrak{n} & \pm 1 \end{pmatrix}.$$

Since this induces an automorphism $X \to X^\tau$ on $\mathfrak{M}_2$, we see that $\det X^\tau = \det X$, and hence the plus signs go together, and so do the minus signs. By Theorem 2 and that part of Theorem 4 which follows from Theorem 2, there exists a matrix $A \in \mathfrak{M}_2$ such that $X^\tau = \pm A X A^{-1}$; here, the plus sign certainly occurs when X is an even element of $\mathfrak{M}_2^+$, and if the minus sign occurs for one odd element of $\mathfrak{M}_2^+$, then it occurs for *every* odd element of $\mathfrak{M}_2^+$. By use of a further inner automorphism using the factor $A^{-1} \dotplus I^{(1)}$, we may assume that

$$\begin{pmatrix} X & \mathfrak{n}' \\ \mathfrak{n} & \pm 1 \end{pmatrix} \in \mathfrak{M}_3^+ \to \begin{pmatrix} \pm X & \mathfrak{n}' \\ \mathfrak{n} & \pm 1 \end{pmatrix}, \tag{8}$$

so that
$$M = \begin{pmatrix} 1 & 0 & 0 \\ 0 & -1 & 0 \\ 0 & 0 & -1 \end{pmatrix} \to M \text{ or } M \to N = \begin{pmatrix} -1 & 0 & 0 \\ 0 & 1 & 0 \\ 0 & 0 & -1 \end{pmatrix}.$$

Since
$$N = \begin{pmatrix} 0 & -1 & 0 \\ 1 & 0 & 0 \\ 0 & 0 & 1 \end{pmatrix} M \begin{pmatrix} 0 & 1 & 0 \\ -1 & 0 & 0 \\ 0 & 0 & 1 \end{pmatrix},$$

we may assume (after a further inner automorphism, if necessary) that I_1, M and N are all invariant under the automorphism (but (8) need not hold).

Thus, after a suitably chosen inner automorphism, we have I_1, M and N invariant. Therefore there exist A, B and $C \in \mathfrak{M}_2$ such that

$$\begin{pmatrix} X & \mathfrak{n} \\ \mathfrak{n}' & \pm 1 \end{pmatrix} \in \mathfrak{M}_3^+ \to \begin{pmatrix} \pm AXA^{-1} & \mathfrak{n} \\ \mathfrak{n}' & \pm 1 \end{pmatrix},$$

$$\begin{pmatrix} \pm 1 & \mathfrak{n}' \\ \mathfrak{n} & X \end{pmatrix} \in \mathfrak{M}_3^+ \to \begin{pmatrix} \pm 1 & \mathfrak{n}' \\ \mathfrak{n} & \pm BXB^{-1} \end{pmatrix},$$

$$\begin{pmatrix} a & 0 & b \\ 0 & \pm 1 & 0 \\ c & 0 & d \end{pmatrix} \in \mathfrak{M}_3^+ \to \begin{pmatrix} \alpha & 0 & \beta \\ 0 & \pm 1 & 0 \\ \gamma & 0 & \delta \end{pmatrix}, \qquad (9)$$

where
$$\begin{pmatrix} \alpha & \beta \\ \gamma & \delta \end{pmatrix} = \pm C \begin{pmatrix} a & b \\ c & d \end{pmatrix} C^{-1}$$

and $\mathfrak{n} = (0,0)'$. Here, the $+1$ on the left goes with the $+1$ on the right always (and the -1's go together); further, when X is an even element of $\mathfrak{M}_2^+$, the plus sign occurs before AXA^{-1}, BXB^{-1} and CXC^{-1}, while if the minus sign occurs before one of these for any odd $X \in \mathfrak{M}_2^+$, it occurs there for every odd $X \in \mathfrak{M}_2^+$.

Now we may assume that at most one of A, B and C has determinant -1; for if both A and B (say) have determinant -1, apply a further inner automorphism (with factor N) which leaves I_1, M and N invariant and changes the signs of det A and det B. Suppose hereafter, without loss of generality, that $\det A = \det B = 1$.

Next, N is invariant, but by (9) goes into
$$\begin{pmatrix} \pm A \begin{pmatrix} -1 & 0 \\ 0 & 1 \end{pmatrix} A^{-1} & \mathfrak{n}' \\ \mathfrak{n} & -1 \end{pmatrix},$$

so that
$$\pm A \begin{pmatrix} -1 & 0 \\ 0 & 1 \end{pmatrix} A^{-1} = \begin{pmatrix} -1 & 0 \\ 0 & 1 \end{pmatrix}.$$

This gives two possibilities:
$$A = I^{(2)} \quad \text{or} \quad \begin{pmatrix} 0 & 1 \\ -1 & 0 \end{pmatrix}.$$

The same holds true for B (but not necessarily for C, since $\det C = \pm 1$). Suppose firstly that either A or B is $I^{(2)}$, say $A = I^{(2)}$. Then

$$T = \begin{pmatrix} 1 & 1 & 0 \\ 0 & 1 & 0 \\ 0 & 0 & 1 \end{pmatrix} \rightarrow \begin{pmatrix} \pm \begin{pmatrix} 1 & 1 \\ 0 & 1 \end{pmatrix} & 0 \\ 0 & 0 & 1 \end{pmatrix}.$$

Case 1 T invariant. Then

$$\begin{pmatrix} 0 & 1 & 0 \\ -1 & 0 & 0 \\ 0 & 0 & 1 \end{pmatrix} \quad \text{and} \quad \begin{pmatrix} 0 & 1 & 0 \\ 1 & 0 & 0 \\ 0 & 0 & -1 \end{pmatrix}$$

are both invariant (the first matrix is invariant in virtue of the remarks after (9); the second is invariant because it is M times the first). For either possible choice of B we find that

$$\begin{pmatrix} -1 & 0 & 0 \\ 0 & 0 & 1 \\ 0 & 1 & 0 \end{pmatrix} \rightarrow \begin{pmatrix} -1 & 0 & 0 \\ 0 & & \\ 0 & \pm \begin{pmatrix} 0 & 1 \\ 1 & 0 \end{pmatrix} \end{pmatrix}.$$

Therefore

$$U = \begin{pmatrix} 0 & 1 & 0 \\ 0 & 0 & 1 \\ 1 & 0 & 0 \end{pmatrix} = \begin{pmatrix} -1 & 0 & 0 \\ 0 & -1 & 0 \\ 0 & 0 & 1 \end{pmatrix} \begin{pmatrix} -1 & 0 & 0 \\ 0 & 0 & 1 \\ 0 & 1 & 0 \end{pmatrix} \begin{pmatrix} 0 & 1 & 0 \\ 1 & 0 & 0 \\ 0 & 0 & -1 \end{pmatrix}$$

is mapped into

$$\begin{pmatrix} -1 & 0 & 0 \\ 0 & -1 & 0 \\ 0 & 0 & 1 \end{pmatrix} \begin{pmatrix} -1 & 0 & 0 \\ 0 & \pm \begin{pmatrix} 0 & 1 \\ 1 & 0 \end{pmatrix} \\ 0 & \end{pmatrix} \begin{pmatrix} 0 & 1 & 0 \\ 1 & 0 & 0 \\ 0 & 0 & -1 \end{pmatrix} = \begin{cases} U, & \text{if } + \text{ is used,} \\ V, & \text{if } - \text{ is used,} \end{cases}$$

where $V = I_1 U I_1^{-1}$. Thus, in this case, $T \to T = I_1 T I_1^{-1}$, and either $U \to U$ or $U \to I_1 U I_1^{-1}$. Since T and U generate[①] $\mathfrak{M}_3^+$, the automorphism is inner.

Case 2
$$T \to \begin{pmatrix} -1 & -1 & 0 \\ 0 & -1 & 0 \\ 0 & 0 & 1 \end{pmatrix}.$$

Then
$$\begin{pmatrix} 0 & 1 & 0 \\ 1 & 0 & 0 \\ 0 & 0 & -1 \end{pmatrix} \to \begin{pmatrix} 0 & -1 & 0 \\ -1 & 0 & 0 \\ 0 & 0 & -1 \end{pmatrix}$$

and one finds in this case that
$$U \to \begin{pmatrix} 0 & -1 & 0 \\ 0 & 0 & 1 \\ -1 & 0 & 0 \end{pmatrix} \text{ or } \begin{pmatrix} 0 & -1 & 0 \\ 0 & 0 & -1 \\ 1 & 0 & 0 \end{pmatrix}.$$

If we set $Z = TU^2$, then
$$\begin{pmatrix} 1 & 0 & 0 \\ 1 & 1 & 0 \\ 0 & 0 & 1 \end{pmatrix} = (UZ^{-1})^2 UZ^2. \tag{10}$$

Now certainly the left side of (10) maps into
$$\begin{pmatrix} -1 & 0 & 0 \\ -1 & -1 & 0 \\ 0 & 0 & 1 \end{pmatrix},$$

whereas, knowing T^τ and U^τ, we can compute Z^τ and thence can find the image of the right side of (10). We readily find (for either value of U^τ) that the right side of (10) maps into
$$\begin{pmatrix} 1 & \cdot & \cdot \\ 3 & \cdot & \cdot \\ \cdot & \cdot & \cdot \end{pmatrix},$$

and hence we have a contradiction.

Therefore Case 2 cannot occur, and so if either A or B equals $I^{(2)}$, the automorphism is inner. Suppose hereafter that
$$A = B = \begin{pmatrix} 0 & 1 \\ -1 & 0 \end{pmatrix}.$$

[①] Hua L K and Reiner I, loc. cit.

In this case we have

$$T \to \pm \begin{pmatrix} 1 & 0 & 0 \\ -1 & 1 & 0 \\ 0 & 0 & 1 \end{pmatrix}.$$

Case 1*

$$T \to \begin{pmatrix} 1 & 0 & 0 \\ -1 & 1 & 0 \\ 0 & 0 & 1 \end{pmatrix}.$$

Then as before

$$\begin{pmatrix} 0 & 1 & 0 \\ -1 & 0 & 0 \\ 0 & 0 & 1 \end{pmatrix} \quad \text{and} \quad \begin{pmatrix} 0 & 1 & 0 \\ 1 & 0 & 0 \\ 0 & 0 & -1 \end{pmatrix}$$

are invariant, and again $U^\tau = U$ or V. After a further inner automorphism by a factor of I_1 (in the latter case) we also have $U \to U$. But then

$$T \to T'^{-1}, \quad U \to U'^{-1}$$

(this automorphism is easily shown to be a non-inner automorphism).

Case 2*

$$T \to \begin{pmatrix} -1 & 0 & 0 \\ 1 & -1 & 0 \\ 0 & 0 & 1 \end{pmatrix}.$$

Then

$$\begin{pmatrix} 0 & 1 & 0 \\ 1 & 0 & 0 \\ 0 & 0 & -1 \end{pmatrix} \to \begin{pmatrix} 0 & -1 & 0 \\ -1 & 0 & 0 \\ 0 & 0 & -1 \end{pmatrix}$$

and again we find that there are two possibilities for U^τ, each of which leads to a contradiction, just as in Case 2. Therefore Theorem 3 holds when $n = 3$.

6. A fundamental lemma

Theorem 3 will be proved by induction on n; the result has already been established for $n = 2$ and 3. In going from $n-1$ to n, the following lemma is basic:

Lemma 2 Let $n \geq 4$, and define $J_1 = (-1) \dotplus I^{(n-1)}$. In any automorphism τ of $\mathfrak{M}_n$, $J_1^\tau = \pm A J_1 A^{-1}$ for some $A \in \mathfrak{M}_n$.

Proof By Corollary 1, $J_1^\tau \in \mathfrak{M}_n^-$, and J_1^τ is an involution. After a suitable inner automorphism, we may assume that $J_1^\tau = W(x, y, z)$ (as defined by (4)), where $2x + y + z = n$ and $x + y$ is odd. Every element of $\mathfrak{M}_n$ which commutes with J_1 maps into an element of $\mathfrak{M}_n$ which commutes with W. Every matrix in $\mathfrak{M}_n^+$ maps into a matrix in $\mathfrak{M}_n^+$. Combining these facts, we see that the group $\mathfrak{G}_1$ consisting of those elements of $\mathfrak{M}_n^+$ which commute with J_1 is isomorphic to $\mathfrak{G}_2$, the corresponding group for W. If we prove that this can happen only for $x = 0$, $y = 1$, $z = n - 1$ or $x = 0$, $y = n - 1$, $z = 1$, the result will follow.

The group $\mathfrak{G}_1$ consists of the matrices in $\mathfrak{M}_n^+$ of the form $(\pm 1) \dotplus X_1$, $X_1 \in \mathfrak{M}_{n-1}$, and so clearly $\mathfrak{G}_1 \cong \mathfrak{M}_{n-1}$.

The group $\mathfrak{G}_2$ is easily found to consist of all matrices $C \in \mathfrak{M}_1^+$ of the form (we illustrate the case where $x = 2$):

$$C = \left(\begin{array}{cccccccccc} a_1 & 0 & a_2 & 0 & 0 & \cdots & 0 & 2\beta_1 & \cdots & 2\beta_z \\ \frac{a_1-d_1}{2} & d_1 & \frac{a_2-d_2}{2} & d_2 & \alpha_1 & \cdots & \alpha_y & \beta_1 & \cdots & \beta_z \\ a_3 & 0 & a_4 & 0 & 0 & \cdots & 0 & 2\delta_1 & \cdots & 2\delta_z \\ \frac{a_3-d_3}{2} & d_3 & \frac{a_4-d_4}{2} & d_4 & \gamma_1 & \cdots & \gamma_y & \delta_1 & \cdots & \delta_z \\ \varepsilon_1 & -2\varepsilon_1 & \zeta_1 & -2\zeta_1 & & & & & & \\ \vdots & \vdots & \vdots & \vdots & & U & & & 0 & \\ \varepsilon_y & -2\varepsilon_y & \zeta_y & -2\zeta_y & & & & & & \\ \eta_1 & 0 & \theta_1 & 0 & & & & & & \\ \vdots & \vdots & \vdots & \vdots & & 0 & & & V & \\ \eta_z & 0 & \theta_z & 0 & & & & & & \end{array}\right) \begin{array}{l} \\ \\ 2x \\ \text{rows} \\ \\ y \\ \text{rows} \\ \\ z \\ \text{rows} \\ \end{array}$$

$$\underbrace{}_{\substack{2x \\ \text{columns}}} \quad \underbrace{}_{\substack{y \\ \text{columns}}} \quad \underbrace{}_{\substack{z \\ \text{columns}}}$$

For the moment put

$$K = \begin{pmatrix} 1 & 0 \\ -1/2 & 1 \end{pmatrix} \dotplus \cdots \dotplus \begin{pmatrix} 1 & 0 \\ -1/2 & 1 \end{pmatrix} \dotplus I^{(n-2x)}.$$

$$(x \text{ terms})$$

Then a simple calculation gives:

$$KCK^{-1} = \begin{pmatrix} a_1 & 0 & a_2 & 0 & 0 & \cdots & 0 & 2\beta_1 & \cdots & 2\beta_z \\ 0 & d_1 & 0 & d_2 & \alpha_1 & \cdots & \alpha_y & 0 & \cdots & 0 \\ a_3 & 0 & a_4 & 0 & 0 & \cdots & 0 & 2\delta_1 & \cdots & 2\delta_z \\ 0 & d_3 & 0 & d_4 & \gamma_1 & \cdots & \gamma_y & 0 & \cdots & 0 \\ 0 & -2\varepsilon_1 & 0 & -2\zeta_1 & & & & & & \\ \vdots & \vdots & \vdots & \vdots & & U & & & 0 & \\ 0 & -2\varepsilon_y & 0 & -2\zeta_y & & & & & & \\ \eta_1 & 0 & \theta_1 & 0 & & & & & & \\ \vdots & \vdots & \vdots & \vdots & & 0 & & & V & \\ \eta_z & 0 & \theta_z & 0 & & & & & & \end{pmatrix}$$

and so C is similar to

$$\begin{pmatrix} a_1 & a_2 & 2\beta_1 & \cdots & 2\beta_z \\ a_3 & a_4 & 2\delta_1 & \cdots & 2\delta_z \\ \eta_1 & \theta_1 & & & \\ \vdots & \vdots & & V & \\ \eta_z & \theta_z & & & \end{pmatrix} \dotplus \begin{pmatrix} d_1 & d_2 & \alpha_1 & \cdots & \alpha_y \\ d_3 & d_4 & \gamma_1 & \cdots & \gamma_y \\ -2\varepsilon_1 & -2\zeta_1 & & & \\ \vdots & \vdots & & U & \\ -2\varepsilon_y & -2\zeta_y & & & \end{pmatrix}$$

$$= \begin{pmatrix} S_1 & 2R_1 & x \\ Q_1 & T_1 & \\ x & z & \end{pmatrix} \dotplus \begin{pmatrix} S_2 & Q_2 & x \\ 2R_2 & T_2 & \\ x & y & \end{pmatrix} \begin{matrix} x \\ y \end{matrix} ,$$

with a fixed similarity factor depending only on W. Therefore $\mathfrak{G}_2 \cong \mathfrak{G}$, where $\mathfrak{G} = \mathfrak{G}(x, y, z)$ is the group of matrices in $\mathfrak{M}_n^+$ of the form

$$\begin{pmatrix} S_1 & 2R_1 \\ Q_1 & T_1 \\ x & z \end{pmatrix} \begin{matrix} x \\ z \end{matrix} \dotplus \begin{pmatrix} S_2 & Q_2 \\ 2R_2 & T_2 \\ x & y \end{pmatrix} \begin{matrix} x \\ y \end{matrix} ,$$

where $S_1 \equiv S_2 \pmod{2}$. Here $2x + y + z = n$ and $x + y$ is odd.

We wish to prove that $\mathfrak{M}_{n-1} \cong \mathfrak{G}(x, y, z)$ only when $x = 0$, $y = 1$, $z = n - 1$ or $x = 0$, $y = n - 1$, $z = 1$. In order to establish this, we shall prove that in all other cases the number of involutions in $\mathfrak{G}$ which are nonsimilar in $\mathfrak{G}$ is greater than the number of involutions in $\mathfrak{M}_{n-1}$ which are nonsimilar in $\mathfrak{M}_{n-1}$; this latter number is, of course, A_{n-1} (given by (6)).

We shall briefly denote the elements of $\mathfrak{G}$ by $A \dotplus B$, where

$$A = \begin{pmatrix} S_1 & 2R_1 \\ Q_1 & T_1 \end{pmatrix} \quad \text{and} \quad B = \begin{pmatrix} S_2 & Q_2 \\ 2R_2 & T_2 \end{pmatrix}.$$

If $A_1 \dotplus B_1$ and $A_2 \dotplus B_2$ are two involutions in $\mathfrak{G}$, where either
$$A_1 \overset{s}{\neq} A_2$$
in $\mathfrak{M}_{x+z}$ or
$$B_1 \overset{s}{\neq} B_2$$
in $\mathfrak{M}_{x+y}$, then certainly
$$A_1 \dotplus B_1 \overset{s}{\neq} A_2 \dotplus B_2$$
in $\mathfrak{G}$ (these may be similar in $\mathfrak{M}_n$, however). Therefore, the matrices $A \dotplus B$, where
$$A = I^{(a_1)} \dotplus (-I)^{(b_1)} \dotplus \underset{(c_1 \text{ terms})}{L \dotplus \cdots \dotplus L},$$
$$B = I^{(a_2)} \dotplus (-I)^{(b_2)} \dotplus \underset{(c_2 \text{ terms})}{L \dotplus \cdots \dotplus L},$$
obtained by taking different sets of values of $(a_1, b_1, c_1, a_2, b_2, c_2)$, if they lie in $\mathfrak{G}$, are certainly nonsimilar in $\mathfrak{G}$. Here we have
$$a_1 + b_1 + 2c_1 = x + z, \quad a_2 + b_2 + 2c_2 = x + y, \quad b_1 + b_2 + c_1 + c_2 \text{ even}.$$
If $x \neq 0$, we impose the further restriction that $c_1 \leqslant (z+1)/2$, $c_2 \leqslant (y+1)/2$, and that in B instead of L we use L'. These conditions will insure that $A \dotplus B \in \mathfrak{G}$. We certainly do not (in general) get all of the nonsimilar involutions of $\mathfrak{G}$ in this way, but instead we obtain only a subset thereof. Call the number of such matrices N.

For $x = 0$, we have $N = B_y B_z + (A_y - B_y)(A_z - B_z)$. Since y is odd, $A_y = 2B_y$, and therefore
$$N = B_y A_z = B_y A_{n-y}.$$

Case 1 n even. Then $N = (y+1)(y+3)(n-y+1)(n-y+3)/32$. If neither y nor $n-y$ is 1 (certainly neither can be zero), then
$$(y+1)(n-y+1) \geqslant 4(n-2) \quad \text{and} \quad (y+3)(n-y+3) \geqslant 6n,$$
so that
$$N \geqslant (24/32)n(n-2).$$
For $n = 4$, $x = 0$, either $y = 1$ or $z = 1$. For $n \geqslant 6$, we have $N > A_{n-1}$. Hence in this case $\mathfrak{G}$ is not isomorphic to $\mathfrak{M}_{n-1}$ (if either y or $n - y = 1$, then $W(x, y, z) = \pm J_1$).

Case 2 n odd. Then $N = (y+1)(y+3)(n-y+2)^2/32$. We find again that $N > A_{n-1}$ for $n \geqslant 5$.

This settles the cases where $x = 0$. Suppose that $x \neq 0$ hereafter. Then N is the number of solutions of

$$a_1 + b_1 + 2c_1 = x + z, \quad a_2 + b_2 + 2c_2 = x + y, \quad b_1 + b_2 + c_1 + c_2 \text{ even,}$$

$$0 \leqslant c_1 \leqslant \frac{z+1}{2}, \quad 0 \leqslant c_2 \leqslant \frac{y+1}{2}.$$

Using $[r]$ to denote the greatest integer less than or equal to r, we readily find that N is given by

$$\frac{1}{2} \left[\frac{z+3}{2} \right] \left[\frac{y+3}{2} \right] \left(x + z + 1 - \left[\frac{z+1}{2} \right] \right) \left(x + y + 1 - \left[\frac{y+1}{2} \right] \right).$$

By considering separately the cases where y and z are both even, one even and one odd, and so on, it is easy to prove that $N \geqslant A_{n-1}$ in all cases except when both y and z are zero. Leaving aside this case for the moment, consider the matrix $A_0 \dotplus I^{(x+y)} \in \mathfrak{G}$, where $A_0 \in \mathfrak{M}_{x+z}$ is given by

$$A_0 = \begin{pmatrix} 1 & 2 & 2 & \cdots & 2 \\ 0 & -1 & 0 & \cdots & 0 \\ 0 & 0 & -1 & \cdots & 0 \\ \vdots & \vdots & \vdots & & \vdots \\ 0 & 0 & 0 & \cdots & -1 \end{pmatrix}.$$

The matrix $A_0 \dotplus I^{(x+y)}$ is certainly an involution in $\mathfrak{G}$. Since, in $\mathfrak{M}_{x+z}$,

$$A_0 \overset{s}{=} \begin{pmatrix} 1 & 0 & \cdots & 0 \\ 0 & -1 & \cdots & 0 \\ \vdots & \vdots & & \vdots \\ 0 & 0 & \cdots & -1 \end{pmatrix} = A_1,$$

$A_0 \dotplus I^{(x+y)}$ can be similar (in $\mathfrak{G}$) only to that matrix (counted in the N matrices) of the form $A_1 \dotplus I^{(x+y)}$. But from

$$A_1 \cdot \begin{pmatrix} a_1 & a_2 & \cdots & a_x & 2b_1 & \cdots & 2b_z \\ \cdots & \cdots & \cdots & \cdots & \cdots & & \cdots \\ \cdots & \cdots & \cdots & \cdots & \cdots & & \cdots \\ \cdots & \cdots & \cdots & \cdots & \cdots & & \cdots \end{pmatrix} = \begin{pmatrix} a_1 & a_2 & \cdots & a_x & 2b_1 & \cdots & 2b_z \\ \cdots & \cdots & \cdots & \cdots & \cdots & & \cdots \\ \cdots & \cdots & \cdots & \cdots & \cdots & & \cdots \\ \cdots & \cdots & \cdots & \cdots & \cdots & & \cdots \end{pmatrix} \cdot A_0$$

we obtain

$$a_1 = a_2 = \cdots = a_x = 2b_1,$$

which is impossible. Hence $\mathfrak{G}$ contains at least $N+1$ nonsimilar involutions, and therefore $\mathfrak{G}$ is not isomorphic to $\mathfrak{M}_{n-1}$ in these cases.

We have left only the case $y = z = 0$, $x = n/2$; then n is singly even. Here we may choose $A = W(c_1, b_1, a_1)$, $B = W(c_1, b_2, a_2)$, where

$$a_1 + b_1 + 2c_1 = x, \quad a_2 + b_2 + 2c_1 = x, \quad b_1 + b_2 \text{ even.}$$

Then $A \dotplus B \in \mathfrak{G}$, and the various matrices are nonsimilar. The number of such matrices is $(x+1)(x+2)(x+3)/12$, which is greater than A_{n-1} for $n \geqslant 14$. For $n = 6$, $\mathfrak{M}_{n-1}$ contains an element of order 5, while $\mathfrak{G}$ does not. For $n = 10$, $\mathfrak{M}_{n-1}$ contains an element of order 7, while $\mathfrak{G}$ does not. This completes the proof of the lemma.

7. Proof of Theorem 3

We are now ready to give a proof of Theorem 3 by induction on n. Hereafter, let $n \geqslant 4$ and suppose that Theorem 3 holds for $n - 1$. If τ is any automorphism of $\mathfrak{M}_n$, by Corollary 1 and Lemma 2 we know that τ takes $\mathfrak{M}_n^+$ into itself, and $J_1^\tau = \pm A J_1 A^{-1}$. If we change τ by a suitable inner automorphism, then we may assume that $J_1 \to \pm J_1$. When n is odd, certainly $J_1 \to J_1$; when n is even, by multiplying τ by the automorphism $X \in \mathfrak{M}_n \to (\det X) \cdot X$ if necessary, we may again assume $J_1 \to J_1$.

Therefore, every $M \in \mathfrak{M}_n^+$ which commutes with J_1 goes into another such element, that is,

$$\begin{pmatrix} \pm 1 & \mathfrak{n}' \\ \mathfrak{n} & X \end{pmatrix}^\tau = \begin{pmatrix} \pm 1 & \mathfrak{n}' \\ \mathfrak{n} & X^\tau \end{pmatrix}.$$

Since this induces an automorphism on $\mathfrak{M}_{n-1}$, we have $\det X^\tau = \det X$, so that the plus signs go together, as do the minus signs. Furthermore, by our induction hypothesis,

$$X^\tau = \pm A X^* A^{-1},$$

where $A \in \mathfrak{M}_{n-1}$ and either $X^* = X$ for all $X \in \mathfrak{M}_{n-1}$ or $X^* = X'^{-1}$ for all $X \in \mathfrak{M}_{n-1}$; here the minus sign can occur only for $X \in \mathfrak{M}_{n-1}^-$, and if it occurs for one such X, it occurs for all $X \in \mathfrak{M}_{n-1}^-$. After changing our original automorphism by a factor of $I^{(1)} \dotplus A^{-1}$, we may assume that $X^\tau = \pm X^*$.

Let J_ν be obtained from $I^{(n)}$ by replacing the νth diagonal element by -1. Then

$$J_1 J_n = \begin{pmatrix} -1 & 0 & \cdots & 0 & 0 \\ 0 & 1 & \cdots & 0 & 0 \\ \vdots & \vdots & & \vdots & \vdots \\ 0 & 0 & \cdots & 1 & 0 \\ 0 & 0 & \cdots & 0 & -1 \end{pmatrix} \rightarrow \begin{pmatrix} -1 & & & \mathfrak{n}' & \\ & 1 & \cdots & 0 & 0 \\ \mathfrak{n} & \vdots & & 0 & 0 \\ & 0 & \cdots & 1 & 0 \\ & 0 & \cdots & 0 & -1 \end{pmatrix}^{*} \pm .$$

The minus sign here is impossible by Lemma 2, since $n \geqslant 4$. Hence $J_1 J_n$ is invariant, and therefore so is J_n. By the same reasoning all of the $J_\nu (\nu = 1, \cdots, n)$ are invariant.

From the above remarks we see that for $X \in \mathfrak{M}_{n-1}^+$,

$$\begin{pmatrix} 1 & \mathfrak{n}' \\ \mathfrak{n} & X \end{pmatrix}^\tau = \begin{pmatrix} 1 & \mathfrak{n}' \\ \mathfrak{n} & A_1 X^* A_1^{-1} \end{pmatrix}, \cdots, \begin{pmatrix} X & \mathfrak{n} \\ \mathfrak{n}' & 1 \end{pmatrix}^\tau = \begin{pmatrix} A_n X^* A_n^{-1} & \mathfrak{n} \\ \mathfrak{n}' & 1 \end{pmatrix},$$

where $A_\nu \in \mathfrak{M}_{n-1}$, and in fact $A_1 = I$. Now suppose that $Z \in \mathfrak{M}_{n-2}^+$, and form $I^{(2)} \dotplus Z$. Since it commutes with both J_1 and J_2, its image must do likewise. But then

$$A_1 \begin{pmatrix} 1 & \mathfrak{n}' \\ \mathfrak{n} & Z \end{pmatrix} A_1^{-1} = \begin{pmatrix} 1 & \mathfrak{n}' \\ \mathfrak{n} & \bar{Z} \end{pmatrix}$$

for every $Z \in \mathfrak{M}_{n-2}^+$. Setting

$$A_1 = \begin{pmatrix} a & \mathfrak{x}' \\ \mathfrak{y} & A \end{pmatrix},$$

we obtain $\mathfrak{x}' Z = \mathfrak{x}'$, $\mathfrak{y} = \bar{Z}\mathfrak{y}$. Since this holds for all $Z \in \mathfrak{M}_{n-2}^+$, we must have $\mathfrak{x} = \mathfrak{y} = \mathfrak{n}$, so that A_1 is itself decomposable. A similar argument (considering the matrices commuting with both J_1 and J_ν, for $\nu = 3, \cdots, n$) shows that A_1 is diagonal. Correspondingly, all of the A_ν are diagonal. It is further clear that all of the $A_\nu (\nu = 1, \cdots, n)$ are sections of a single diagonal matrix $D^{(n)}$. Using the further inner automorphism factor D^{-1}, we may henceforth assume that $X^\tau = X^*$ for every decomposable $X \in \mathfrak{M}_n^+$, where either $X^* = X$ always or $X^* = X'^{-1}$ always. Since $\mathfrak{M}_n^+$ is generated by the set of decomposable elements of $\mathfrak{M}_n^+$, the theorem is proved.

Automorphisms of the projective unimodular group*

Notation Let $\mathfrak{M}_n$ denote the group of $n \times n$ integral matrices of determinant ± 1 (the unimodular group). By $\mathfrak{M}_n^+$ we denote that subset of $\mathfrak{M}_n$ where the determinant is $+1$; $\mathfrak{M}_n^-$ is correspondingly defined. Let $\mathfrak{P}_{2n}$ be obtained from $\mathfrak{M}_{2n}$ by identifying $+X$ and $-X$, $X \in \mathfrak{M}_{2n}$ (this is the same as considering the factor group of $\mathfrak{M}_{2n}$ by its centrum). We correspondingly obtain $\mathfrak{P}_{2n}^+$ and $\mathfrak{P}_{2n}^-$ from $\mathfrak{M}_{2n}^+$ and $\mathfrak{M}_{2n}^-$. Let $I^{(n)}$ (or briefly I) be the identity matrix in $\mathfrak{M}_n$, and let X' denote the transpose of X. The direct sum of A and B is represented by $A \dotplus B$, while

$$A \stackrel{s}{=} B$$

means that A is similar to B.

In this paper we shall find explicitly the generators of the group $\mathfrak{B}_{2n}$ of all automorphisms of $\mathfrak{P}_{2n}$, thereby obtaining a complete description of these automorphisms. This generalizes the result due to Schreier[1] for the case $n = 1$.

We shall frequently refer to results of an earlier paper: *Automorphisms of the unimodular group* (L. K. Hua and I. Reiner. *Trans. Amer. Math. Soc.*, 1951, 71: 331-348). We designate this paper by AUT.

1. The commutator subgroup of $\mathfrak{P}_{2n}$

The following useful result is an immediate consequence of the corresponding theorem for $\mathfrak{M}_{2n}$ (AUT, Theorem 1).

Theorem 1 *Let $\mathfrak{S}_{2n}$ be the commutator subgroup of $\mathfrak{P}_{2n}$. Then clearly $\mathfrak{S}_{2n} \subset \mathfrak{P}_{2n}^+$. For $n = 1$, $\mathfrak{S}_{2n}$ is of index 2 in $\mathfrak{P}_{2n}^+$, while for $n > 1$, $\mathfrak{S}_{2n} = \mathfrak{P}_{2n}^+$.*

Theorem 2 *In any automorphism of $\mathfrak{P}_{2n}$, always $\mathfrak{P}_{2n}^+$ goes into itself.*

Proof This is a corollary to Theorem 1 when $n > 1$, since the commutator subgroup goes into itself under any automorphism. For $n = 1$, suppose that $\pm S \to \pm S_1$ and $\pm T \to \pm T_1$, where

---

* Received by the editors May 18, 1951. Reprinted from *Transactions of the American Mathematical Society*, 1952, **72**(3): 467–473. By Hua L K and Reiner I.
 [1] *Abh. Math. Sem. Hamburgischen Univ.*, 1924, **3**: 167.

$$S = \begin{pmatrix} 0 & 1 \\ -1 & 0 \end{pmatrix}, \quad T = \begin{pmatrix} 1 & 1 \\ 0 & 1 \end{pmatrix}. \tag{1}$$

Since S and T generate $\mathfrak{M}_2^+$, it follows that $\pm S$ and $\pm T$ generate $\mathfrak{P}_2^+$, and hence so must $\pm S_1$ and $\pm T_1$. It is therefore sufficient to prove that $\det S_1 = \det T_1 = +1$. From $(ST)^3 = I$ we deduce $S_1 T_1 = \pm T_1^{-1} S_1^{-1} T_1^{-1} S_1^{-1}$, so that $\det S_1 T_1 = 1$. Hence either S_1 and T_1 are both in $\mathfrak{P}_2^+$ or both in $\mathfrak{P}_2^-$; we shall show that the latter alternative is impossible.

Suppose that $\det S_1 = \det T_1 = -1$. From $S^2 = I$ we deduce $S_1^2 = \pm I$; if $S_1^2 = -I$, then $S_1^2 + I = 0$ and the characteristic equation of S_1 is $\lambda^2 + 1 = 0$, from which it follows that $\det S_1 = 1$; this contradicts our assumption that $\det S_1 = -1$, so of necessity $S_1^2 = I$. But if this is the case, then it is easy to show that there exists a matrix $A \in \mathfrak{M}_2$ such that $A S_1 A^{-1}$ takes one of the two canonical forms

$$\begin{pmatrix} 1 & 0 \\ 0 & -1 \end{pmatrix} \quad \text{and} \quad \begin{pmatrix} 1 & 0 \\ 1 & -1 \end{pmatrix}.$$

By considering instead of the original automorphism τ, a new automorphism τ' defined by: $X^{\tau'} = A X^\tau A^{-1}$, we may hereafter assume that

$$S_1 = \pm \begin{pmatrix} 1 & 0 \\ 0 & -1 \end{pmatrix} \quad \text{or} \quad \pm \begin{pmatrix} 1 & 0 \\ 1 & -1 \end{pmatrix}.$$

Let

$$T_1 = \pm \begin{pmatrix} a & b \\ c & d \end{pmatrix},$$

then $ad - bc = -1$.

Now we observe that $J = (1) \dotplus (-1)$ is distinct from $\pm I$ and $\pm S$, that it commutes with S, and that JT is an involution. Hence there exists a matrix $M \in \mathfrak{P}_2$ distinct from $\pm I$ and $\pm S_1$, such that M commutes with S_1, and MT_1 is an involution.

Case 1

$$S_1 = \pm \begin{pmatrix} 1 & 0 \\ 0 & -1 \end{pmatrix}.$$

Since $(S_1 T_1)^3 = \pm I$, we find that $a - d = \pm 1$. The only matrices commuting with S_1 which are distinct from $\pm I$ and $\pm S_1$ are

$$\pm \begin{pmatrix} 0 & 1 \\ 1 & 0 \end{pmatrix} \quad \text{and} \quad \pm \begin{pmatrix} 0 & 1 \\ -1 & 0 \end{pmatrix}.$$

If M is either of the first two matrices, then the condition that MT_1 be an involution yields $b + c = 0$. Thus $a = d \pm 1$, $b = -c$, and $ad - bc = -1$. Combining these, we obtain $d(d \pm 1) + c^2 = -1$, which is impossible. The other two choices for M imply $b = c$, and therefore $d(d \pm 1) - c^2 = -1$. Hence $1 - 4(1 - c^2)$ is a perfect square; but $4c^2 - 3 = f^2$ implies $(2c + f)(2c - f) = 1$, whence $c = \pm 1$. But then $ad = 0$; from $a - d = \pm 1$ we deduce that $a^2 - d^2 = \pm 1$, whence $(S_1 T_1^2)^3 = \pm I$, which is impossible.

Case 2
$$S_1 = \pm \begin{pmatrix} 1 & 0 \\ 1 & -1 \end{pmatrix}.$$

From $(S_1 T_1)^3 = \pm I$ we obtain $a - d + b = \pm 1$. For M there are the four possibilities

$$\pm \begin{pmatrix} 1 & -2 \\ 0 & -1 \end{pmatrix} \quad \text{and} \quad \pm \begin{pmatrix} 1 & -2 \\ 1 & -1 \end{pmatrix}.$$

Since MT_1 is an involution, in the first two cases we have $a - 2c - d = 0$, whence

$$ad - bc = \{(a+d)^2 + (a - d \pm 1)^2 - 1\}/4 \neq -1.$$

In the second two cases we find that $a - 2c + b - d = 0$, so that $2c = a + b - d = \pm 1$, which is again a contradiction. This completes the proof of Theorem 2.

2. Automorphisms of $\mathfrak{P}_2^+$

Let us now determine all automorphisms of $\mathfrak{P}_2$. Since every such automorphism takes $\mathfrak{P}_2^+$ into itself, we begin by considering all automorphisms of $\mathfrak{P}_2^+$.

Theorem 3 *Every automorphism of $\mathfrak{P}_2^+$ is of the form $X \in \mathfrak{P}_2^+ \to AXA^{-1}$ for some $A \in \mathfrak{M}_2$; that is, all automorphisms of $\mathfrak{P}_2^+$ are "inner" (with $A \in \mathfrak{M}_2$ rather than $A \in \mathfrak{P}_2^+$).*

Proof Let τ be any automorphism of $\mathfrak{P}_2^+$, and define S and T as before; let $S_0 \in \mathfrak{M}_2$ be a fixed representative of $\pm S^\tau$. By Theorem 2, $S_0 \in \mathfrak{M}_2^+$, and therefore $S_0^2 = -I$. Let T_0 be that representative of $\pm T^\tau$ for which $(S_0 T_0)^3 = I$ is valid. Then $S \to S_0$, $T \to T_0$ induces a mapping from $\mathfrak{M}_2^+$ onto itself. The mapping is one-to-one, for although an element of $\mathfrak{M}_2^+$ can be expressed in many different ways as a product of powers of S and T, these expressions can be gotten from one another by use of $S^2 = -I$, $(ST)^3 = I$; since S_0 and T_0 satisfy these same relations, the mapping is one-to-one. It is an automorphism because τ is one. Therefore (AUT, Theorem 2) there exists an $A \in \mathfrak{M}_2$ such that $S_0 = \pm ASA^{-1}$, $T_0 = \pm ATA^{-1}$. This proves the result.

Corollary *Every automorphism of $\mathfrak{P}_2$ is of the form $X \in \mathfrak{P}_2 \to AXA^{-1}$ for some $A \in \mathfrak{M}_2$.*

(This corollary is a simple consequence of Theorem 3, as is shown in AUT by the remarks following the statement of Theorem 4.)

3. The generators of $\mathfrak{B}_{2n}$

Our main result may be stated as follows:

Theorem 4 *The generators of $\mathfrak{B}_{2n}$ are*

(i) *The set of all inner automorphisms:*

$$\pm X \in \mathfrak{P}_{2n} \to \pm AXA^{-1} \quad (A \in \mathfrak{M}_{2n}),$$

and

(ii) *The automorphism $\pm X \in \mathfrak{P}_{2n} \to \pm X'^{-1}$.*

Remark For $n = 1$, the automorphism (ii) is a special case of (i).

In the proof of Theorem 4 by induction on n, the following lemma (which has already been established for $n = 1$) will be basic:

Lemma 1 *Let $J_1 = (-1) \dotplus I^{(2n-1)}$. In any automorphism τ of $\mathfrak{P}_{2n}$, $J_1^\tau = \pm AJ_1A^{-1}$ for some $A \in \mathfrak{M}_{2n}$.*

Proof The result is already known for $n = 1$. Hereafter let $n \geqslant 2$. Certainly $(J_1^\tau)^2 = \pm I$ and $\det J_1^\tau = -1$. If $(J_1^\tau)^2 = -I$, then the minimum function of J_1^τ is $\lambda^2 + 1$, and its characteristic function must be some power of $\lambda^2 + 1$, whence $\det J_1^\tau = 1$. Therefore $(J_1^\tau)^2 = I$ is valid in $\mathfrak{M}_{2n}$. After a suitable inner automorphism, we may assume that

$$J_1^\tau = W(x, y, z) = L \dotplus \cdots \dotplus L \dotplus (-I)^{(y)} \dotplus I^{(z)},$$

where

$$L = \begin{pmatrix} 1 & 0 \\ 1 & -1 \end{pmatrix}$$

occurs x times, $2x + y + z = 2n$, and $x + y$ is odd (this follows from AUT, Lemma 1). Let $\mathfrak{G}_1$ be the group consisting of all elements of $\mathfrak{P}_{2n}$ which commute with J_1, and $\mathfrak{G}_2$ the corresponding group for J_1^τ. The lemma will be proved if we can show that $\mathfrak{G}_1$ is not isomorphic to $\mathfrak{G}_2$ unless $J_1^\tau = \pm J_1$. The group $\mathfrak{G}_1$ consists of the matrices $\pm(1 \dotplus X_1) \in \mathfrak{P}_{2n}$, so that $\mathfrak{G}_1 \cong \mathfrak{M}_{2n-1}$. The number of nonsimilar involutions in $\mathfrak{G}_1$ is therefore $n(n+1)$ (see AUT, §4). We shall prove that $\mathfrak{G}_2$ contains more than $n(n+1)$ involutions which are nonsimilar in $\mathfrak{G}_2$, except when $x = 0$, $y = 1$, $z = 2n - 1$ or $x = 0$, $y = 2n - 1$, $z = 1$.

Those elements $\pm C \in \mathfrak{P}_{2n}$ which commute with W must satisfy one of the two equations: $CW = WC$ or $CW = -WC$. The solutions of the first of these equations form a subgroup of $\mathfrak{G}_2$, and this subgroup is known (see AUT, proof of Lemma 2) to be isomorphic to $\mathfrak{G}_0 = \mathfrak{G}_0(x, y, z)$ consisting of all matrices in $\mathfrak{P}_{2n}$ of the form

$$\begin{pmatrix} S_1 & 2R_1 \\ Q_1 & T_1 \end{pmatrix} \dotplus \begin{pmatrix} S_2 & Q_2 \\ 2R_2 & T_2 \end{pmatrix},$$

where S_1, S_2, T_1, and T_2 are square matrices of dimensions x, x, z, and y respectively, and where $S_1 \equiv S_2 \pmod{2}$, $2x + y + z = 2n$, and $x + y$ and $x + z$ are both odd.

Next we prove that $\bar{C}W = -W\bar{C}$ is solvable only when $y = z$. The space $\mathfrak{U}$ of vectors $\mathfrak{u}$ such that $W\mathfrak{u} = \mathfrak{u}$ is of dimension $x + z$, while the space $\mathfrak{B}$ of vectors $\mathfrak{b}$ for which $W\mathfrak{b} = -\mathfrak{b}$ has dimension $x + y$. But if $\bar{C}W = -W\bar{C}$, then $W\bar{C}\mathfrak{u} = -\bar{C}\mathfrak{u}$ and $W\bar{C}^{-1}\mathfrak{b} = \bar{C}^{-1}\mathfrak{b}$, so the dimensions of $\mathfrak{U}$ and $\mathfrak{B}$ must be the same, whence $y = z$. Hence if $y \neq z$, there are no solutions of $\bar{C}W = -W\bar{C}$, $\bar{C} \in \mathfrak{M}_{2n}$.

We may now proceed to find a lower bound for the number of nonsimilar matrices in $\mathfrak{G}_0(x, y, z)$. We briefly denote the elements of $\mathfrak{G}_0$ by $A \dotplus B$, where

$$A = \begin{pmatrix} S_1 & 2R_1 \\ Q_1 & T_1 \end{pmatrix} \quad \text{and} \quad B = \begin{pmatrix} S_2 & Q_2 \\ 2R_2 & T_2 \end{pmatrix}.$$

If $A_1 \dotplus B_1$ and $A_2 \dotplus B_2$ are two distinct involutions in $\mathfrak{G}_0$, where either

$$A_1 \overset{s}{\neq} A_2 \text{ in } M_{x+z} \quad \text{or} \quad B_1 \overset{s}{\neq} B_2 \text{ in } M_{x+y},$$

then certainly

$$A_1 \dotplus B_1 \overset{s}{\neq} A_2 \dotplus B_2 \text{ in } \mathfrak{G}_0.$$

Now let

$$A = I^{(a_1)} \dotplus (-I)^{(b_1)} \dotplus L \dotplus \cdots \dotplus L,$$

$$B = I^{(a_2)} \dotplus (-I)^{(b_2)} \dotplus L \dotplus \cdots \dotplus L,$$

where L occurs c_1 times in A and c_2 times in B; the various elements $A \dotplus B$ gotten by taking different sets of values of $(a_1, b_1, c_1, a_2, b_2, c_2)$, if they lie in $\mathfrak{G}_0$, are certainly nonsimilar in $\mathfrak{G}_0$, except that $A \dotplus B$ and $(-A) \dotplus (-B)$ are the same element of $\mathfrak{G}_0$. Hence the number N of nonsimilar involutions of $\mathfrak{G}_0$ is at least half of the number N_1 of solutions of

$$a_1 + b_1 + 2c_1 = x + z,$$

$$a_2 + b_2 + 2c_2 = x + y,$$

where if $x \neq 0$ we impose the restrictions that $c_1 \leqslant (z+1)/2$, $c_2 \leqslant (y+1)/2$, and that in B instead of L we use L' (these conditions insure that $A \dotplus B \in \mathfrak{G}_0$). As in the previous paper, one readily shows that $N > n(n+1)$ unless $J_1^\tau = \pm J_1$. We omit the details.

This leaves only the case where $y = z$. If $\bar{C}W = -W\bar{C}$, then $\bar{C}^k W = (-1)^k W \bar{C}^k$; therefore no odd power of $\bar{C}$ can be $\pm I$. Let p be a prime such that $n < p < 2n$. Since $x + y = n$, certainly n is odd, and $p \geqslant n+2$. Now $\mathfrak{G}_1$ (being isomorphic to $\mathfrak{M}_{2n-1}$) contains infinitely many elements of order p. However, $\mathfrak{G}_2$ contains only two such elements, since $\bar{C}^p \neq \pm I$ by the above argument, while if $C \in \mathfrak{G}_0$ and $C^p = \pm I$, then setting $C = A^{(n)} \dotplus B^{(n)}$ shows that $A^p = \pm I$ and $B^p = \pm I$. However, $A \in \mathfrak{M}_n$, and if $A^p = \pm I$, then the minimum function of A must divide $\lambda^p \mp 1$. But the degree of the minimum function is at most n, and therefore is less than $p - 1$, whereas $\lambda^p \mp 1$ is the product of a linear factor $\lambda \mp 1$ and an irreducible factor of degree $p - 1$; thence the minimum function of A is $\lambda \mp 1$, so $A = \pm I$. In the same way $B = \pm I$. Hence the only solutions are $C = I^{(n)} \dotplus I^{(n)}$ and $C = -I^{(n)} \dotplus I^{(n)}$. This completes the proof of the lemma. We remark that the use of the existence of the prime p could have been avoided, but the proof is much quicker this way.

4. Proof of the main theorem

We are now ready to prove Theorem 4 by induction on n. Hereafter, let $n \geqslant 2$ and assume that Theorem 4 holds for $n-1$. Let τ be any automorphism of $\mathfrak{P}_{2n}$; then by Lemma 1, $J_1^\tau = \pm A J_1 A^{-1}$ for some $A \in \mathfrak{M}_{2n}$. If we change τ by a suitable inner automorphism, we may assume that $J_1^\tau = \pm J_1$.

Therefore, every $M \in \mathfrak{P}_{2n}$ which commutes with J_1 goes into another such element, that is,

$$\pm \begin{bmatrix} 1 & \mathfrak{n}' \\ \mathfrak{n} & X \end{bmatrix}^\tau = \pm \begin{bmatrix} 1 & \mathfrak{n}' \\ \mathfrak{n} & Y \end{bmatrix},$$

where $\mathfrak{n}$ denotes a column vector all of whose components are zero, and $X \in \mathfrak{M}_{2n-1}$. Thus, τ induces an automorphism on $\mathfrak{M}_{2n-1}$. Consequently (AUT, Theorem 4) there exists a matrix $A \in \mathfrak{M}_{2n-1}$ such that $Y = AX^*A^{-1}$ for all $X \in \mathfrak{M}_{2n-1}$, where either $X^* = X$ for all $X \in \mathfrak{M}_{2n-1}$ or $X^* = X'^{-1}$ for all $X \in \mathfrak{M}_{2n-1}$. After a further inner automorphism by a factor of $(1) \dotplus A^{-1}$, we may assume that $J_1^\tau = \pm J_1$ and also that $X^\tau = Y = X^*$ for all $X \in \mathfrak{M}_{2n-1}$.

Let J_ν be obtained from $I^{(2n)}$ by replacing the νth diagonal element by -1.

Then

$$(J_1 J_{2n})^\tau = \pm \begin{pmatrix} 1 & 0 & \cdots & 0 & 0 \\ 0 & -1 & \cdots & 0 & 0 \\ \vdots & \vdots & & \vdots & \vdots \\ 0 & 0 & \cdots & -1 & 0 \\ 0 & 0 & \cdots & 0 & 1 \end{pmatrix}^\tau = \pm \begin{pmatrix} 1 & & & & \mathfrak{n}' \\ & -1 & \cdots & 0 & 0 \\ & \vdots & & \vdots & \vdots \\ \mathfrak{n} & 0 & \cdots & -1 & 0 \\ & 0 & \cdots & 0 & 1 \end{pmatrix}^*$$

$$= \pm J_1 J_{2n},$$

so that $\pm J_{2n}$ is invariant. Similarly, all of the matrices $\pm J_\nu$ ($\nu = 1, \cdots, 2n$) are invariant. Therefore for any $X \in \mathfrak{M}_{2n-1}$ we have

$$\pm \begin{pmatrix} 1 & \mathfrak{n}' \\ \mathfrak{n} & X \end{pmatrix}^\tau = \pm \begin{pmatrix} 1 & \mathfrak{n}' \\ \mathfrak{n} & A_1 X^* A_1^{-1} \end{pmatrix}, \cdots, \pm \begin{pmatrix} X & \mathfrak{n} \\ \mathfrak{n}' & 1 \end{pmatrix}^\tau = \pm \begin{pmatrix} A_{2n} X^* A_{2n}^{-1} & \mathfrak{n} \\ \mathfrak{n}' & 1 \end{pmatrix},$$

with $A_\nu \in \mathfrak{M}_{2n-1}$, and in fact $A_1 = I$.

Now suppose that $Z \in \mathfrak{M}_{2n-2}$, and consider $\pm(Z + I^{(2)})$; since it commutes with J_{2n-1} and J_{2n}, so does its image. But therefore

$$A_{2n} \begin{pmatrix} Z & \mathfrak{n} \\ \mathfrak{n}' & 1 \end{pmatrix} A_{2n}^{-1} = \begin{pmatrix} \bar{Z} & \mathfrak{n} \\ \mathfrak{n}' & 1 \end{pmatrix},$$

where $\bar{Z}$ denotes some matrix in $\mathfrak{M}_{2n-2}$. From this one easily deduces that A_{2n} must be of the form $B \dotplus (1)$, with $B \in \mathfrak{M}_{2n-2}$. By considering the matrices commuting with J_ν and J_{2n} for $\nu = 1, \cdots, 2n-2$ we see that A_{2n} must be diagonal. Furthermore, it is clear that all of the $A_\nu (\nu = 1, \cdots, 2n)$ must be diagonal, and all are sections of one diagonal matrix $D^{(2n)}$. Using the further inner automorphism factor D^{-1}, we find that $\pm X^\tau = \pm X^*$ for every decomposable matrix $\pm X \in \mathfrak{P}_{2n}$. Since $\mathfrak{P}_{2n}$ is generated by the set of its decomposable matrices, the theorem is proved.

《华罗庚文集》已出版书目

(按出版时间排序)

1 华罗庚文集数论卷 I　王元　审校　2010 年 5 月
2 华罗庚文集数论卷 II　贾朝华　审校　2010 年 5 月
3 华罗庚文集数论卷 III　王元　潘承彪　贾朝华　编译　2010 年 5 月
4 华罗庚文集代数卷 I　万哲先　审校　2010 年 5 月
5 华罗庚文集多复变函数论卷 I　陆启铿　审校　2010 年 5 月
6 华罗庚文集应用数学卷 I　杨德庄　主编　2010 年 5 月
7 华罗庚文集应用数学卷 II　杨德庄　主编　2010 年 5 月
8 华罗庚文集代数卷 II　李福安　审校　2011 年 2 月